SIGNAL TRANSDUCTION
IN PLANTS

SIGNAL TRANSDUCTION IN PLANTS
CURRENT ADVANCES

Edited by

S. K. Sopory
International Centre for Genetic Engineering and Biotechnology
New Delhi, India

Ralf Oelmüller
Friedrich-Schiller University
Jena, Germany

and

S. C. Maheshwari
International Centre for Genetic Engineering and Biotechnology
New Delhi, India

Kluwer Academic / Plenum Publishers
New York, Boston, Dordrecht, London, Moscow

Library of Congress Cataloging-in-Publication Data

Signal transduction in plants: current advances/edited by S.K. Sopory, Ralf Oelmüller, and S.C. Maheshwari.
   p. cm.
   Includes bibliographical references (p.).
   ISBN 0-306-46671-6
   1. Plant cellular signal transduction.   2. Plant molecular biology.   I. Sopory, S. K.   II. Oelmüller, Ralf, 1957–   III. Maheshwari, S. C.   IV. ICGEB Symposium on Plant Signal Transduction (1st: 1999: New Delhi, India)

QK725 .S5474 2001
571.6'2—dc21

2001038774

Proceedings of the First ICGEB Symposium on Plant Signal Transduction, held 4–6 October, 1999, in New Delhi, India

ISBN 0-306-46671-6

©2001 Kluwer Academic/Plenum Publishers, New York
233 Spring Street, New York, New York 10013

http://www.wkap.nl/

10 9 8 7 6 5 4 3 2 1

A C.I.P. record for this book is available from the Library of Congress

All rights reserved

No part of this book may be reproduced, stored in a retrieval system, or transmitted in any form or by any means, electronic, mechanical, photocopying, microfilming, recording, or otherwise, without written permission from the Publisher

Printed in the United States of America

Preface

"Signal Transduction" is an area of intensive scientific research at the present time. The current research focus in many laboratories is towards unraveling the mode by which environment and various physical and chemical signals influence gene expression, synthesis of proteins, and enzyme activity. Although much work has been done on animals, until recently molecules that perceive external signals and the signaling cascades operating in plants remain largely unknown. A year before relinquishing charge as Head of the New Delhi component of the International Centre for Genetic Engineering and Biotechnology, Prof. K. K. Tewari conceived the plan of organizing an international symposium on Plant Signal Transduction to enable younger researchers in India and elsewhere to have an idea of this new field of research. The symposium was finally held between October 4 and 6, 1999. This volume presents updated contributions that were made in the meetings.

Because, unlike in animals, growth and development of plants are controlled so profoundly by light and other external factors, the mechanisms of perception and transduction of signals are in several respects rather unique to plants. This meeting began by discussion of the various kinds of novel photoreceptors that exist in plants, sensing red, far-red and blue light and proceeded to a consideration of how various signaling cascades operate to alter cellular metabolism and gene expression, bringing about the various developmental effects. One of the more memorable highlights of the meeting was the remarkable full-screen portrayal of photoreceptors moving in the nucleus from the cytoplasm upon exposure of cell to light and their exit when light was turned off. Plant growth and morphogenesis are also affected by biotic and abiotic stresses and hormones. Several contributions focussed on the role of protein phosphorylation and dephosphorylation, light activated and other kinases and phosphatases that seem special to plants. G proteins, second messengers such as Ca^{++} and phosphoinositides, alicyclic acid and nitric oxide. Because of the special role of chloroplasts in the life of a plant, an entire session was devoted to their biogenesis and interactions with the nucleus. Finally, since all higher plants are multicellular organisms and various external or internal signals must bring their effects through enhancement or suppression of cell division, and sometimes cell death, the concluding session concerned itself with a discussion of roles of various newly discovered molecules, such as cyclins, cyclin-dependent kinases, inhibitors like Rb (retinoblastoma) proteins affecting the cell cycle, as also proteases and caspases that cause cell death.

Publishing proceedings of a conference like this has its usual pitfalls and disappointments. Despite our announcement in the meeting itself of the advance arrangement with Plenum Press to publish the proceedings and extending the deadline originally set, several participants failed to send their manuscripts in time. However, we have included the abstracts that were submitted to give an idea to readers of the full range of contributions made.

We hope that the volume will be an important addition to the literature of modern plant biology and inspire many younger workers to expand the horizons of knowledge. At the end, the Editors would like to acknowledge the help received from the present Director of ICGEB, Prof. V. S. Chauhan, and the staff of this Institute as well as at the Botanisches Institut of the University at Jena in Germany.

S. K. Sopory
R. Oelmüller
S. C. Maheshwari

Contents

1. Differential Perception of Environmental Light by Phytochromes 1
   MASAKI FURUYA

2. Functions and Actions of *Arabidopsis* Phytochromes 9
   KAREN J. HALLIDAY, UTA M. PRAEKELT, MICHAEL G. SALTER,
   AND GARRY C. WHITELAM

3. Signal Transduction in Photomorphogenesis: Intracellular Partitioning
   of Factors and Photoreceptors 19
   EBERHARD SCHÄFER, STEFAN KIRCHER, PATRICIA GIL,
   KLAUS HARTER, LANA KIM, FRANK WELLMER,
   LAZLO KOZMA-BOGNAR, EVA ADAM, AND FERENC NAGY

4. Molecular Genetic Analysis of Constitutively Photomorphogenic
   Mutants of *Arabidopsis* 25
   JITENDRA P. KHURANA, AKHILESH K. TYAGI, PARAMJIT KHURANA,
   ANJU KOCHHAR, PRADEEP K. JAIN, ANIRUDDHA RAYCHAUDHURI,
   REKHA CHAWLA, ARVIND K. BHARTI, ASHVERYA LAXMI,
   AND UJJAINI DASGUPTA

5. Cytosolic pH as a Secondary Messenger During Light Activation of
   Phosphoenolpyruvate Carboxylase in Mesophyll Cells of C_4 Plants ... 39
   Ch. BHASKAR RAO, NASSER SYED, JHADESWAR MURMU,
   AND A. S. RAGHAVENDRA

6. A Role for Ectoapyrases and Extracellular ATP in Plant Growth and
   Development 49
   STANLEY J. ROUX, COLLIN THOMAS, AND ASHA RAJAGOPAL

7. Auxin- and GTP-binding Proteins and Protein Kinases from
   the Protonema of the Moss *Funaria hygrometrica* 59
   M. M. JOHRI, KISHORE C. PANIGRAHI, J. S. D'SOUZA, AND D. MITRA

8. Myoinositol Phosphates as Implicated in Metabolic Signaling and
   Calcium Homeostasis in Plants 71
   SHASHIPRABHA DASGUPTA, DIPAK DASGUPTA, SUSWETA BISWAS,
   AND BIRENDRA B. BISWAS

9. The Phosphoinositide (PI) Pathway and Signaling in Plants 83
   IMARA Y. PERERA, INGO HEILMANN, AND WENDY F. BOSS

10. Phytochrome Signal Transduction in Cucumber Cotyledons 93
    DOLORS VIDAL, CARMEN BERGARECHE, M. TERESA GIL,
    AND ESTHER SIMON

11. Calcium, Calmodulin and Phosphoinositides in Leaflet Movements
    Mediated by Phytochrome: Nyctinastic and Rhythmic Movements 103
    LUISA MOYSSET, EVA FERNANDEZ, LUIS A. GOMEZ,
    AND ESTHER SIMON

12. Calmodulin and Plant Responses to the Environment: Modulation
    of Plant Tolerance to Toxic Metals by a Plasma Membrane Calcium/
    Calmodulin Binding Channel 113
    RAMANJULU SUNKAR, TZAHI ARAZI, BOAZ KAPLAN,
    DVORA DOLEV, AND HILLEL FROMM

13. Calcium Signaling: Downstream Components in Plants 125
    GIRDAR K. PANEY, VEENA, RENU DESWAL, SONA PANDEY,
    S. B. TEWARI, W. TYAGI, VANGA SIVA REDDY,
    ALOK BHATTACHARYA, AND SUDHIR K. SOPORY

14. Apoplast Calmodulin: The Identification, Functions and
    Transmembrane Mechanism 137
    DAYE SUN AND LIGENG MA

15. CDPKs in Plant Signaling Networks: Progress in Research on
    a Groundnut CDPK 145
    MAITRAYEE DAS GUPTA AND SUBHO CHAUDHURI

16. The Function of the Maize CRINKLY4 Receptor-like Kinase in
    a Growth Factor Like Signaling System 157
    PHILIP W. BECRAFT, MEENA R. CHANDOK, PING JIN, TAO GUO,
    YVONNE ASUNCION-CRABB, AND YUAN ZHANG

17. Novel Calcium/Calmodulin-modulated Proteins: Chimeric Protein
    Kinase and Small Auxin Up-regulated RNA 167
    B.W. POOVAIAH, WUYI WANG, AND TIANBAO YANG

18. A Novel Ca^{2+}/CaM-regulated Microtubule Motor Protein from Plants:
    Role in Trichome Morphogenesis and Cell Division 177
    A. S. N. REDDY

19. Progress Towards the Identification of Cytokinin Receptors 193
    RICHARD HOOLEY

20. Salicylic Acid- and Nitric Oxide-mediated Signal Transduction in
    Disease Resistance 201
    DANIEL F. KLESSIG, JÖRG DURNER, ROY NAVARRE,
    DHIRENDRA KUMAR, JYOTI SHAH, JUN MA ZHOU, SHUQUN ZHANG,
    DAVID WENDEHENNE, PRADEEP KACHROO, HERMAN SILVA,
    KEIKO YOSHIOKA, YOUSSEF TRIFA, DOMINIQUE PONTIER,
    ERIC LAM, ZHIXIANG CHEN, MARC ANDERSON, AND HE DU

21. Involvement of ROS and Caspase-like Proteases During Cell Death Induction by Plant Pathogens: Signaling in Plant Cell Death 209
   ERIC LAM, OLGA DEL POZO, AND DOMINIQUE PONTIER

22. The Role of HSF in Heat Shock Signal Transduction and Heat Shock Response in Plants 217
   RALF PRÄNDL, CHRISTIAN LOHMANN, STEFANIE DÖHR, AND FRITZ SCHÖFFL

23. Towards Understanding the Recognition and Signal Transduction Processes in the Soybean-Phytophthora Sojae Interaction 227
   MADAN K. BHATTACHARYYA, BONNIE G. ESPINOSA, TAKAO KASUGA, YONGQING LIU, SHANMUKHASWAMI S. SALIMATH, MARK GIJZEN, VAINO POISA, AND RICHARD BUZZELL

24. Elements of Signal Transduction Involved in Thylakoid Membrane Dynamics 241
   PETRA WEBER, ANNA SOKOLENKO, SAID ESHAGHI, HRVOJE FULGOSI, ALEXANDER V. VENER, BERTIL ANDERSSON, ITZHAK OHAD, AND REINHOLD G. HERRMANN

25. Novel Aspects in Photosynthesis Gene Regulation 259
   RALF OELMÜLLER, TATJANA PESKAN, MARTIN WESTERMANN, IRENA SHERAMETI, MEENA CHANDOK, SUDHIR K. SOPORY, ANKE WÖSTEMEYER, VICTOR KUSNETSOV, STAVER BEZHANI, AND THOMAS PFANNSCHMIDT

26. Regulation of rDNA Transcription in Spinach Plastids by Transcription Factor CDF2 279
   SILVA LERBS-MACHE

27. Plastid Ribosome Biogenesis During the Early Steps of Chloroplast Differentiation: Elements Controlling the Activation of Nuclear Genes Encoding Plastid Ribosomal Proteins 287
   REGIS MACHE, JEAN-LUC GALLOIS, AND PATRICK ACHARD

28. Mechanism of Regulation of Gene Expression for Chloroplast Proteins 297
   AKHILESH K. TYAGI, JITENDRA P. KHURANA, ARUN K. SHARMA, AMITABH MOHANTY, AMIT DHINGRA, SAURABH RAGHUVANSHI, AND TRIPTI GAUR

29. Functional Analysis of Pea Chloroplast DNA Polymerase and its Accessory Proteins 309
   AMOS GAIKWAD, NASREEN EHTESHAM, D. V. HOP, SHAOJ CHEN, AND SUNIL KUMAR MUKHERJEE

30. Signal Transduction by Mitogen-activated Protein Kinases (MAPKs) .. 321
   ERWIN HEBERLE-BORS, ORNELLA CALDERINI, VIKTOR VORONIN, AND CATHAL WILSON

Index 331

Differential Perception of Environmental Light by Phytochromes

MASAKI FURUYA
Hitachi Advanced Reseach Laboratory, Hatoyama, Saitama 350-0395, Japan

1. **INTRODUCTION**

Plants monitor wavelength, fluence and timing of light irradiation in order to adjust their development and reproduction to the seasonal and daily changes of environment. In the early 20th century botanists were already aware that development and reproduction in plants are influenced significantly by light (Klebs 1910). Evidence for photoregulation in plants so accumulated that Mohr (1962) operationally classified light-dependent reactions in plants into two categories, low and high energy reactions; the former reaction was induced by a pulse of light irradiation, whereas the latter required a continuous irradiation for long period of time. The former was further divided into low irradiance response and very low irradiance response in terms of light energy required and photoreversibility of effects by Blaauw *et al* (1968). Later, Briggs *et al* (1984) renamed them as very low fluence response (VLFR), low fluence response (LFR) and high irradiance response (HIR). As far as photoreceptor pigments for these three categories of photoperception in plants were concerned, it had been an open question for long time except that phytochrome was believed as the photoreceptor for red/far-red reversible LFR. Red/far-red reversible LFR (Borthwick *et al* 1952) was consistent with the photoreversible spectral changes of phytochrome between red light absorbing form, Pr, and far-red light absorbing form, Pfr (Butler *et al* 1959). The central dogma for phytochrome action that Pfr is only the active form was long supported by workers in this field, but evidence has accumulated against the dogma (Smith 1983). The discovery that phytochrome is a multi-gene family (Sharrock and Quail

1989, Clack *et al* 1994) has opened new approaches to solve this problem by addressing which member of the phytochrome family is responsible for which phytochrome-mediated responses (Furuya 1993).

2. DISTINCT ACTION SPECTRA FOR PHYTOCHROME - DEPENDENT RESPONSES IN PLANTS

Action spectra for red/far-red photoreversible effects on lettuce seed germination and photoperiodic floral induction were first determined by Borthwick *et al* (1952); this finding showed that environmental light is used as a signal controlling plant development rather than merely as an energy source for photosynthesis. HIR showing a peak in the FR region was called FR-HIR, and the involvement of phytochrome in the mediation of the FR-HIR was speculated based upon experimental results using prolonged dichromatic irradiation with continuous R and FR (Hartmann 1966). It, however, had been obscure until recently as to which phytochrome species would work for which effect. An indispensable tool for this purpose has been the isolation of mutants for each type of phytochrome, leading to the finding that FR-HIR resulted from PhyA while R-HIR from PhyB (Quail *et al* 1995).

In recent years we have determined the action spectra for VLFR, LFR and HIR, using phytochrome A (PhyA)- and phytochrome B (PhyB)-deficient mutants in *Arabidopsis thaliana* at the Okazaki large spectrograph facility (Watanabe *et al* 1982). We discovered that PhyA regulates VLFR (Shinomura *et al* 1996) and FR-HIR (Shinomura *et al* 2000) by essentially different manners. Thus, PhyA induces seed germination upon perception of VLF lights of broad spectral range from 300 to 770 nm (Fig. 1), in which no photoreversible effect was observed. The action spectrum for PhyA-dependent VLFR coincides quite well with the absorption spectrum of purified PhyA in the P_r form ($PhyA_r$), indicating that absorption of the light by $PhyA_r$ is the primary action of the response. The light capture by PhyA is a very sensitive photoperception response in plants, and it occurs at approximately 3-orders of magnitude less light energy than that for the elementary process of FR-HIR and the traditionally known R/FR reaction. This was the first evidence that PhyA is the photoreceptor for VLFR.

By contrast, Shinomura *et al* (2000) reported that in the cyclic treatment with far-red light pulse (pFR) for HIR, both the newly synthesized $PhyA_r$ as also the photoconverted $PhyA_{fr}$ do not appear to be active in inducing downstream signals, but the signal is produced somewhere in phototransformation from $PhyA_{fr}$ to $PhyA_r$. The pFR is required every few minutes for many hours to induce the HIR effect, suggesting that the signal is short-lived. The effect of pFR is repeatedly photoreversible by red light pulse (pR) if given subsequently after pFR. This mode of photoperception for FR-HIR is totally different from that of VLFR by PhyA, indicating that PhyA acts by two essentially distinct mechanisms in photoperception, depending upon the developmental stages of plants and environmental light condition.

PhyB photoreversibly switches responses to on or off upon alternating irradiation with red and far-red light (Tab. 1). PhyB photoreversibly induces germination by irradiation with red light (550-690 nm) of 10-1,000·mmol m^{-2}, and the induction is prevented photoreversibly by far-red light (700-800 nm) of similar fluence. Considering that this action spectrum for the PhyB response (Shinomura *et al* 1996) is essentially the same as that for the red/far-red reversible effect on lettuce seed germination reported by Borthwick *et al* (1952), we guess that the first action spectrum for lettuce seed germination resulted from a PhyB effect.

In addition, very similar action spectra for PhyA- and PhyB-specific induction for *Cab* gene expression were obtained in the etiolated seedlings of *Arabidopsis* mutants (Hamazato *et al* 1997), suggesting that differential regulation by different phytochromes is a rather general property of plants. It is now evident that structurally similar but functionally distinct phytochromes perceive light as environmental signals by different modes of their photoperception (Tab. 1).

Table 1. Different modes of photoperception by phytochromes A and B => Phototransformation, -> Dark transformation , ( ) Intermittent treatment, X, Y and Z, Interacting partner molecules to phytochromes.

Fluence (Mode)	Effective Wavelength	Phytochrome species, their signaling and Response
Switch	R / FR	PhyBr <=> PhyBfr + X--> Stable signal --> LFR (off)　　　(on)
Trigger	UV-Vis-FR	PhyAr => ? + Y --> Short-lived signal --> VLFR
Cycle	FR (B) / R	PhyAr <=> (PhyAfr <=> ?) -> ? + Z -> Signal --> HIR

3. PHYTOCHROME-DEPENDENT GENE EXPRESSION

One of the major down-stream responses to phytochrome signals is the photoregulation of gene expression. In the past two decades, the photoregulation of gene expression in higher plants was studied in respect of a limited number of genes in the genome such as *Lhcb* (Thompson and

White 1991, Tobin and Kehoe 1994). Hence we have developed a fluorescent differential display (FDD) technique to perform a genome-wide screening for identification of a set of phytochrome-regulated genes in *Arabidopsis thaliana* (Kuno *et al* 2000). Approximately 30,000 bands of cDNA were displayed by FDD, and northern blot (or RT-PCR) analysis revealed that thirteen up-regulated genes included nine photosynthetic proteins, two enzymes involved in biosynthesis of chlorophyll and one DNA damage repair/toleration-related protein besides one unknown gene, and two down-regulated genes such as the xyloglucan endotransglycosylase-related gene and a novel member of the ASK protein kinase family. Thus, the majority of up-regulated genes detected 4 hours after FR irradiation were identical to the nuclear genes for plastid proteins involved in photosynthesis and biosynthesis of chlorophyll. In contrast to the up-regulated genes, the two down-regulated genes encode proteins functionally distinct from plastid proteins.

The expression of the above-described 15 PhyA-inducible genes was examined to determine whether phytochromes other than PhyA are involved in their regulation. Kuno *et al* (2000) clearly demonstrated that PhyB and phytochromes other than PhyA and PhyB also photoreversibly regulate the expression of the 15 genes. In other words, none of these genes was regulated exclusively in a PhyA-specific manner. Although the modes of photoperception by PhyA and PhyB are essentially different in terms of fluence required, effective wavelengths and photoreversibility, PhyA and PhyB share the photoregulation of transcription of the same genes. This is consistent with the results of previous studies of phytochrome-regulation of *Lhcb1*3* gene expression (Reed *et al* 1994, Anderson *et al* 1997, Hamazato *et al* 1997). Overlapping effects of phytochromes have also been observed in several phytochrome-related responses, such as regulation of the photoinduction of seed germination (Shinomura *et al* 1996), hypocotyl elongation, promotion of cotyledon expansion and flowering (Reed *et al* 1994). Thus, the overlapping effects of phytochromes might be a general phenomenon in photoregulation of plants, suggesting that the signals derived from different modes of photoperception (Tab. 1) would transmit into a mutual transduction pathway. To uncover the mechanisms of the overlapping effects of different phytochromes, further analysis of their signal transduction pathways is obviously necessary.

4. FUTURE PROSPECT

Most of the spectrophotometric and molecular properties known in literature were centering around only PhyA, but very little with other phytochromes (Furuya and Song 1994). Hence, first of all we would like to know what is the crucial difference of molecular properties between PhyA and PhyB that results in the above-described differential photoperception. In this connection, Wagner *et al* (1996) reported that chromophore-bearing N-terminal domains of PhyA and PhyB determine their photosensory specificity and differential light lability. Synthesis *in vitro* of variously designed PhyA and PhyB apoproteins and chemically synthesized chromophores could provide us the material to answer this question. Protein-protein interactions are central to signal transduction cascades in cells. If we could find any unique interaction between PhyA or PhyB and some of the molecules directly interacting with phytochromes (Ni *et al* 1998, Fankhauser *et al* 1999, Choi *et al* 1999), we might explain the different modes of photoperception (X, Y and Z, Tab. 1).

At this moment, we know only a fragmentary idea of the nature of signals from PhyA and PhyB particularly derived through the work on phytochrome-signaling mutants (Neff *et al* 2000). A long-distance signaling for gene expression in leaves was demonstrated using microbeam irradiation method in *Sinapis* (Nick *et al* 1993) and in tobacco (Bischoff *et al* 1998), whereas cell-autonomously short-lived signal is transmitted from a type II phytochrome in inducing actomyosin-dependent movement of cytoplasmic particles (Takagi and Furuya 2000). These scattered evidences must be systematized into a central dogma eventually.

ACKNOWLEDGEMENT

This work was partly supported by grants to M.F. from the Hitachi Advanced Research Laboratory (B2023) and the Program for Promotion of Basic Research Activities for Innovative Biosciences.

REFERENCES

Anderson, S.L., Somers, D.E., Millar, A.J., Hanson, K., Chory, J., and Kay, S.A., 1997, Attenuation of phytochrome A and B signaling pathways by the *Arabidopsis* circadian clock. *Plant Cell* **9**: 1727-1743.

Bischoff, F., Millar, A.J., Kay, S.A., and Furuya, M., 1997, Phytochrome-induced intercellular signalling activates *cab::luciferase* gene expression. *Plant J.* **12** (10): 839-849.

Blaauw, O.H., Blaauw-Jansen, G., and Van Leeuwen, W.J., 1968, An irreversible red-light-induced growth response in *Avena. Planta* **82**: 87-104.

Briggs, W.R., Mandoli, D.F., Shinkle, J.R., Kaufman, .L.S., Watson, J.C., and Thompson, W.F., 1984, Phytochrome regulation of plant development at the whole plant, physiological, and molecular levels. *In* G Colombetti, F Lenci, P-S Song, eds, Sensory Perception and Transduction in Aneural Organisms, *Plenum, New York*, pp 265-280.

Bothwick, H.A., Hendrick, S.B., Parker, E.H., Toole, E.H., and Toole, V.K., 1952, A reversible photoreaction controlling seed germination. *Proc. Natl. Acad. Sci. USA* **38**: 662-666.

Butler, W.L., Norris, K. H., Siegelman, H.W., and Hendricks, S.B., 1959, Action spectra of phytochrome *in vitro*. *Photochem. Phtotobiol.* **3**: 521-528.

Clack, T., Mathews, S., and Sharrock, R.A., 1994, The phytochrome apoprotein family in *Arabidopsis* is encoded by five genes: the sequences and expression of *PHYD* and *PHYE*. *Plant Mol. Biol.* **25**: 413-427.

Furuya, M., 1993, Phytochromes: Their molecular species, gene families and functions. *Annu. Rev. Plant Physiol. Plant Mol. Biol.* **44**: 617-645.

Furuya, M., and Song, P.S., 1994, Assembly and properties of holophytochrome. In "Photomorphogenesis in Plants (2nd edition)" (R.E. Kendrick & G.H.M. Kronenberg eds), p.105-140, *Kluwer Academic Publ.*, Dordrecht, Boston, London.

Hamazato, F., Shinomura, T., Hanzawa, H., Chory, J., and Furuya, M., 1997, Fluence and wavelength requirements for *Arabidopsis CAB* gene induction by different phytochromes. *Plant Physiol.* **115**: 1533-1540.

Hartmann, K.M., 1966, A general hypothesis to interpret "high energy phenomena" of photomorphogenesis on the basis of phytochrome. *Photochem. Photobiol.* **5**: 349-366.

Klebs, G., 1910, Alterations in the development and forms of plants as a result of environment. *Proc. R. Soc. Lond.* B **82**: 547-558.

Kuno, N., Muramatsu, T., Hamazato, F., and Furuya, M., 2000, Phytochrome-Regulated Genes in Etiolated Seedlings of *Arabidopsis thaliana* using a Fluorescent Differential Display Technique. *Plant Physiol.* **122**: 15-24.

McCormac, A.C., Wagner, D., Boylan, M.T., Quail, P.H., Smith, H., and Whitelam, G.C., 1993, Photoresponses of transgenic *Arabidopsis* seedlings expressing introduced phytochrome B-encoding cDNAs: evidence that phytochrome A and phytochrome B have distinct photoregulatory functions. *Plant J.* **4**: 19-27.

Mohr, H., 1962, Primary effects of light on growth. *Annu. Rev. Plant Physiol.* **13**: 465-488.

Neff, M. M., Fankhauser, C., and Chory, J., 2000, Light: an indicator of time and place. *Genes Dev.* **14**: 257-271.

Ni, M., Tepperman, M., and Quail, P.H., 1998, PIF3, a phytochrome-interacting factor necessary for normal photoinduced signal transduction, is a novel basic helix-loop-helix protein. *Cell* **95**: 657-667.

Nick, P., Ehmann, B., Furuya, M., and Schäfer, E., 1993, Cell communication, stochastic cell responses, and anthocyanin pattern in mustard cotyledons. *Plant Cell* **5**, 541-552.

Quail, P.H., Boylan, M.T., Parks, B.M., Short, T.W., Xu, Y., and Wagner, D., 1995, Phytochromes: photosensory perception and signal transduction. *Science* **268**: 675-680.

Reed, J.W., Nagatani, A., Elich, T.D., Fagan, M., and Chory, J., 1994, Phytochrome A and phytochrome B have overlapping but distinct functions in *Arabidopsis* development. *Plant Physiol.* **104**: 1139-1149.

Sharrock, R.A., and Quail, P.H., 1989, Novel phytochrome sequences in *Arabidopsis thaliana*: structure, evolution, and differential expression of a plant regulatory photoreceptor family. *Genes Dev.* **3**: 1745-1757.

Shinomura, T., Nagatani, A., Hanzawa, H., Kubota, M., Watanabe, M., and Furuya, M., 1996, Action spectra for phytochrome A- and B-specific photoinduction of seed germination in *Arabidopsis thaliana*. *Proc. Natl. Acad. Sci. USA* **93**: 8129-8133.

Shinomura, T., Uchida, K., and Furuya, M., 2000, Elementary processes of photoperception by phytochrome A for high irradiance response of hypocotyl eleongation in *Arabidopsis thaliana*. *Plant Physiol.* **122**: 147-156.

Smith, H., 1983, Is Pfr the active form of phytochrome? *Phil. Trans. R. Soc. Lond.* B **303**: 443-452.

Takagi, S., Mineyuki, Y., and Furuya, M., 2000, A rapid, cell-autonomous response of agitation in Vallisneria epidermal cells is under control of type II phytochrome. *Plant Cell Physiol.* **41**, suppl. S55, 116.

Wagner, D., Fairchild, C.D., Kuhn, R.M., and Quail, P.H., 1996, Chromophore-bearing NH_2H terminal domains of phytochromes A and B determine their photosensory specificity and differential light lability. *Proc. Natl. Acad. Sci. USA* **93**: 4011-4015.

Watanabe, M., Furuya, M., Miyoshi, Y., Inoue, Y., Iwahashi, I., and Matsumoto, K., 1982, Design and performance of the Okazaki large spectrograph for photobiological research. *Photochem. Photobiol.* **36**: 491-498.

Functions and Actions of *Arabidopsis* Phytochromes

KAREN J. HALLIDAY, UTA M. PRAEKELT, MICHAEL G. SALTER, AND GARRY C. WHITELAM
Department of Biology, University of Leicester, University Road, Leicester LE1 7RH, UK

1. INTRODUCTION

Environmental cues play a crucial role in regulating plant growth and development. Given the sedentary nature of plants, this close coupling between environment and development is essential for survival under conditions that are ever changing. The light environment is a particularly important determinant of plant development and the periodicity, direction, quantity and quality of incident light are continuously monitored by plants. Regulatory light signals are perceived by a number of specialised photoreceptor systems, each of which displays maximal light absorption in a specific region of the light spectrum. The photoreversible phytochromes absorb predominantly in the red (R) and far-red (FR) region of the spectrum, the cryptochromes and phototropin absorb light in the blue/UV-A region of the spectrum, and other, as yet unknown, photoreceptors are involved in perception of UV-A and UV-B light. The phytochromes are perhaps the best characterised of these photoreceptor systems. In the higher plants, phytochromes comprise a number of closely related photoreceptor proteins, the apoproteins of which are encoded by a small family of genes. In *Arabidopsis thaliana* there are five apophytochrome-encoding genes, *PHYA*, *PHYB*, *PHYC*, *PHYD* and *PHYE* (Sharrock and Quail 1989, Clack et al 1994). In the last few years genetic approaches, mainly focussed on *Arabidopsis*, have had a major impact in increasing our understanding of the functions of individual members of the phytochrome family and in identifying components of the signal transduction networks that couple light reception to alteration in gene expression and changes in growth and development.

2. PHYTOCHROMES A AND B PLAY MAJOR ROLES IN SEEDLING PHOTOMORPHOGENESIS

Analysis of genetic mutants, null for one, two or multiple phytochrome species has provided a window into the complex network that comprises the phytochrome signalling system. These analyses indicate that there is extensive overlap in function, interaction and redundancy among members of the phytochrome family. Although there may be some degree of shared function for all the phytochromes, phytochrome A (phyA) and phytochrome B (phyB) appear to be the major determinants of seedling photomorphogenesis. Loss of either phyA or phyB in *Arabidopsis* has a dramatic affect on plant phenotype. For example, mutants that are null for phyA lack completely all of the de-etiolation responses shown by wild type seedlings to prolonged FR and representing the High Irradiance Response (HIR) mode of phytochrome action. Thus, in contrast with the wild type, *phyA* seedlings fail to display inhibition of hypocotyl elongation and the promotion of cotyledon opening and expansion in response to prolonged FR (see Whitelam *et al* 1998). Under natural canopy shade conditions, a light environment enriched in FR, *phyA* mutants fail to survive. This conditional lethality of the *phyA* mutation indicates the essential nature of the FR HIR (Yanovsky *et al* 1995). In addition to defects in the HIR, *phyA* mutants lack all phytochrome responses that reflect operation of the Very Low Fluence Response (VLFR) mode of phytochrome action (e.g. Botto *et al* 1996, Shinomura *et al* 1996). Despite these defects in seedling photomorphogenesis under FR or in FR-rich environments, *phyA* seedlings grown under continuous white light or unfiltered daylight are virtually indistinguishable from the wild type. *Arabidopsis* is a facultative long day plant and *phyA* mutants grown under short photoperiods show reduced sensitivity to both night-breaks and low fluence rate incandescent day extensions (Johnson *et al* 1994, Reed *et al* 1994). This indicates that phyA plays a role in photoperiodic control of flowering time. Although it is not known precisely how phyA exercises this photoperiodic control, phyA has been implicated in the entrainment of the circadian oscillator (Somers *et al* 1999).

In contrast to *phyA*, the phenotype of *phyB* mutants grown under white light is markedly different from the wild type. Thus *phyB* is characteristically elongated, pale and early flowering compare with wild type (Koornneef *et al* 1980, Reed *et al* 1993). The gross phenotypic differences between *phyA* and *phyB* mutant seedlings suggest that these two phytochromes fulfil quite different principal roles. Seeds from *phyB* null mutants show reduced germination, and whilst de-etiolating, *phyB* seedlings demonstrate reduced hypocotyl inhibition and cotyledon expansion in response to brief or prolonged R light (Koornneef *et al* 1980, Reed *et al* 1993, Shinomura *et al* 1994). In the wild type these responses are all R/FR reversible and represent action of phytochrome *via* a Low Fluence Response (LFR) mode.

The elongated, early flowering phenotype of *phyB* is reminiscent of the syndrome of responses shown by wild type seedlings exposed to a low R/FR ratio and collectively known as the shade avoidance response. The constitutive shade avoidance phenotype of the *phyB* mutant is thought to reflect a substantial role for phyB in this respect. Indeed, the severely attenuated response of *phyB* mutants to low R/FR ratio light suggests that this is the case (Whitelam *et al* 1998). Such adaptive growth strategies are often essential to maximise chances of reproductive success, providing a means to both anticipate a potential shade situation and escape from direct shade.

The *phyB* mutant flowers early under both long and short photoperiod, but retains a degree of photoperiodic sensitivity. This constitutive early flowering phenotype of *phyB* suggests that phyB may act as a repressor of flowering in a clock-independent manner. Nevertheless, phyB has been shown to play a role in entrainment of the circadian oscillator under red light conditions (Somers *et al* 1999), suggesting the possibility that phyB may also exert control of flowering time through pathways that do signal *via* the circadian clock.

2.1 phyA and phyB Act Redundantly

A notable feature of the phytochromes and other photoreceptors is the degree of redundancy and cross-talk that exists among them. Phytochromes A and B act redundantly in the control of seedling de-etiolation. For example, under R light, hypocotyls of the *phyAphyB* double mutant are longer than those of the elongated *phyB* mutant (Reed *et al* 1994). This reveals a role for phyA in the inhibition of hypocotyl elongation that is not evident from analysis of the monogenic *phyA* mutant, which responds in wild-type manner to R. Phytochromes A and B both act redundantly in the induction of *CAB* gene expression in response to R treatments (Reed *et al* 1994). Thus, although monogenic *phyA* or *phyB* mutants respond reasonably well to R, the *phyAphyB* double mutant shows a greatly attenuated response. This indicates that in the absence of either phyA or phyB, the remaining phytochrome is capable of mediating these responses to R treatment. Likewise, Devlin *et al* (1996) have shown that the biomass of individual monogenic *phyA* or *phyB* mutant plants is not significantly different from that of wild type plants, yet *phyAphyB* double mutants have a greatly reduced stature with only about 30-40% of the biomass of the wild type. This indicates that the action of either phyA or phyB is sufficient for normal biomass production.

3. THE ROLES OF PHYTOCHROMES D AND E

Analysis of *phyAphyB* double mutants has provided clear evidence that some photoresponses are controlled by other phytochromes. For example, although phyB has been shown to play the major role in shade avoidance responses to low R/FR ratio the *phyB* mutant and the *phyA phyB* double mutant still respond to these light treatments (Halliday *et al* 1994, Devlin *et al* 1996). The *phyA phyB* double mutant responds to low R/FR ratio, and end-of-day FR treatments, by increased elongation growth and early flowering (Devlin *et al* 1996). However, unlike wild type plants that show elongation of petioles, the *phyAphyB* double mutant displays an elongation of internodes in response to low R/FR ratio (Fig. 1), giving the appearance of a caulescent plant. Since the monogenic *phyB* mutant does not respond in this way to low R/FR ratio, this indicates that whilst phyA plays little part in shade avoidance responses, it normally acts in conjunction with phyB to maintain the rosette habit of *Arabidopsis*. The action of phytochromes other than phyA and phyB is also indicated by the observation that seeds from the *phyAphyB* mutant display R-FR reversible promotion of germination (Devlin *et al* 1996, Poppe and Schäfer 1997).

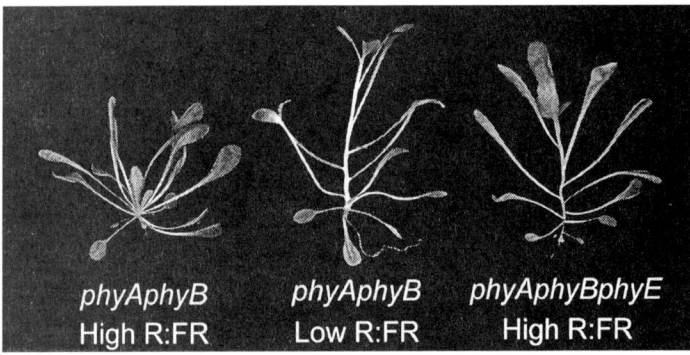

Figure 1. phyAphyB double mutants respond to low R/FR ratio and in doing so phenocopy the *phyAphyBphyE* triple mutant.

The finding that accessions of the Ws ecotype of *Arabidopsis* carry a deletion in the *PHYD* gene provided the first insights into the precise role of phyD (Aukerman *et al* 1997). Following introgression experiments in which a wild type *PHYD* gene has been introduced into Ws, or in which the phyD allele from Ws has been introduced into other ecotypes, it became apparent that monogenic phyD mutants are essentially indistinguishable from the wild type (Devlin *et al* 1999). However, the effects of phyD deficiency are very obvious in a *phyB* background, where *phyBphyD* double mutants have more elongated hypocotyls and petioles and are earlier flowering than *phyB* mutants (Aukerman *et al* 1997, Devlin *et al* 1999). The *phyBphyD* double mutant, when grown under high R/FR ratio light resembles the *phyB* mutant grown under low R/FR ratio conditions. This, coupled with the attenuation of further responses to low R/FR ratio in the *phyBphyD* double mutant, indicates that phyD acts redundantly with phyB in the shade avoidance response. These findings are not entirely unsurprising as PHYB and PHYD are the most related of the apophytochromes with ≈80% amino acid sequence identity.

The *phyAphyBphyD* triple mutant retains rosette habit and a response to low R/FR ratio and to end-of-day FR in terms of flowering time and internode elongation. This indicates the action of phytochromes other than phyA, phyB and phyD in the control of these responses.

From a screen utilising M2 seedlings derived following irradiation-induced mutagenesis of *phyAphyB* double mutants, several early flowering seedlings with elongated internodes were isolated (Fig. 1). One of these proved to define the fourth phytochrome mutant, *phyE* (Devlin *et al* 1998). In fact, the *phyAphyBphyE* triple mutant not only resembles the *phyAphyB* mutant when grown under low R/FR ratio (Fig. 1), it exhibits only the most marginal of responses to low R/FR or end-of-day FR, implicating phyE in the control of these responses. The small residual responses retained in the *phyAphyBphyE* triple mutant reflect the roles of phyD and phyC.

Thus, it appears that in addition to phyB, phyD and phyE and perhaps phyC also mediate changes in growth and reproductive strategy in response to low R/FR. However, although there is some overlap in phyD and phyE function, their roles are not identical. The constitutive elongated internode and early flowering phenotype of *phyAphyBphyE,* coupled with the very severely attenuated responses of this triple mutant to low R/FR ratio, compared with the retained responses seen in the *phyAphyBphyD* mutant point to a more prominent role for phyE in the control of these responses.

4. PHYTOCHROME SIGNAL TRANSDUCTION

The subtle and discrete ways in which different phytochromes act to control photomorphogenesis indicate that there is a complex and highly specific network of signalling pathways. The light signalling pathway is also required to interact with signalling pathways which monitor other environmental factors such as temperature, pathogen attack or physical stresses which may at any one time be exerting either conflicting or complementary pressure on the developmental process. There must, therefore, be a flexible network of signalling pathways that allow the importance of a specific environmental input to be assessed. Considerable effort is now being focussed on the dissection of these signalling pathways using both genetic and transgenic means to identify the key players.

A genetic approach has been successfully applied to identify mutants that act downstream of the phytochromes. Because their obvious phenotypes make mutants more readily recognisable, much of this work has concentrated on identifying signalling mutants for phyA and phyB. Mutants that define genes encoding positive intermediates of photomorphogenesis, such as *HY5*, *FHY1* and *FHY3* (Whitelam *et al* 1998), and those defining genes for repressors of photomorphogenesis, such as the *COP/DET/FUS* group have been identified (Chamovitz and Deng 1996, von Armin and Deng 1996). Some mutants define genes for signalling molecules that act close to the phytochromes and are highly specific to photomorphogenesis, whereas others define genes for components that act downstream of the initial signal perception and which may not be specific for light signal transduction.

4.1 The phyA Signal Transduction Pathway is Branched

Given the overlapping functions of phyA and phyB in seedling photomorphogenesis there is question of whether these two photoreceptors share common signalling pathway components. The recent observation that both phyA and phyB undergo light-induced nuclear translocation (*Kircher et al* 1999) together with the finding that both phyA and phyB interact physically with PIF3, a protein that has the features of a transcriptional regulator (Ni *et al* 1998) and which seems to be involved in mediating seedling responses to R and FR (Halliday *et al* 1999), would seem to indicate the operation of a common pathway. Nevertheless, several mutants have been isolated that affect specifically the action of either phyA or phyB. The first of these were the *fhy1* and *fhy3* mutants (Whitelam *et al* 1993, Barnes *et al* 1996) These mutants, in common with *phyA* display an elongated hypocotyl in FR, but show a wild type phenotype in R. Neither mutant affects the levels of phyA, indicating that their wild type gene products act specifically in the phyA signalling pathway.

Although *fhy1* resembles phyA in having an elongated hypocotyl under FR, it differs from *phyA* in that the promotion of *fhy1* seed germination by FR is unaffected (Johnson *et al* 1994). This infers that FHY1 is necessary for only a discrete subset of phyA-mediated morphological responses. Further analysis of *fhy1* has shown that, whereas the induction of expression of the chalcone synthase (*CHS*) gene by FR, acting through phyA, is significantly reduced in *fhy1*, the induction of expression of the *CAB* gene is hardly affected (Barnes *et al* 1996). These findings are consistent with the notion that the phyA signalling pathway is branched and that FHY1 acts in one of the branches. Interestingly, it had previously been postulated that the phyA signal transduction pathway was branched, and that *CHS* and *CAB* define different branches, as a result of microinjection studies (Barnes *et al* 1997).

Several other mutants, defining genes for proteins that act specifically in a subset of the pathways, have been identified. These include mutants that lead to specific defects in the phyA and phyB pathways (see Whitelam and Devlin 1998). This raises the possibility that the phyA and phyB signalling pathways involve both shared and unique intermediates.

4.2 Interaction Between Light- and Gravity-Sensing Pathways

A clear example of interaction between phytochrome signal transduction pathways and pathways involved transducing other environmental cues is provided by the effect of R and FR on the gravitropic response of *Arabidopsis* seedlings. For seedlings growing in the dark, the orientation of hypocotyls is determined solely by the gravitational vector. This gravitropic default pathway is modulated by light. Irradiation of etiolated seedlings with either R or FR, in the absence of a phototropic stimulus, leads to complete randomisation of hypocotyl growth orientation. A detailed study of this phenomenon with mutants and transgenic plants has revealed that the response is mediated by phyA acting via the VLFR mode, and by phyB acting via the LFR mode (Poppe *et al* 1996). Whilst the physiological function of hypocotyl randomisation by R and FR is not clear, it is possible that it is related to the R enhancement of phototropic curvature.

A recently identified mutant, *gil1* (for gravitropic in light), may be of great value in defining the point of cross-talk between the phytochrome and gravitropic signalling pathways. The hypocotyls of mutant seedlings are not fully randomised by either very low fluences of FR or by continuous high fluence rate FR or R. Identification of the wild type *GIL1* gene suggests that it encodes a novel protein with a predicted size of 54 kDa. The *gil1* mutant has no other obvious phenotypes, suggesting that GIL1 may represent a component that is specific to the interaction between phytochrome and gravity signaling.

5. CONCLUSIONS

Understanding of the components involved in the perception and transduction of regulatory light signals has advanced significantly in recent years, principally through the application of genetic methods. *Arabidopsis* mutants that define four (of the five) apophytochrome genes have been identified and characterised. Although *phyA* and *phyB* play the predominant role in seed germination and seedling de-etiolation, they are involved in the perception of quite discrete light signals. In the established plant, phytochrome A is involved in the detection of daylength, whilst phytochromes B, D and E act redundantly in the perception of R/FR ratio signals.

Several of the components that act downstream of phyA and phyB are also being identified. Mutants that define signalling components that are specific to particular phytochromes and those that define components that are common to both phytochromes have been identified.

REFERENCES

Aukerman, M.J., Hirschfeld, M., Wester, L., Weaver, M., Clack, T., Amasino, R.M., and Sharrock, R.A., 1997, A deletion in the *PHYD* gene of the *Arabidopsis* Wassilewskija ecotype defines a role for phytochrome D in red/far-red light sensing. *Plant Cell* **9**: 1317-1326.

Barnes, S.A., Quaggio, R.B., Whitelam, G.C., and Chua, N-H., 1996, *fhy1* defines a branchpoint in phytochrome A signal transduction pathways for gene expression. *Plant J.* **10**: 1155-1161.

Botto, J.F., Sánchez, R.A., Whitelam, G.C., and Casal, J.J., 1996, Phytochrome A mediates the promotion of seed germination by very low fluences of light and canopy shade-light in *Arabidopsis*. *Plant Physiol.* **110**: 439-444.

Chamovitz, D.A., and Deng, X-W., 1996, Light signalling in plants. *Crit. Rev. Plant Sci.* **15**: 455-478.

Clack, T., Mathews, S., and Sharrock, R.A., 1994, The phytochrome apoprotein family in *Arabidopsis* is encoded by five genes: The sequences and expression of *PHYD* and *PHYE*. *Plant Mol. Biol.* **25**: 413-427.

Devlin, P.F., Halliday, K.J., Harberd, N.P., and Whitelam, G.C., 1996, The rosette habit of *Arabidopsis thaliana* is dependent upon phytochrome action: novel phytochromes control internode elongation and flowering time. *Plant J.* **10**: 1127-1134.

Devlin, P.F., Patel, S., and Whitelam, G. C., 1998, Phytochrome E influences internode elongation and flowering time in *Arabidopsis*. *Plant Cell* **10**: 1479-1488.

Devlin, P.F., Robson, P.R.H., Patel, S., Goosey, L., Sharrock, R.A., and Whitelam, G.C., 1999, Phytochrome D acts in the shade avoidance syndrome in *Arabidopsis thaliana* controlling elongation growth and flowering time. *Plant Physiol.* **119**: 909-915.

Halliday, K.J., Hudson, M., Ni, M., Qin, M.M., and Quail, P.H., 1999, *poc1*: An *Arabidopsis* mutant perturbed in phytochrome signaling because of a T-DNA insertion in the promoter of *PIF3*, a gene encoding a phytochrome-interacting bHLH protein. *Proc. Natl. Acad. Sci. USA* **96**: 5832-5837.

Halliday, K.J., Koornneef, M., and Whitelam, G.C., 1994, Phytochrome B, and at least one other phytochrome, mediate the accelerated flowering response of *Arabidopsis thaliana* L. to low red:far-red ratio. *Plant Physiol.* **104**: 1311-1315.

Johnson, E., Bradley, J.M., Harberd, N.H., and Whitelam, G.C., 1994, Photoresponses of light-grown *phyA* mutants of *Arabidopsis*: phytochrome A is required for the perception of daylength extensions. *Plant Physiol.* **105**: 141-149.

Kircher, S., Kozma-Bognar, L., Kim, L., Adam, E., Harter, K., Schäfer, E., and Nagy, F., 1999, Light quality–dependent nuclear import of the plant photoreceptors phytochrome A and B. *Plant Cell* **11**: 1445–1456.

Koornneef, M., Rolf, E., and Spruit, C.J.P., 1980, Genetic control of light-inhibited hypocotyl elongation in *Arabidopsis thaliana* (L.) Heynh. *Z. Pflanzenphysiol.* **100**: 147-160.

Ni, M., Tepperman, J.M., and Quail, P.H., 1998, PIF3, a phytochrome-interacting factor necessary for normal photoinduced signal transduction, is a novel basic helix-loop-helix protein. *Cell* **95**: 657-667.

Poppe, C., Hangarter, R.P., Sharrock, R.A., Nagy, F., and Schäfer, E., 1996, The light-induced reduction of the gravitropic growth-orientation of seedlings of *Arabidopsis thaliana* (L) Heynh is a photomorphogenic response mediated synergistically by the far-red-absorbing forms of phytochromes A and B. *Planta* **199**: 511-514.

Poppe, C., and Schäfer, E., 1997, Seed germination of *Arabidopsis thaliana phyA/phyB* double mutants is under phytochrome control. *Plant Physiol.* **114**: 1487-1492.

Reed, J.W., Nagatani, A., Elich, T.D., Fagan, M., and Chory, J., 1994 Phytochrome A and phytochrome B have overlapping but distinct functions in *Arabidopsis* development. *Plant Physiol.* **104**: 1139-1149.

Reed, J.W., Nagpal, P., Poole, D.S., Furuya, M., and Chory, J., 1993, Mutations in the gene for the red/far-red light receptor phytochrome B alter cell elongation and physiological responses throughout *Arabidopsis* development. *Plant Cell* **5**: 147-157.

Shinomura, T., Nagatani, A., Chory, J., and Furuya, M., 1994, The induction of seed-germination in *Arabidopsis thaliana* is regulated principally by phytochrome B and secondarily by phytochrome A. *Plant Physiol.* **104**: 363-371.

Shinomura, T., Nagatani, A., Hanzawa, H., Kubota, M., Watanabe, M., and Furuya, M., 1996, Action spectra for phytochrome A-specific and B-specific photoinduction of seed-germination in *Arabidopsis thaliana*. *Proc. Natl. Acad. Sci. USA* **93**: 8129-8133.

Somers, D.E., Devlin, P.F., and. Kay, S.A., 1998, Phytochromes and cryptochromes in the entrainment of the *Arabidopsis* circadian clock. *Science* **282**: 1488-1490.

von Arnim, A., and Deng, X-W., 1996, A role for transcriptional repression during light control of plant development. *BioEssays* **18**: 905-910.

Whitelam, G.C., Johnson, E., Peng, J., Carol, P., Anderson, M.L., Cowl, J., and Harberd, N.P., 1993, Phytochrome A null mutants of *Arabidopsis* display a wild-type phenotype in white light. *Plant Cell* **5**: 757-768.

Whitelam, G.C., Patel, S., and Devlin, P.F., 1998, Phytochromes and photomorphogenesis in *Arabidopsis*. *Phil. Trans. Roy. Soc. Lond. B* **353**: 1445-1453.

Whitelam, G.C., and Devlin, P.F., 1998, Light signalling in *Arabidopsis*. *Plant Physiol. Biochem.* **36**: 125-133.

Yanovsky, M.J., Casal, J.J., and Whitelam, G.C., 1995, Phytochrome A, phytochrome B and HY4 are involved in hypocotyl growth responses to natural radiation in *Arabidopsis*: weak de-etiolation of the *phyA* mutant under dense canopies. *Plant Cell Environ.* **18**: 788-794.

Signal Transduction in Photomorphogenesis: Intracellular Partitioning of Factors and Photoreceptors

EBERHARD SCHÄFER, STEFAN KIRCHER, PATRICIA GIL, KLAUS HARTER, LANA KIM, FRANK WELLMER, LAZLO KOZMA-BOGNAR, EVA ADAM AND FERENC NAGY

Albert-Ludwigs-Universität Freiburg, Institut für Bologie II, Botanik, Schänzlestr. 1, D-79104 Freiburg i.Br., Germany

1. INTRODUCTION

Plants are sessile organisms which have evolved a fascinating capacity to adapt to changes in natural environment. One of the most important variable environmental factors is light. To monitor light quality, light quantity as well as temporal and spatial patterns of light at least three different photoreceptor classes are used: i) the red far-red reversible photoreceptors, phytochromes; ii) the various classes of blue/UVA photoreceptors: cryptochromes and phototropin (Briggs and Huala 1999); iii) UVB photoreceptors which, so far, have only been characterised by action spectroscopy (Wellmann 1983).

In this chapter we will concentrate on the role of phytochromes in the control of photomorphogenesis. In *Arabidopsis* five genes encoding for phytochrome apoproteins (phyA-E) have been identified (Sharrock and Quail 1989, Clack *et al* 1994). Phytochromes are dimers with one open chain tetrapyrole covalently attached to each monomer. The molecular mass of each monomer is about 120 kDa. The unique feature of phytochromes is their photoreversibility. They are synthesized in the red-absorbing form P_r. After absorption of light it is transformed to its physiologically active far-red absorbing form, P_{fr}. Phytochrome in its P_{fr} form will be reverted back to P_r after light absorption. Therefore, monochromatic or polychromatic irradiation will establish a photoequilibrium (P_{fr}/P_r) characteristic for the spectral distribution of the light. This allows phytochromes to be powerful light quantity and light quality detectors, especially in the red/far-red spectral region.

Until recently it was generally believed that phytochromes are localised in the cytoplasm either as soluble proteins or associated with the plasmalemma. Predictions based on the amino acid sequences, biochemical assays and cell fractionation studies indicated a cytosolic localization of phytochromes (cf. Kendrick and Kronenberg 1994). This view was supported by immunocytochemical studies (Speth *et al* 1986 and 1987). Analysis of chloroplast movement in *Mougeotia* and phototropism and polarotropism studies in mosses and ferns showed an action dichroism indicative of a structured organisation of phytochromes either within the plasmalemma or originally attached to it (Wada *et al* 1993).

Because of the well established statement that phytochromes control the transcription rate of many genes, a light-dependent flow of information between cytosol (photoreceptor) and nucleus (genes) must be predicted (Fig. 1). Some probably cytosolic elements of the phytochrome signaling pathway like G-proteins and Ca^{2+}/Calmodulin (Bowler *et al* 1994), or the phytochrome kinase substrate PKS1 (Fankhauser *et al* 1999) were characterised. In this review we are focussing on recent insights in the regulation of the intracellular localization of signal transduction components. In the first part we will describe our progress analysing light-dependent intracellular partitioning of bZIP transcription factors in parsley and in the second part recent advances in the analysis of cytosol/nuclear partitioning of phytochromes.

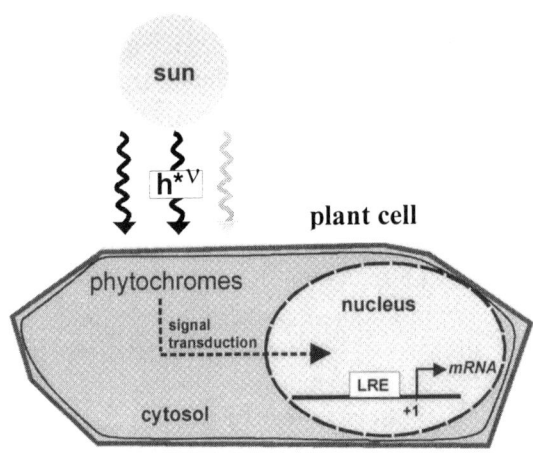

Figure 1. Phytochrome signal transduction includes communication between cytosolic and nuclear compartments.

2. LIGHT DEPENDENT NUCLEAR TRANSPORT OF CPRF2 IN PARSLEY CELLS

In parsley cell cultures, intensive studies demonstrated the light-dependent transcriptional regulation of chalcone synthase, the key enzyme of flavonoid biosynthesis. For its light regulation a promoter element (LRE) consistent of one binding site for bZIP factors and one for myb factors has been shown to be sufficient and necessary for its light regulation (see Feldbrügge *et al* 1997 and references therein). Previous studies showed a rapid red light-dependent protein phosphorylation (Harter *et al* 1994a), a partial cytosolic localisation of bZIP factors and a light-dependent nuclear transport (Harter *et al* 1994b). Recent studies using monospecific antibodies against CPRF2, one member of the bZIP factor family in parsley, demonstrated by indirect immunostaining a primarily cytosolic localisation in dark-adapted cells and a strong and fast stimulation of nuclear transport after red light treatment (Kircher *et al* 1999a). We recently addressed the problem of cytosolic retention of CPRF2 and the possible role of phosphorylation in the control of intracellular distribution. A rapid red-light induced phosphorylation of CPRF2 *in vivo* an *in vitro* could be shown which does not affect the DNA-binding efficiency of the factor (Wellmer *et al* 1999). This is in contrast to CPRF4 where strong phosphorylation dependence of protein/DNA interaction could be shown (Wellmer *et al* unpublished). The phosphorylation site of CPRF2 was identified as a serine residue and was mapped primarily to the C-terminal half, whereas the N-terminal half seems to be responsible for cytosolic retention (Wellmer *et al* 1999, Kircher *et al* 1999a). Further studies using gel filtration and in-gel phosphorylation assays showed that CPRF2 is retarded in the dark in the cytosol in a high molecular weight complex. This complex contains, besides CPRF2, also a CPRF2 specific kinase with a 40 kDa active subunit (Wellmer *et al* 1999).

3. LIGHT CONTROL OF INTRACELLULAR LOCALISATION OF PHYTOCHROMES

As indicated in the introduction, it was generally believed that phytochromes are soluble cytosolic proteins. The observation that in purified oat nuclei run-on transcription could be controlled in a red/far-red reversible manner (Mösinger and Schäfer 1984, Mösinger *et al* 1987) was ignored or believed to be an artefact. These doubts were challenged by the observation of a nuclear localisation of truncated phyB fused to the GUS reporter gene after transient transformation (Sakamato and Nagatani 1996). More recently, Yamaguchi *et al* (1999) and Kircher *et al* (1999b), analysing *Arabidopsis* or tobacco stably transformed with phyB fused to the *in vivo* reporter GFP (green fluorescent protein), demonstrated a light-dependent nuclear transport of this fusion protein. In dark-grown seedlings or dark-adapted cells of light-grown plants phyB-GFP seems to be exclusively cytosolic. Nuclear import and intranuclear spot formation could be induced by continuous red light treatment or sequential red light pulses. The effect of the red light pulses were far-red reversible indicating that nuclear transport of phyB-GFP is regulated by a typical low fluence response of phytochrome, mediated by phyB itself (Kircher *et al* 1999b). Because overexpression of phyB-GFP results in a phyB overexpression phenotype in tobacco and can complement a phyB mutant of *Arabidopsis*, it is assumed that the localisation pattern of the fusion protein reflects that of the endogenous molecule (Kircher *et al* 1999b, Yamaguchi *et al* 1999).

Expression of rice phyA-GFP in tobacco also showed a cytosolic localisation of the fusion protein in darkness. Nuclear localisation and, again, spot formation within the nucleus could even be induced with a far-red light pulse (Kircher *et al* 1999b). This indicates that nuclear localisation of phyA-GFP is a phyA-mediated very low fluence rate response showing, therefore, no red/far-red reversibility. The light-induced nuclear localisation is preceeded by a very rapid cytosolic spot formation reminiscent of the previously described light-induced SAP formation (Speth *et al* 1986, 1987). Nuclear localisation of phyA-GFP can also be induced by continuous far-red and blue light (Kircher *et al* 1999b, Kim *et al* unpublished). Thus, light-dependent nuclear localisation of rice phyA-GFP reflects aspects of phyA-dependent photomorphogenetic responses in monocotyledonous plants.

Recent studies about *Arabidopsis* phyA-GFP expressed in a phyA mutant of *Arabidopsis* demonstrated that the mutant phenotype could be rescued. The light-dependent nuclear import reflects in this case the light regulations of dicotyledonous phyA, i.e. a very low fluence response and a far-red HIR (Kim *et al* unpublished).

4. CONCLUSIONS

Signal transduction in photomorphogenesis includes as an important step light dependent translocation of phytochromes into the nucleus. Spectral dependences and irradiance dependence of this step is different for phyA and phyB. Thus, light dependent nuclear transport allows already at this initial step a qualitative differentiation between *phyA* and *phyB* mediated responses. Within the nucleus both phytochromes seem to share some common interaction partners (i. e. PIF 3, Ni *et al* 1999) but probably again specific mediators like FAR1 and SPA1 (see Fig. 2, Hudson *et al* 1999, Hoecker *et al* 1999).

The role of cytosolic phytochrome remains obscure. Besides its function to control its own nuclear transport, light-mediated control of transcription factors like CPRF2 seems to be an important function. Other responses like chloroplast orientation, microtubule orientation *etc* can probably also be mediated by cytosolic phytochrome but detailed analysis of both cytosolic and nuclear transduction events remains to be done.

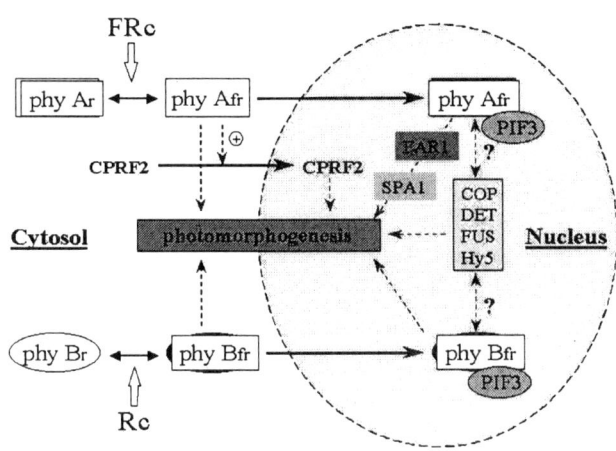

Figure 2. Within the nucleus both phytochromes seem to share some common interaction partners (i.e. PIF 3, Ni *et al* 1999) but probably again specific mediators like FAR1 and SPA1.

ACKNOWLEDGEMENT

The work was supported by grants of DFG (SFB388) to E.S. and K.H.; Volkswagen Foundation and HFSPO to F.N. and E.S. and by OMFB (1998/E132) to F.N.

REFERENCES

Bowler, C., Neuhaus, G., Yamagata, H., and Chua, N.-H., 1994, Cyclic GMP and calcium mediate phytochrome phototransduction. *Cell* **77**: 73-81.
Briggs, W.R., and Huala, E., 1999, Blue-light photoreceptors in higher plants. *Annu. Rev. Cell Dev. Biol.* **15**: 33-62.

Clack, T., Mathews, S., and Sharrock, R.A., 1994, The phytochrome apoprotein family in *Arabidopsis* is encoded by five genes – the sequences and expression of *PHYD* and *PHYE*. *Plant Mol. Biol.* **25**: 413-427.

Fankhauser, C., and Chory, J., 1997, Light control of plant development. *Annu. Rev. Cell Dev. Biol.* **13**: 203-229.

Fankhauser, C., Yeh, K.-C., Lagarias, J.C., Zhang, H., Elich, T.D., and Chory, J., 1999, PKS1, a substrate phosphorylated by phytochrome that modulates light signaling in *Arabidopsis*. *Science* **284**: 1539-1541.

Feldbrügge, M., Sprenger, M., Hahlbrock, K., and Weisshaar, B., 1997, PcMYB1, a novel plant protein containing a DNA-binding domain with one MYB repeat, interacts *in vivo* with a light-regulatory promoter unit. *Plant J.* **11**: 1079-1093.

Harter, K., Frohnmeyer, H., Kircher, S., Kunkel, T., Mühlbauer, S., and Schäfer, E., 1994a, Light induces rapid changes of the phosphorylation pattern in the cytosol of evacuolated parsley protoplasts. *Proc. Natl. Acad. Sci. USA* **91**: 5038-5042.

Harter, K., Kircher, S., Frohnmeyer, H., Krenz, M., Nagy, F., and Schäfer, E., 1994b, Light-regulated modification and nuclear translocation of cytoplasmic G-box binding factors (GBFs) in parsley (*Petroselinum crispum* L.). *Plant Cell* **6**: 545-559.

Hoecker, U., Tepperman, J.M., and Quail, P.H., 1999, SPA1, a WD-repeat protein specific to phytochrome A signal transduction. *Science* **284**: 496-499.

Hudson, M., Ringli, C., Boylan, M.T., and Quail, P.H., 1999, The *FAR1* locus encodes a novel nuclear protein specific to phytochrome A signaling. *Genes Dev.* **13**: 2017-2027.

Kendrick, R. E. and Kronenberg, G. H. M., eds., 1994, Photomorphogenesis in Plants. Dordrecht, The Netherlands: Kluwer Academic Publishers.

Kircher, S., Wellmer, F., Nick, P., Rügner, A., Schäfer, E., and Harter, K., 1999a, Nuclear import of the parsley bZIP transcription factor CPRF2 is regulated by phytochrome photoreceptors. *J. Cell Biol.* **144**: 201-211.

Kircher, S., Kozma-Bognar, L., Kim, L., Adam, E., Harter, K., Schäfer, E., and Nagy, F., 1999b, Light quality-dependant nuclear import of the plant photoreceptors phytochromes A and B. *Plant Cell* **11**: 1445-1456.

Mösinger, E., and Schäfer, E., 1984, In vivo phytochrome control of in vitro transcription rates in isolated nuclei from oat seedlings. *Planta* **161**: 444-450.

Mösinger, E., Batschauer, A., Viestra, R., Apel, K., and Schäfer, E., 1987, Comparison of the effects of exogenous native phytochrome and in vivo irradiation on in vitro transcription in isolated nuclei from barley (*Hordeum vulgare*). *Planta* **170**: 505-514.

Ni, M., Tepperman, J.M., and Quail, P.H., 1999, Binding of phytochrome B to its nuclear signalling partner PIF3 is reversibly induced by light. *Nature* **400**: 781-784.

Sakamoto, K., and Nagatani, A., 1996, Nuclear localization activity of phytochrome B. *Plant J.* **10**: 859-868.

Sharrock, R.A., and Quail, P.H., 1989, Novel phytochrome sequences in *Arabidopsis thaliana*: structure, evolution, and differential expression of a plant regulatory photoreceptor family. *Genes Dev.* **3**: 1745-57.

Speth, V., Otto, V., and Schäfer, E., 1986, Intracellular localization of phytochrome in oat coleoptiles by electron microscopy. *Planta* **168**: 299-304.

Speth, V., Otto, V., and Schäfer, E., 1987, Intracellular localization of phytochrome and ubiquitin in red light-irradiated oat coleoptiles by electron microscopy. *Planta* **171**: 332-338.

Wada, M., Grolig, F., and Haupt, W., 1993, Light-orientated chloroplast positioning: contribution to progress in photobiology. *J. Photochem. Photobiol. B Biol.* **17**: 3-25.

Wellmer, F., Kircher, S., Rügner, A., Frohnmeyer, H., Schäfer, E., and Harter, K., 1999, Phosphorylation of the parsley bZIP transcription factor CPRF2 is regulated by light. *J. Biol. Chem.* **274**: 29476-29482.

Yamaguchi, R., Nakamura, M., Mochizuki, N., Kay, S.A., and Nagatani, A., 1999, Light-dependent translocation of a phytochrome B-GFP fusion protein to the nucleus in transgenic *Arabidopsis*. *J. Cell Biol.* **3**: 437-445.

Molecular Genetic Analysis of Constitutively Photomorphogenic Mutants of *Arabidopsis*

JITENDRA P. KHURANA, AKHILESH K. TYAGI, PARAMJIT KHURANA, ANJU KOCHHAR, PRADEEP K. JAIN, ANIRUDDHA RAYCHAUDHURI, REKHA CHAWLA, ARVIND K. BHARTI, ASHVERYA LAXMI, AND UJJAINI DASGUPTA
Department of Plant Molecular Biology, University of Delhi South Campus, New Delhi-110 021, India

1. INTRODUCTION

Light is one of the most important environmental signals that profoundly influence growth and development in plants all through their life cycle. Plants have thus evolved many diverse sensory photoreceptors that scan spectral regions of the solar spectrum essential for regulating photomorphogenic development. The family of red/far-red reversible phytochromes and the blue light-sensing cryptochromes and phototropin principally absorb radiation in the visible range; the latter class also perceives biologically active UV-A radiation (*see* Quail *et al* 1995, Cashmore *et al* 1999, Christie *et al* 1999, Khurana and Poff 1999). Some morphogenic responses, e.g. cotyledon curling and anthocyanin production, are also elicited by UV-B radiation but the identity of the receptors(s) remains elusive (*see* Stapleton 1992, Bharti and Khurana 1997). Among these sensory pigment systems, phytochromes are the best studied class primarily because they regulate important plant processes like onset of seed germination, inhibition of stem growth, cotyledon expansion and chloroplast development, and induction of flowering. Moreover, phytochrome was the first photosensory pigment identified in the 1950s (*see* Sage 1992 for a historical account). The phytochrome apoprotein is encoded by a multigene family in most species examined, but all of them harbor an open-chain tetrapyrrole chromophore towards the N-terminal region. In *Arabidopsis*, which represents one of the smallest genomes amongst angiosperms, the phytochrome apoprotein is encoded by five divergent genes (*PHYA* to *PHYE*).

The primary mechanism of action of these sensory photoreceptors has been the subject of intensive investigations (*see* Khurana 1999). Earlier studies on the overexpression of deletion constructs of phytochromes in transgenics defined that the signal output domain, essential for biological activity, resides largely in the C-terminal region (*see* Quail *et al* 1995). Recent studies on the molecular analyses of *Synechocystis* and *Avena* phytochromes have revealed that the higher plant phytochromes belong to a class of serine/threonine kinases with histidine kinase ancestry (Yeh *et al* 1997, Yeh and Lagarias 1998, *see also* Maheshwari *et al* 1999). The yeast two-hybrid screens have identified a negative regulator of phytochrome function, PKS1 (*Phytochrome Kinase Substrate*), that binds to phytochromes and gets phosphorylated in a phytochrome-dependent manner (Fankhauser *et al* 1999). Phytochrome has also been shown to regulate changes in gene expression and the analysis of the upstream promoter sequences has defined several *cis*-acting elements and *trans*-acting proteins that confer light responsiveness (*see* Terzaghi and Cashmore 1995). Studies with agonists and antagonists of classical signalling molecules have provided evidence that the phytochrome-regulated responses, including changes in gene expression, are mediated through G-proteins, cGMP, Ca^{2+}/calmodulin and phosphatidylinositols (*see* Sopory and Munshi 1998, Grover *et al* 1998 1999). However, the scenario has changed rapidly with the demonstration that phytochromes translocate to the nucleus upon photoactivation (Kircher *et al* 1999, Yamaguchi *et al* 1999). The yeast two-hybrid screens have identified more *p*hytochrome *i*nteracting *f*actors, such as PIF3 (Ni *et al* 1998), and the molecular analysis of a *far*-red impaired response mutant has identified a protein FAR1 (Hudson *et al* 1999), which are nuclear localized and may act as transcription factors.

Genetic screens have also been very powerful to dissect the phytochrome signal transduction pathway (*see* Khurana *et al* 1998, Whitelam and Devlin 1998). Essentially, two types of mutations have been analyzed. One group of mutations results in reduced responsiveness of the seedlings to light. With the exception of HY5 (which is nuclear localized), these strains represent mutations either in the phytochromes *phyA* and *phyB*, or their specific downstream signalling component. The other group includes mainly the mutants that display constitutive photomorphogenesis in dark and has resulted in identification of novel proteins of COP/DET/FUS class, most of which are nuclear localized and act as negative regulators of photomorphogenesis in dark (*see* von Arnim and Deng 1996). Some of the less pleiotropic *COP/DET/FUS* loci encode products involved in regulating either the cellular levels of auxins, cytokinins and brassinosteroids, or their activity, and have been implicated in light-hormone interaction (*see* Khurana *et al* 1998, Khurana and Chawla 1998).

As a part of a larger effort, our group has been carrying out investigations to understand the mechanisms of photosensory perception and signal transduction, using *Arabidopsis* as a model system. Several mutants defective in phototropic response to unilateral blue light, anthocyanin production, or phytochrome-mediated photomorphogenic responses have been isolated and are being characterized. The work on phototropism mutants in fact began in the laboratory of Professor Kenneth L. Poff at the Michigan State University (Khurana and Poff 1989, Khurana *et al* 1989, Konjevic *et al* 1992) and is being pursued here. At least four *transparent testa* (*tt*) mutants defective in anthocyanin production, both in the seed and the seedlings, have been characterized at the molecular level (Bharti and Khurana 1999). In this article, however, the focus is on another class of *Arabidopsis* mutants that display constitutive photomorphogenesis in dark. They have been designated *pho*, for their *p*lumular *h*ook *o*pen phenotype in dark. Initial results indicate that some of these mutants define novel regulatory components that act downstream to phytochromes and play an important role in photomorphogenic development. Some of the salient findings of our work are discussed in this chapter.

2. ISOLATION OF MUTANTS AND GENETIC ANALYSIS

The young wild-type *Arabidopsis* seedlings, when grown in dark, display characteristic features of an etiolated dicot seedling, i.e. a long hypocotyl with a typical apical hook and unexpanded cotyledons. An exposure to light causes dramatic changes in seedling morphology and the hypocotyl growth is arrested, plumular hook opens completely and cotyledons expand to develop photosynthetic competence. To obtain mutants defective in photomorphogenic responses, 3-d-old dark-grown seedlings derived from mutagenized (EMS or fast-neutron) M_2 seed population were examined. Several variable phenotypic aberrations were recorded (Fig. 1). Some seedlings displayed a completely open plumular hook, with complete or partially expanded cotyledons, or a partially open hook, and others had a *ctr* phenotype, i.e. they exhibited an exaggerated constitutive triple response, characteristic of ethylene responses, even in the absence of exogenous ethylene. Some of these *p*lumular *h*ook *o*pen (*pho*) mutants have been characterized in more detail.

Conventional genetic analysis revealed that all the *pho* mutants represent single recessive mutations of nuclear origin. Complementation analysis allocated them to five different groups: *pho1* to *pho5*. Further complementation tests, with already known mutants with essentially a similar phenotype, revealed that *pho1* is allelic to *cop3/hls1* and *pho3* is allelic to *cop2/amp1*, whereas others define novel loci involved in photomorphogenesis (for more details, *see* Khurana *et al* 1996 and 1998). Despite their allelic nature, the *pho* mutants exhibit certain unique features not described earlier for *cop/hls/amp* mutants.

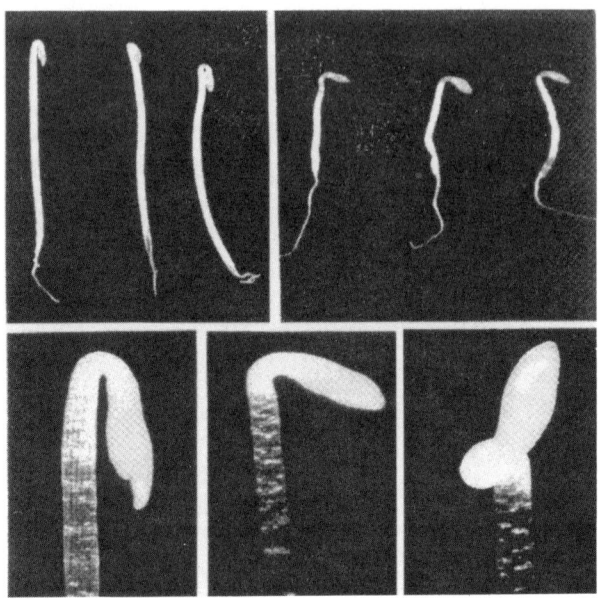

Figure 1. Phenotype of the young 3-d-old, dark-grown seedlings of wild-type *Arabidopsis* and some mutants. The upper panel shows the wild-type (left) and *pho5* mutant (right) seedlings, with tight apical hook and partially open hook, respectively. The lower panel shows a magnified view (from left to right) of the apical portion of the wild-type, *pho5* and *ctr* mutant seedlings.

3. PHYSIOLOGICAL CHARACTERIZATION OF *pho* MUTANTS

As stated above, *pho* mutants show aberrant phenotype at the young seedling stage. The seedlings of mutants *pho1, pho2* and *pho3* (EMS-mutagenized) display a completely open hook and expanded cotyledons with long petioles, whereas *pho4* and *pho5* mutant (fast-neutron bombarded) seedlings have partially open hooks. To find out how pleiotropic is the effect of *pho* mutants, various other growth and developmental processes that are specifically phytochrome regulated were examined in detail.

3.1 Seed Germination without Photoactivation of Phytochrome

The hallmark of typical phytochrome responses, i.e. the red/far-red reversibility, was first reported for seed germination in *Lactuca sativa* (Borthwick *et al* 1952). Since then many other photoblastic seeds, including *Arabidopsis*, have shown their dependence on phytochrome (*see* Smith 1975, Shinomura *et al* 1996). The germination in wild-type *Arabidopsis* seeds is promoted substantially by 1 min red light irradiation and its effect is reversed completely by a subsequent far-red (2 min) irradiation. Although germination behaviour of *pho4* and *pho5* was comparable to wild-type, the seeds of *pho1*, *pho2* and *pho3* mutants exhibited more than 90% germination in total darkness and were completely insensitive to red or far-red light. This is in contrast to studies on other *cop/det/fus* mutants, some of which exert more pleiotropic effects than do *pho* mutants, but still retain light requirement for the onset of seed germination (Deng *et al* 1991), except *det1* (Chory *et al* 1989); however, the dark germination response of *det1* is conditional (Shinomura *et al* 1994).

3.2 Leaf Differentiation and Flowering in Dark

Although the 3-d-old dark-grown seedlings displaying *pho* phenotype had normal seedling height, the hypocotyl growth was arrested completely in *pho1*, *pho2* and *pho3* after 4 or 5 days, as if they have perceived some light signal, whereas the wild-type hypocotyls continued to elongate, in dark. However, in *cop3* and *cop2*, which are allelic to *pho1* and *pho3*, respectively, the hypocotyl growth behaviour was essentially similar to that of wild-type (Hou *et al* 1993). The hypocotyl growth was retarded considerably in *pho5* but it was affected only partially in *pho4*. Besides inhibition of hypocotyl growth, primary leaf primordia were visible in seedlings of *pho1*, *pho2* and *pho3* grown in dark for more than 6 days. Prolonged growth in dark not only allowed further leaf differentiation but even bolting and flowering between 30 to 50 days of growth. In some instances, a silique-like structure was also visible but it was too minute to analyze anatomically. These dark-differentiated leaves were unusual in that they did not form a typical rosette-like structure, analogous to that of light-grown *Arabidopsis*. This was because of the increased internode length, a feature reminiscent of the elongated internodes visible in *phyB* and *phyE* mutants of *Arabidopsis* grown in light (Devlin *et al* 1998). In comparison to other *pho* mutants, although *pho5* mutant developed a rosette-like structure, it did not bolt and, in strain *pho4*, only one of the alleles permitted differentiation of a few leaves in dark.

The cotyledons and the dark-differentiated leaves of *pho1* and *pho2* were subjected to transmission electron microscopy. The wild-type plastid appeared spherical in shape and the paracrystalline structure (prolamellar body) was quite distinct. In contrast, a large percentage of plastids of *pho1* and *pho2* were elliptical in shape, resembling more the differentiated chloroplast of light-grown leaves. However, a greatly reduced prolamellar body with parallel strands of unstacked membranes radiating out of it were clearly visible. This suggests that the leaf differentiation is probably not coupled tightly with plastid differentiation.

3.3 Precocious Flowering

The pleiotropic changes observed in the dark-grown phenotype of *pho* mutant seedlings necessitated a close examination of their light-grown morphology. Phenotypically, the adult *pho1* and *pho2* plants appear normal, like wild-type, but *pho3*, *pho4* and *pho5* have considerably altered morphology. All these mutants have reduced fertility. The stature of *pho3* plants is tall but that of *pho5* is quite short, whereas *pho4* plants are semi-dwarf and have other abnormalities too, including decreased apical dominance, plageotropic branching and twisted siliques as well as delayed flowering. Probably the most striking feature of *pho1*, *pho2* and *pho3* mutants is their precocious flowering behaviour, which may be attributed to their altered developmental programme and not due to the altered photoperiodic sensitivity. This is because the number of leaves in a rosette of *pho* mutants at the time of flowering is not significantly different from that in the wild-type. The *pho* mutants share this early-flowering trait with PHYB mutant *hy3*, and *amp1* and *hls1* mutants, allelic to *pho3* and *pho1*, respectively (Chaudhury *et al* 1993, Reed *et al* 1993, Lehman *et al* 1996). This investigation adds at least one more, i.e. *pho2*, to this category.

4. MOLECULAR CHARACTERIZATION OF *pho* MUTANTS

4.1 De-repression of Gene Expression in Dark

During de-etiolation process induced by light through phytochrome, several genes are either up- or down-regulated. The genes encoding polypeptides of the photosynthetic apparatus, i.e. *CAB* and *RBCS*, have been extensively studied in this regard. It was, thus, imperative to find out if the transcript abundance of the light-regulated genes is affected in some way in these constitutively photomorphogenic *pho* mutants.

4.1.1 Photosynthesis-related genes

The steady-state transcript levels of several light-responsive genes, both of nuclear (e.g. *CAB, RBCS*) and plastidic (e.g. *PSBA, RBCL*) origin, were monitored by northern analysis in the seedlings grown in dark or light from 2 to 7 days. The expression pattern of the genes examined, in at least dark-grown *pho1* and *pho2* seedlings, was development-dependent and, like wild-type, the maximal transcript levels were recorded between 3 to 4 days. However, as compared to wild-type, the transcript levels of all these genes were relatively higher in both the mutants, on any given day, and this increase was more distinct in early stages of seedling development. This de-repression of gene expression in dark was also observed for two other genes encoding polypeptides of the oxygen-evolving complex, i.e. *PSBO* and *PSBP* (Kochhar *et al* 1996, Jain *et al* 1998).

Whether these gene expression changes are also reflected at the protein level was checked by western analysis using specific antibodies to D1 (product of *PSBA* gene), RBCL, cyt-b_{559}, and 33 and 16 kDa polypeptides (products of *PSBO* and *PSBQ* genes, respectively). Although the changes observed at the transcript levels of some of these genes and their representative polypeptides did not correspond with each other completely, at least the 33 and 16 kDa polypeptides could be detected in *pho1* and *pho2* seedlings grown in dark, but not in the wild-type.

4.1.2 Genes of terpenoid biosynthesis pathway

Terpenoids probably constitute the largest class of natural plant products and play diverse functional roles in plants (McGarvey and Croteau 1995). The growth regulators, gibberellins and abscisic acid, and the photosynthetic pigments, carotenoids and phytol, also belong to this class. To find whether the *pho* mutants may have affected the terpenoid biosynthetic pathway, thereby altering the endogenous status of these important hormones and pigments which is also controlled by light, transcript levels of some of the genes encoding enzymes of the terpenoid pathway were analyzed. These include *GGPS, PDS* and *PSY*, encoding geranylgeranyl pyrophosphate synthase, phytoene desaturase and phytoene synthase, respectively. Since these genes express at quite low levels, RT-PCR analysis was done using gene specific primers. The levels of *GGPS* transcript were low in dark and were not affected by light in 3-d-old wild-type seedlings, whereas those of *PSY* and *PDS* were up-regulated; this is in contrast to an earlier report (Lintig *et al* 1997) where light was shown to up-regulate the expression of only *PSY*, and not of *PDS*. However, the transcript levels of *GGPS, PSY* and *PDS* in all the mutants that differentiate to flowering stage, i.e. *pho1, pho2* and *pho3*, were much higher. This suggests that the *PHO* gene products play a diverse role in regulating metabolism in plants, including changes in the level of hormones and photosynthetic pigments.

4.2 Up-regulation of Auxin-responsive Genes

The detailed analysis of *pho* mutants performed in this investigation clearly shows that there may be some aberration in hormone responses too and that light does interact with hormones for photomorphogenic development. Pursuing this observation further, the expression of two of the auxin-responsive genes, *SAUR-AC1* and *TCH*4 (Gil *et al* 1994, Xu *et al* 1995), was monitored in the mutant *pho5* which displays certain traits of auxin overproduction or oversensitivity. For this purpose, *GUS* reporter gene fused with the promoter of *SAUR-AC1* and *TCH*4, and expressed in transgenic *Arabidopsis,* was mobilized into *pho5* background by conventional breeding method. The F2 segregating seedlings, displaying a mutant phenotype in dark, were allowed to grow further in light and tested for tissue distribution of GUS activity. Microscopic examination showed that although the *GUS* expression driven by *SAUR-AC1* or *TCH*4 was not ectopic and the distribution pattern in the intact seedling leaves, hypocotyl and roots was fairly comparable to wild-type, the GUS activity was substantially higher in the *pho5* mutant plants. Whether this increased responsiveness of *SAUR* and *TCH*4 promoters in *pho5* mutant background is because of oversensitivity to auxin or its overproduction, remains to be investigated.

4.3 Cloning and Sequencing of *PHO1/HLS1* Gene and Regulation of its Expression

The *HOOKLESS1* (*HLS1*) gene was cloned based upon the analysis of the mutant *hls1* (Lehman *et al* 1996) which we found to be allelic to *pho1*. Using the primers from the coding region of *HLS1* gene, PCR amplification was done with the genomic DNA isolated from *pho1* mutant seedlings and its wild-type parent, ecotype Estland of *A. thaliana*; *hls1* mutation is in Columbia ecotype background. The PCR amplified products were cloned and processed for sequencing. The analysis of the DNA sequences not only revealed the occurrence of ecotype-specific changes but also a C to T transition at 962 position (Fig. 2), thus creating a stop codon. This may result in the formation of a truncated product of *HLS1* gene in *pho1/hls1* mutant, thus leading to abortion of its function.

```
           F   R   T   P   S   I   L   V   N   P   V   Y   A   H   R   V   N   V   S   R
902 TTTCGTACACCGTCGATTTTGGTTAACCCGGTTTACGCTCATCGAGTTAATGTTTCGCGG

       *   V   T   V   I   K   L   E   P   V   D   A   E   T   L   Y   R   I   R   F
962 TGAGTCACGGTTATCAAGTTAGAGCCGGTTGATGCTGAGACGTTGTACCGAATCCGGTTT

        S   T   T   E   F   F   P   R   D   I   D   S   V   L   N   N   K   L   S   L
1022 AGCACAACAGAGTTTTTCCCGCGGGATATTGATTCGGTACTTAATAACAAACTCTCGCTT
```

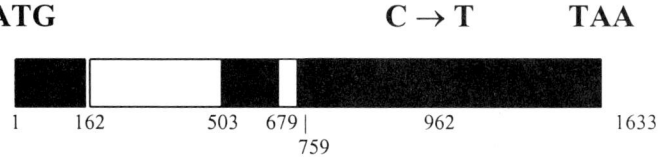

Figure 2. Diagrammatic sketch of the *HLS1/PHO1* genomic clone (1633 bp) and partial nucleotide/amino acid sequence showing the site of lesion (C→T) in the *pho1* mutant of *Arabidopsis thaliana*, ecotype Estland. Black and open boxes show exons and introns, respectively.

Northern analysis using *HLS1/PHO1* as probe confirmed that the expression of this gene is down-regulated by light (Lehman *et al* 1996). However, contrary to our expectation, its transcript levels, as determined by RT-PCR, were not affected significantly in the dark-grown seedlings of the constitutively photomorphogenic mutants, *pho1, pho2* and *pho3*. The light signal transduction pathway operative in regulating the expression of *HLS1/PHO1* is probably different from that involved in photosynthesis-related or terpenoid metabolic pathway genes.

5. CONCLUSION

The pleiotropic effects displayed by the constitutively photomorphogenic mutants described here clearly suggest that the products of the *PHO* genes play significant role in the photomorphogenic development of young seedlings as well as the adult plants. Although *pho1* and *pho3* have been found to be allelic to *cop3/hls1* and *cop2/amp1*, respectively, some of the traits described are unique and could simply be ascribed to the genetic background of the ecotype (Estland) used in this investigation. The uniqueness of *pho1, pho2* and *pho3* lies in their dark germination, differentiation followed by flowering in dark, de-repression of terpenoid pathway genes, and precocious flowering in adult light-grown plants, and that too collectively in individual mutant backgrounds. The occurrence of early-flowering trait in three of these *pho* mutants is quite exciting and their molecular analysis may enrich the already growing list of components that regulate flowering time (*see* Pieiro and Coupland 1998).

Although the young seedlings of *pho4* and *pho5* show partial de-etiolation in dark, the adult plants display many features characteristic of auxin response mutants (Leyser *et al* 1996, Kim *et al* 1998). However, the complementation tests have not detected any allelism between *pho4* and *pho5* and other known mutants. A detailed analysis of these at the physiological as well as molecular level is in progress and should prove useful in illuminating the interaction of light and hormones in photomorphogenic development.

Coming to the functional significance, the recessive nature of all the *pho* mutants described here implies that they represent negative regulators of photomorphogenesis. Close phenotypic similarity between *pho1*, *pho2* and *pho3* indicates that their wild-type gene products interact closely, whereas products of *pho4* and *pho5* may act further downstream or in an independent branched pathway. Although details have not been given in this text, we have evidence to support that *PHO* gene products act downstream to both PHYA and PHYB forms of phytochrome. Comparing the degree of pleiotropy in the mutant phenotypes, it is conceivable that early signalling components like COP1 and DET1 also act upstream of PHOs. As of now, it is difficult to ascribe a precise role to *PHO* gene products and more input is obviously required to decipher their molecular nature; but it is apparent that they represent light signalling components critical for regulating photomorphogenic development.

ACKNOWLEDGEMENTS

The work in our laboratory was financially supported by the Department of Biotechnology of the Government of India, and the University Grants Commission, New Delhi. AL and UD thank the CSIR for providing Research Fellowship. The antibodies specific to photosynthetic apparatus polypeptides were generously provided by Dr Autar K. Mattoo, USDA, Beltsville, USA.

REFERENCES

Bharti, A.K., and Khurana, J.P., 1997, Mutants of *Arabidopsis* as tools to understand the regulation of phenylpropanoid pathway and UVB protection mechanisms. *Photochem. Photobiol.* **65**: 765-776.

Bharti, A.K., and Khurana, J.P., 1999, Characterization of *transparent testa* (*tt*) mutants of *Arabidopsis* impaired in flavonoid pathway. In *Abstracts, Int. Symp. Plant Signal*

Transduction, International Centre for Genetic Engineering and Biotechnology, New Delhi, Abstract No.2, p. 90.

Borthwick, H.A., Hendricks, S.B., Parker, M.W., Toole, E.H., and Toole, V.K., 1952, A reversible photoreaction controlling seed germination. *Proc. Natl. Acad. Sci. USA* **38**: 662-666.

Cashmore, A.R., Jarillo, J.A., Wu, Y.-J., and Liu, D., 1999, Cryptochromes: Blue light receptors for plants and animals. *Science* **284**: 760-765.

Chaudhury, A.M., Letham, S., Craig, S., and Dennis, E.S., 1993, *amp1*-a mutant with high cytokinin levels and altered embryonic pattern, faster vegetative growth, constitutive photomorphogenesis and precocious flowering. *Plant J.* **4**: 907-916.

Chory, J., Peto, C., Feinbaum, R., Pratt, L., and Ausubel, F., 1989, *Arabidopsis thaliana* mutant that develops as a light-grown plant in the absence of light. *Cell* **58**: 991-999.

Christie, J.M., Salomon, M., Nozue, K., Wada, M., and Briggs, W.R., 1999, LOV (light, oxygen, or voltage) domains of the blue-light photoreceptor phototropin (nph1): binding sites for the chromophore flavin mononucleotide. *Proc. Natl. Acad. Sci. USA* **96**: 8779-8783.

Deng, X.-W., Caspar, T., and Quail, P.H., 1991, *cop1*: a regulatory locus involved in light-controlled development and gene expression in *Arabidopsis*. *Genes Dev.* **5**: 1172-1182.

Devlin, P. F., Patel, S. R., and Whitelam, G. C., 1998, Phytochrome E influences internode elongation and flowering time in *Arabidopsis*. *Plant Cell* **10**: 1479-1487.

Fankhauser, C., Yeh, K.-C., Lagarias, J.C., Zhang, H., Elich, T.D., and Chory, J., 1999, PKS1, a substrate phosphorylated by phytochrome that modulates light signalling in *Arabidopsis*. *Science* **284**: 1539-1541.

Gil, P., Liu, Y., Orbovic, V., Verkamp, E., Poff, K.L., and Green, P.J., 1994, Characterization of auxin inducible *SAUR-AC1* gene for use as a molecular genetic tool in *Arabidopsis*. *Plant Physiol.* **104**: 777-784.

Grover, M., Dhingra, A., Sharma A.K., Maheshwari, S.C., and Tyagi, A.K., 1999, Involvement of phytochrome(s), Ca^{2+} and phosphorylation in light-dependent control of transcript levels for plastid genes (*psbA, psaA* and *rbcL*) in rice (*Oryza sativa*). *Physiol. Plant.* **105**: 701-707.

Grover, M., Sharma A.K., Dhingra, A., Maheshwari, S.C., and Tyagi, A.K., 1998, Regulation of plastid gene expression in rice involves calcium and protein phosphatase/kinases for signal transduction. *Plant Sci.* **137**: 185-190.

Hou, Y., von Arnim, A.G., and Deng, X.-W., 1993, A new class of *Arabidopsis* constitutive photomorphogenic genes involved in regulating cotyledon development. *Plant Cell* **5**: 329-339.

Hudson, M., Ringli, C., Boylan, M. T., and Quail, P.H., 1999, The *FAR1* locus encodes a novel nuclear protein specific to phytochrome A signalling. *Genes Dev.* **13**: 2017-2027.

Jain, P. K., Kochhar, A., Khurana, J. P., and Tyagi, A. K., 1998, The *psbO* gene for 33-kDa precursor polypeptide of the oxygen-evolving complex in *Arabidopsis thaliana* and control of its expression. *DNA Res.* **5**: 221-228.

Khurana, J.P., 1999, Light signal transduction in plants: The emerging trends. *Ind. Photobiol. Soc. Newslett.* **38**: 28-32.

Khurana, J.P., and Chawla, R., 1998, Molecular genetic analysis of integration of light and hormonal signals in plant development. In *Plant Tissue Culture and Molecular Biology – Applications and Prospects* (P.S. Srivastava, ed.), Narosa Publishing House, New Delhi, pp. 642-669.

Khurana, J.P., Kochhar, A., and Jain, P.K., 1996, Genetic and molecular analysis of light-regulated plant development. *Genetica* **97**: 349-361.

Khurana, J.P., Kochhar, A., and Tyagi, A.K., 1998, Photosensory perception and signal transduction in higher plants-molecular genetic analysis. *Crit. Rev. Plant Sci.* **17**: 465-539.

Khurana, J.P., and Poff, K.L. 1989, Mutants of *Arabidopsis thaliana* with altered phototropism. *Planta* **178**: 400-406.

Khurana, J.P., and Poff, K.L., 1999, Blue light perception and signal transduction in higher plants. In *Concepts in Photobiology: Photosynthesis and Photomorphogenesis* (G.S. Singhal, G. Renger, S.K. Sopory, K.D. Irrgang and Govindjee, eds.), Narosa Publishing House, New Delhi, pp. 796-820.

Khurana, J.P., Ren, Z., Steinitz, B., Parks, B., Best, T., and Poff, K.L., 1989, Mutants of *Arabidopsis thaliana* with decreased amplitude in their phototropic response. *Plant Physiol.* **91**: 685-689.

Kim, B.C., Soh, M.S., Hong, S.H., Furuya, M., and Nam, H.G., 1998, Photomorphogenic development of the *Arabidopsis shy2-1D* mutation and its interaction with phytochromes in darkness. *Plant J.* **15**: 61-68.

Kircher, S., Kozma-Bognar, L., Kim, L., Adam, E., Harter, K., Schäfer, E., and Nagy, F., 1999, Light quality-dependent nuclear import of the plant photoreceptors phytochrome A and B. *Plant Cell* **11**: 1445-1456.

Kochhar, A., Khurana, J. P., and Tyagi, A.K., 1996, Nucleotide sequence of the *psb*P gene encoding precursor of 23-kDa polypeptide of oxygen-evolving complex in *Arabidopsis thaliana* and its expression in the wild-type and a constitutively photomorphogenic mutant. *DNA Res.* **3**: 277-285.

Konjevic, R., Khurana, J.P., and Poff, K.L., 1992, Analysis of multiple photoreceptor pigments for phototropism in a mutant of *Arabidopsis thaliana*. *Photochem. Photobiol.* **55**: 789-792.

Lehman, A., Black, R., and Ecker, J.R., 1996, *Hookless*1, an ethylene response gene, is required for differential cell elongation in the *Arabidopsis* hypocotyl. *Cell* **85**: 183-194.

Leyser, H.M.O., Pickett, F.B., Dharmasiri, S., and Estelle, M., 1996, Mutations in the *AXR3* gene of *Arabidopsis* result in altered auxin response including ectopic expression from the *SAUR-AC1* promoter. *Plant J.* **10**: 403-413.

Lintig, J.V., Welsch, R., Bonk, M., Giuliano, G., Batschauer, A., and Kleinig, H., 1997, Light-dependent regulation of carotenoid biosynthesis occurs at the level of phytoene synthase expression and is mediated by phytochrome in *Sinapis alba* and *Arabidopsis thaliana* seedlings. *Plant J.* **12**: 625-634.

Maheshwari, S.C., Khurana, J.P., and Sopory, S.K., 1999, Novel light-activated protein kinases as key regulators of plant growth and development. *J. Biosci.* **24**: 499-514.

McGarvey, D.J., and Croteau, R., 1995, Terpenoid metabolism. *Plant Cell* **7**: 1015-1026.

Ni, M., Tepperman, J. M., and Quail P. H., 1998, PIF3, a phytochrome-interacting factor necessary for normal photoinduced signal transduction, is a novel basic helix-loop-helix protein. *Cell* **95**: 657-667.

Pieiro, M., and Coupland, G., 1998, The control of flowering time and floral identity in *Arabidopsis*. *Plant Physiol.* **117**: 1-8.

Quail, P. H., Boylan, M. T., Parks, B. M., Short, T. W., Xu, Y., and Wagner, D., 1995, Phytochromes: photosensory perception and signal transduction. *Science* **268**: 675-680.

Reed, J.W., Nagpal, P., Poole, D.S., Furuya, M., and Chory, J., 1993, Mutations in the gene for the red/far-red light receptor phytochrome B alter cell elongation and physiological responses throughout *Arabidopsis* development. *Plant Cell* **5**: 147-157.

Sage, L.C., 1992, *Pigment of the Imagination. A History of Phytochrome Research*. Academic Press Inc., San Diego, California.

Shinomura, T., Nagatani, A., Chory, J., and Furuya, M., 1994, The induction of seed germination in *Arabidopsis thaliana* is regulated principally by phytochrome B and secondarily by phytochrome A. *Plant Physiol.* **104**: 363-371.

Smith, H., 1975, *Phytochrome and photomorphogenesis*. Mcgraw Hill, London.

Shinomura, T., Nagatani, A., Hanzawa, H., Kubota, M., Watanabe, M., and Furuya, M., 1996, Action spectra for phytochrome A- and B-specific photoinduction of seed germination in *Arabidopsis thaliana*. *Proc. Natl. Acad. Sci. USA* **93**: 8129-8133.

Sopory, S.K., and Munshi, M., 1998, Protein kinases and phosphatases and their role in signalling in plants. *Crit. Rev. Plant Sci.* **17**: 245-318.

Stapleton, A.E., 1992, Ultraviolet radiation and plants: burning questions. *Plant Cell* **4**: 1353-1358.

Terzaghi, W. B., and Cashmore, A.R., 1995, Light regulated transcription. *Annu. Rev. Plant Physiol. Plant Mol. Biol.* **46**: 445-474.

von Arnim, A.G., and Deng, X.-W., 1996, Light control of seedling development. *Annu. Rev. Plant Physiol. Plant Mol. Biol.* **47**: 215-243.

Whitelam, G.C., and Devlin, P.F., 1998, Light signalling in plants. *Plant Physiol. Biochem.* **36**: 125-133.

Xu, W., Purugganan, M.M., Polisensky, D.H., Antosiewicz, D.M., Fry, S.C., and Bramm, J., 1995, *Arabidopsis TCH4*, regulated by hormones and the environment, encodes a xyloglucan endotransglycosylate. *Plant Cell* **7**: 1555-1567.

Yamaguchi, R., Nakamura, M., Mochizuki, N., Kay, S. A., and Nagatani, A., 1999, Light-dependent translocation of a phytochrome B-GFP fusion protein to the nucleus in transgenic *Arabidopsis*. *J. Cell Biol.* **145**: 437-445.

Yeh, K.-C., and Lagarias, J. C., 1998, Eukaryotic phytochromes: Light-regulated serine/threonine protein kinases with histidine kinase ancestry. *Proc. Natl. Acad. Sci. USA* **95**: 13976-13981.

Yeh, K.-C., Wu, S., Murphy, J.T., and Lagarias, J.C., 1997, A cyanobacterial phytochrome two-component light sensory system. *Science* **277**: 1505-1508.

Cytosolic pH as a Secondary Messenger During Light Activation of Phosphoenolpyruvate Carboxylase in Mesophyll Cells of C_4 Plants

CH. BHASKAR RAO, NASSER SYED, JHADESWAR MURMU, AND
A. S. RAGHAVENDRA
Department of Plant Sciences, School of Life Sciences, University of Hyderabad, Hyderabad 500 046, India

1. INTRODUCTION

The primary carbon fixation in C_4 and CAM plants is catalyzed by phosphoenolpyruvate carboxylase (PEPC, EC 4.1.1.31). With a pH optimum of about 8.0, the activity of PEPC is modulated markedly by pH. The enzyme, PEPC, is localized in cytosol of mesophyll cells higher plants. Obviously the cytosolic pH would be a very important factor in determining the activity of PEPC.

Being a key enzyme of C_4 photosynthesis or CAM, the regulation of PEPC has been of considerable interest. PEPC is feed-back inhibited by L-malate and is activated by glucose-6-phosphate (Glc-6-P) (Andreo *et al* 1987, Rajagopalan *et al* 1994, Chollet *et al* 1996, Vidal and Chollet 1997). The sensitivity of PEPC to malate is further influenced by various factors like light or pH. Malate inhibition is competitive at pH 7.0, and non-competitive at pH 8.0 and Glc-6-P protects the enzyme from malate inhibition. On illumination, the concentration of malate in mesophyll cells can rise up to 30 mM and can lead to a strong feedback inhibition of PEPC. However, when leaves are illuminated, there is a marked decrease in sensitivity of PEPC to malate besides an increase in activity. The light activation of PEPC is therefore considered to be an adaptive feature to sustain enzyme activity in presence of high malate concentrations.

Signal Transduction in Plants: Current Advances.
Edited by Sopory *et al.*, Kluwer Academic/Plenum Publishers, 2001.

The present article attempts to review the observations on the light induced increase in the activity of PEPC and relate them to changes in pH of cytosol in mesophyll cells. We propose that the cytosolic pH could be a secondary messenger during such light activation of PEPC, particularly in leaves of C_4 plants.

Readers interested in the properties and regulation of PEPC in leaves of C_4 and CAM plants are referred to several recent reviews (Andreo et al 1987, Rajagopalan et al 1994, Chollet et al 1996, Vidal and Chollet 1997, Lepiniec et al 1994, Toh et al 1994, Raghavendra et al 1998, Nimmo 2000).

2. MARKED LIGHT ACTIVATION OF PEPC IN LEAVES OF C_4 PLANTS

A common feature of PEPC in leaves of C_3 and C_4 plants is that the light-form exhibits less sensitivity to malate inhibition than the dark form (Huber and Sugiyama 1986, Doncaster and Leegood 1987). The activity of PEPC is enhanced 2-4 fold on illumination in leaves of C_4 plants, compared to much smaller and generally <50% stimulation in case of C_3 plants (Rajagopalan et al 1993, Gupta et al 1994). Both kinetic and regulatory properties of C_4-PEPC in leaves are modulated markedly by light/dark transitions in vivo (Andreo et al 1987, Rajagopalan et al 1994, Chollet et al 1996, Vidal and Chollet 1997).

Light-activation is a feature of key photosynthetic enzymes in C_3 plants (Buchanan 1992). Therefore, the light activation of PEPC is not surprising. However, two aspects of such light activation are unique in case of PEPC. The first is that the extent of light activation is much more pronounced in leaves of C_4 plants than that in C_3 or CAM plants (Tab. 1). The second is that the enzyme is located in the cytoplasm and not the chloroplast. It has even been suggested that the light activation pattern can be used as a criterion to establish the C_4 form of PEPC (Rajagopalan et al 1993). The reasons for such pronounced activation in leaves of C_4 plants are not clear. PEPC falls into the category of at least two more cytosolic enzymes: nitrate reductase and sucrose phosphate synthase, which are light activated. A common factor among these three cytosolic and light-regulated enzymes is that the enzyme undergoes phosphorylation-dephosphorylation cascade. PEPC is phosphorylated at a serine residue near its N-terminal by a PEPC protein-serine kinase and is dephosphorylated by a PEPC-protein phosphatase. Further, the phosphorylation of enzyme in vitro by mammalian protein kinase mimics the changes caused by illumination, namely the increase in enzyme activity and decrease in the malate sensitivity (Jiao and Chollet 1991, Huber et al 1994, Rajagopalan et al 1994, Chollet et al 1996).

Table 1. The extent of light-induced changes in two important properties of PEPC from a range of C_3 and C_4 plants.

Photosynthetic type/species	Increase in PEPC activity	Decrease in malate sensitivity	Reference
	(*fold*)		
C_4 species			
Alternanthera pungens	2.4	4.0	Rajagopalan *et al* 1993
Sorghum bicolor	2.2	5.7	Gupta *et al* 1994
Zea mays	2.0	2.1	Jiao and Chollet 1988
C_3 species			
Alternanthera sessilis	1.6	1.0	Rajagopalan *et al* 1993
Triticum aestivum	1.1	1.2	Gupta *et al* 1994
Flaveria pringlei	1.2	1.1	Gupta *et al* 1994

3. ILLUMINATION RESULTS IN ALKALIZATION OF CYTOSOL IN MESOPHYLL CELLS

The activity of PEPC is maximal at alkaline pH. Since PEPC is located in cytosol of mesophyll cells, the cytosolic pH would be an important factor during light activation of PEPC. That the illumination induces marked cytosolic alkalization in mesophyll cells of C_4 plants is shown by two lines of evidence: one involving the use of pH-sensitive fluorescent dyes and the other by determining the pH of cell sap extracted from leaves (Tab. 2).

Table 2. Light-induced cytosolic alkalization in either cell sap or cells of leaves from a range of C_3 and C_4 plants.

System	Rise in cytosolic pH on illumination
C_4 species	
Alternanthera pungens	0.19
Amaranthus hypochondriacus	0.19
Gomphrena procumbens	0.18
Zea mays	0.45
Amaranthus caudatus	0.34
C_3 species	
Alternanthera sessilis	0.07
Pisum sativum	0.06
Pelargonium zonale	0.01
Vicia faba	0.02
Spinacia oleracea	0.03

The data were collected from the reports of Rajagopalan *et al* (1993) and Yin *et al* (1993).

Information on cytosolic and vacuolar pH can be obtained by monitoring changes in the fluorescence of pH-indicating dyes which were introduced into the leaves (Yin *et al* 1990). Pyranine, a trivalent sulfonate anion (pK 7.3), records pH changes in slightly alkaline leaf compartments such as cytosol. The observed fluorescence changes in pyranine are attributable to mostly cytosolic emission. In contrast, 5-carboxy-2',7'-dichlorofluorescein (CDCF) has a much lower pK (4.8) than that of pyranine. Fluorescence microscopy has shown that CDCF fluorescence originates from vacuoles.

Light dependent alkalization was observed in the cytosol of mesophyll cells of C_3 plants, whereas vacuolar pH decreased. The phenomena were dependent on the irradiance and chloroplast activity. The light-dependent increase in pyranine fluorescence (reflecting cytosolic alkalization) was several times greater in C_4 plants than that in C_3 species (Raghavendra *et al* 1993, Yin *et al* 1993). The kinetics of the fluorescence changes were also remarkably different in C_3 and C_4 plants. In contrast to pyranine, the fluorescence of CDCF did not increase but rather decreased on illumination, indicating acidification of vacuole of leaf cells. The pattern of the acidification was similar in C_4 and C_3 leaves.

There is a biochemical basis for the cytosolic alkalization on illumination, presumably perceived by chloroplasts. The high degree of proton influx into the thylakoid lumen on illumination leads to marked alkalization of the stroma and subsequently the cytoplasm. The uptake of 3-PGA into chloroplasts during C_4 photosynthesis may lead to further elevation of cytosolic pH in mesophyll of C_4 leaves (Yin *et al* 1993).

Cytosol-enriched cell sap can be prepared from leaf discs, kept in either darkness or light, using 0.3 M sorbitol. Illumination caused a measurable increase in pH of cytosol-enriched cell sap prepared from the leaves of *Alternanthera pungens*, a C_4 species (Rajagopalan *et al* 1998). The pH of cytosol-enriched cell sap increased during illumination and decreased after transfer to darkness. The extent of light-dependent alkalization of cytosol-enriched cell sap was three-fold higher in leaves of the four C_4 species evaluated than in the four C_3 plants. The degree of cell sap alkalization in C_3-C_4 intermediates was similar to that of C_3 plants. Separate experiments with isolated protoplasts of maize (Devi and Raghavendra 1992) or sorghum (Pierre *et al* 1992) indicated that the pH is important during the light activation of PEPC and in the reduction in sensitivity to malate (due to phosphorylation of the enzyme).

4. CORRELATION BETWEEN THE pH OF CYTOSOL AND PEPC ACTIVITY

A correlation appears to exist between the patterns of change in cytosolic pH and PEPC activity in leaves particularly of C_4 plants. Such relationship is suggested from experiments involving measurements of pH of cell sap in leaves and acid/base-modulation of cytosolic pH in leaves as well as mesophyll protoplasts.

4.1 Light Induced Increase in pH of Cell Sap and Decrease on Transfer to Darkness or after Extraction

The PEPC activity, as well as the pH of the cytosol-enriched cell sap, increased on a similar course during illumination and decreased during subsequent darkness. The course of deactivation of PEPC and acidification of the cell sap in darkness were strikingly similar.

The slow pace of light activation/dark deactivation of PEPC in C_4 species is strikingly similar to the slow but steep rate of light-dependent cytosolic alkalization/dark-acidification in leaves of C_4 plants, compared to the quick but small cytosolic pH changes in C_3 plants (Yin *et al* 1990, 1993, Raghavendra *et al* 1993).

4.2 Acid-Base Modulation

Acid/base-loading has been used as a strategy for studying cytosolic pH regulation in plants (Kurkdjian and Guern 1989). Weak lipophilic acids (e.g. acetic, propionic, butyric or benzoic or salicylic acid) penetrate the cells and acidify the cytosol. PEPC activity was stimulated when the cytosol was alkalized and decreased upon acidification of the cytosol due to base- or acid-loading of leaf discs, respectively (Rajagopalan *et al* 1998).

When leaf discs of a C_4 species, *Alternanthera pungens* or *Amaranthus hypochondriacus* were preincubated in 7.5 mM NH_4Cl, the pH of cell sap increased by nearly 0.3 unit, while the activity of PEPC about doubled compared to the cell sap from control leaf discs (Tab. 3). The sensitivity of PEPC to L-malate (a feedback inhibitor) decreased marginally due to cytosolic alkalization. The pH of cell sap and PEPC activity decreased by nearly 0.4 unit and 50%, respectively, when leaf discs were incubated in weak organic acids (propionic, butyric or salicylic acid). These results demonstrate a marked modulation *in vivo* of cell sap pH and PEPC activity in leaf discs from C_4 plants by external alkalizing/acidifying agents.

The acid/base-effects on PEPC activity in leaf discs were reversible *in vivo*. For example when the base-loading leaf discs were transferred to medium containing weak acid (5 mM propionic or butyric acid in 5 mM acetate buffer, pH 4.5), PEPC activity and the pH of the cell sap returned to the levels in control samples (Rajagopalan *et al* 1998).

The addition of NH_4Cl or methylamine to an illuminated suspension of mesophyll cell protoplasts from *Digitaria sanguinalis* was required for the *in situ* stimulation of PEPC kinase activity and PEPC phosphorylation (Pierre *et al* 1992). These two weak bases permeate cells in their neutral form and therefore tend to increase pH of cytosol. Corresponding increases in phosphorylation of PEPC occurred along with decrease in malate sensitivity particularly on exposure to light and the presence of weak bases. The final *in situ* level of PEPC-kinase activity and the apparent phosphorylation state of its target protein depend on the cytosolic pH value of illuminated mesophyll cell protoplasts.

Table 3. Modulation of cell sap pH and/or PEPC activity by NH_4Cl in leaves of C_3 and C_4 plants.

Plant Species	Increase in	
	Cell sap pH	PEPC activity
	Over control	*% of control*
C_4 species		
Alternanthera pungens	0.26	215
Amaranthus hypochondriacus	0.29	264
Gomphrena globosa	0.26	204
Zea mays	0.32	170
Digitaria sanguinalis (Protoplasts)	0.90	--
C_3 species		
Alternanthera sessilis	0.08	103
Pisum sativum	0.09	105
Lycopersicon esculentum	0.11	119
Tagetes erecta	0.09	114
Tridax procumbens	0.11	114

The data were adapted from the reports of Rajagopalan *et al* (1998), Giglioli-Guivarc'h *et al* (1996) as well as the Ph.D thesis of Rajagopalan (1997).

Examination of protoplasts preloaded with the specific pH fluorescent probe 2',7'-bis-(2-carboxyethyl)-5(and 6)carboxyfluorescein acetoxymethyl ester (BCECF-AM) demonstrated that these two compounds rapidly and efficiently alkalinized pH of cytosol. In these experiments, an increase of ~ 0.6 to 1 pH unit in the cytosol was required to obtain a high induction of PEPC kinase activity. Whether such a large variation in cytosolic pH occurs in mesophyll cells of an illuminated intact C_4 leaf remains to be determined (Giglioli-Guivarc'h et al 1996).

5. CONSEQUENCES OF RISE IN CYTOSOLIC pH LEADING TO MODULATION OF PEPC ACTIVITY

The cytosolic pH is obviously an important factor in regulating the light-induced increase in PEPC activity. The importance of cytosolic pH has been shown during studies on mesophyll protoplasts of maize (Devi et al 1992), sorghum (Pierre et al 1992) and *Digitaria sanguinalis* (Giglioli-Guivarc'h et al 1996) as well as leaf discs of several C_4 and C_3 plants (Rajagopalan et al 1993, 1998).

The mechanism of regulation of PEPC by cytosolic pH is not completely elucidated. However, several possible explanations can be given. Since PEPC responds to alkaline pH the enzyme may be activated directly by any rise in pH. The change in pH may modulate PEPC directly or PEPC-protein kinase or both. Further, changes in cytosolic pH lead to marked changes in the levels of free calcium, which in turn can modulate the status of PEPC-phosphorylation. Cytosolic alkalization leads to an increase in free Ca^{2+} levels, in several animal systems. Similar situation may occur in plant cells. These aspects are further discussed below.

A major regulation of C_4 PEPC in the mesophyll cell cytosol involves opposing photosynthesis related metabolite effectors, that is, Glc-6-P (positive) and L-malate (negative) (Doncaster and Leegood 1987) and a light dependent, reversible phosphorylation process (Jiao and Chollet 1991). The latter post-translational mechanism targets a single serine residue of the holoenzyme subunit in a highly conserved N-terminal domain shared by all plant PEPCs sequenced to date (Lepiniec et al 1994). Within the C_4 leaf, phosphorylation of PEPC is completed ~1 hr after the onset of illumination and causes gradual changes in the enzyme's functional properties, namely, an increase in catalytic activity and apparent affinity for Glc-6-P and a decrease in end-product inhibition by L-malate (as determined at suboptimal but near-physiological conditions of pH and PEP concentration).

C_4 plants maintain soluble Ca^{2+} at low levels in their leaves and are called as calciophobes. It is however not clear if low levels of Ca^{2+} is a prerequisite for the smooth functioning of C_4 pathway (Gavalas and Manetas 1980). Therefore the effects of Ca^{2+} on the activity of PEPC and pattern of PEPC photophosphorylation in leaf extracts were examined and are of great interest.

PEPC is phosphorylated in leaf extracts by an endogenous protein kinase and this process can further be stimulated by the addition of ATP to the medium. The involvement of calcium was suggested by the stimulation by

Ca^{2+} and marked inhibition by EGTA or PEPC-PK activity. However, Ca^{2+} promoted PEPC-PK activity, but was not essential, since PEPC-PK activity was seen even in its absence (Parvathi *et al* 2000). Our observations on the regulation of PEPC activity by Ca^{2+} are important in view of the conflicting reports on the role of Ca^{2+} in the regulation of PEPC-PK activity.

Cytosolic Ca^{2+} acts as a secondary messenger in a variety of physiological responses. PEPC is also localized in the cytosol of mesophyll cells. The interaction of Ca^{2+} and PEPC within the cytosol is therefore also a topic of interest. It has been recently shown that cytosolic Ca^{2+} can act as a relay in the light signal transduction pathway of C_4 PEPC (Giglioli-Guivarc'h *et al* 1996). The changes in the intracellular levels of Ca^{2+} are perceived and the signal is transduced downstream by other secondary messengers to elicit the final response. Protein kinases are among such secondary messengers of signal transduction.

6. A MODEL OF REGULATION OF PEPC IN LEAVES OF C4 PLANTS

On illumination, chloroplasts tend to cause marked alkalization of their own stroma as well as surrounding cytoplasm. The extent of cytoplasmic alkalization in mesophyll chloroplasts of C_4 plants appears to be both due to the photochemical activity as well as the uptake of PGA from cytoplasm. The alkalization of cytoplasm leads to an increase in free calcium of cytosol. Such rise in calcium may stimulate the Ca^{2+}-dependent PEPC-protein kinase (PEPC-PK) or another upstream calcium (or camodulin) dependent protein kinase (CDPK). Thus, the phosphorylation of PEPC catalyzed by PEPC-PK may be modulated by calcium directly or through CDPK indirectly (Fig. 1).

As an end result the phosphorylated form of PEPC is more active and less sensitive to both inhibitors (e.g. L-malate) or activators (e.g. Glc-6-P) than the dephosphorylated form. In light, PEPC in C_4 leaves is predominantly in phosphorylated form while in darkness being in a dephosphorylated form.

We therefore propose that the cytoplasmic pH can be an important secondary messenger (as is the case of free Ca^{2+}) during light activation of PEPC in mesophyll cells of C_4 leaves.

ACKNOWLEDGMENTS

Work in our laboratory and preparation of this manuscript is supported by grants from Council of Scientific and Industrial Research, New Delhi [38(0949)/99/EMR-II].

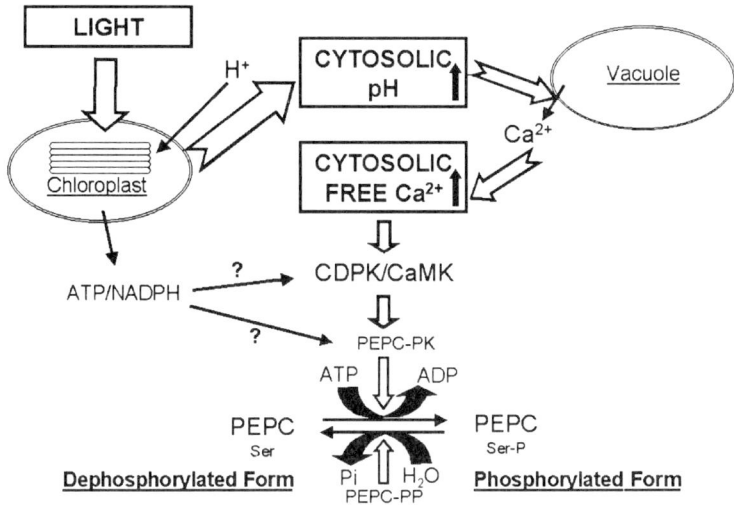

Figure 1. The mediation of light activation of PEPC in mesophyll cells of C_4 plants by cytosolic pH as well as cytosolic Ca^{2+}. The model is described in the text. The steps where no evidences are available are indicated by question-marks. The phosphorylated form of PEPC is more active and less sensitive to malate and GLc-6-P than the dephosphorylated form.

REFERENCES

Andreo, C.S., Gonzalez, D.H., and Iglesias, A.A., 1987, Higher plant phosphoenolpyruvate carboxylase. Structure and regulation. *FEBS Lett.* **213**: 1-8.

Buchanan, B.B., 1992, Carbon dioxide assimilation in oxygenic and anoxygenic photosynthesis. *Photosynth. Res.* **33**: 147-162.

Chollet, R., Vidal, J., and O'Leary, M.H., 1996, Phosphoenolpyruvate carboxylase: A ubiquitous, highly regulated enzyme in plants. *Annu. Rev. Plant Physiol. Plant Mol. Biol.* **47**: 273-298.

Devi, M.T., and Raghavendra, A.S., 1992, Light activation of phosphoenolpyruvate carboxylase in maize mesophyll protoplasts. *J. Plant Physiol.* **138**: 435-439.

Devi, M.T., Rajagopalan, A.V., Raghavendra, A.S., 1992, Structure, regulation and biocsynthesis of phosphoenolpyruvate carboxylase from C_4 plants. *J. Plant Biochem. Biotechnol.* **1**: 73-80.

Doncaster, H.D., and Leegood, R.C., 1987, Regulation of phosphoenolpyruvate carboxylase activity in maize leaves. *Plant Physiol.* **84**: 82-94.

Gavalas, N.A., and Manetas, Y., 1980, Calcium inhibition of phosphoenolpyruvate carboxylase: Possible physiological consequences for C_4 photosynthesis. *Z. Pflanzenphysiol.* **100**: 179-184

Giglioli-Guivarc'h, N., Pierre, J.N., Brown, S., Chollet, R., Vidal, J., and Gadal, P., 1996, The light dependent transduction pathway controlling the regulatory phosphoenolpyruvate carboxylase in protoplasts from *Digitaria sanguinalis*. *Plant Cell* **8**: 573-586.

Gupta, S.K., Ku, M.S.B., Lin, J-H., Zhang, D., and Edwards, G.E., 1994, Light/dark modulation of phophosenolpyruvate carboxylase in C_3 and C_4 species. *Photosynth. Res.* **42**: 133-143.

Huber, S.C., and Sugiyama, T., 1986, Changes in sensitivity to effectors of maize leaf phosphoenolpyruvate carboxylase during light/dark transition. *Plant Physiol.* **81**: 674-677.

Huber, S.C., Huber, J.L., McMichael, R.W. Jr., 1994, Control of plant enzyme activity by reversible protein phosphorylation. *Int. Rev. Cytol.* **149**: 47-98.

Jiao, J.A., and Chollet, R., 1988, Light/dark regulation of maize leaf phosphoenolpyruvate carboxylase by in vivo phosphorylation. *Arch. Biochem. Biophys.* **261**: 409-417.

Jiao, J.A., and Chollet, R., 1991, Posttranslational regulation of phosphoenolpyruvate carboxylase in C_4 and Crassulacean acid metabolism plants. *Plant Physiol.* **95**: 981-985.

Kurkdjian, A., and Guern, J., 1989, Intracellular pH: Measurement and importance in cell activity. *Annu. Rev. Plant Physiol. Plant Mol. Biol.* **40**: 271-303.

Lepiniec, L., Vidal, J., Chollet, R., Gadal, P., and Crétin, C., 1994, Phosphoenolpyruvate carboxylase: structure, regulation and evolution. *Plant Sci.* **99**: 111-124.

Nimmo, H.G., 2000, The regulation of phosphoenolpyruvate carboxylase in CAM plants. *Trends Plant Sci.* **5**: 75-80.

Parvathi, K., Gayathri, J., Gururaj, G., Bhagwat, A.S., and Raghavendra, A.S., 2000, Modulation of phosphoenolpyruvate carboxylase phosphorylation in leaves of *Amaranthus hypochondriacus*, a NAD-ME type of C_4 plant. *Photosynthetica* **38**: In press.

Pierre, J.N., Pacquit, V., Vidal, J., and Gadal, P., 1992, Regulatory phosphorylation of phosphoenolpyruvate in protoplasts from sorghum mesophyll cells and the role of pH and Ca^{2+} as possible components of the light-transduction pathway. *Eur. J. Biochem.* **210**: 531-538.

Raghavendra, A.S., Parvathi, K., and Gayathri, J., 1998, Modulation by calcium of PEP carboxylase and PEPC-protein kinase from leaves of *Amaranthus hypochondriacus*, an NAD-ME type C_4 plant. Kluwer Academic Publishers, Netherlands.

Raghavendra, A.S., Yin, Z.-H., and Heber, U., 1993, Light-dependent pH changes in leaves of C_4 plants. Comparison of the pH response to carbon dioxide and oxygen with that of C_3 plants. *Planta* **189**: 278-287.

Rajagopalan, A.V., 1997, PhD Thesis. Light activation of phosphoenolpyruvate carboxylase in leaf discs of C4 plant species in relation to alkalinization of cell sap. University of Hyderabad, India.

Rajagopalan, A.V., Gayathri, J., and Raghavendra, A.S., 1998, Modulation by weak bases or weak acids of the pH of cell sap and phosphoenolpyruvate carboxylase activity in leaf discs of C_4 plants. *Physiol. Plant.* **104**: 456-462.

Rajagopalan, A.V., Tirumala Devi, M., and Raghavendra, A.S., 1993, Patterns of phosphoenolpyruvate carboxylase activity and cytosolic pH during light activation and dark deactivation in C_3 and C_4 plants. *Photosynth. Res.* **38**: 51-60.

Rajagopalan, A.V., Tirumala Devi, M., and Raghavendra, A.S., 1994, Molecular biology of C_4 phosphoenolpyruvate carboxylase: Structure and regulation and genetic engineering. *Photosynth. Res.* **39**: 115-135.

Toh, H., Kawamura, T., and Izui, K., 1994, Molecular evolution of phosphoenolpyruvate carboxylase. *Plant Cell Environ.* **17**: 31-43.

Vidal, J., and Chollet, R., 1997, Regulatory phosphorylation of C_4 PEP carboxylase. *Trends Plant Sci.* **2**: 230-237.

Yin, Z.-H., Heber, U., and Raghavendra, A.S., 1993, Light-independent pH changes in leaves of C_4 plants. Comparison of cytosolic alkalization and vacuolar acidification with that of C_3 plants. *Planta* **189**: 267-277.

Yin, Z.-H., Neimanis, S., Wagner, U., and Heber, U., 1990, Light-dependent pH changes in leaves of C_3 plants. I. Recording pH changes in various cellular compartments by fluorescent probes. *Planta* **182**: 244-252.

A Role for Ectoapyrases and Extracellular ATP in Plant Growth and Development

STANLEY J. ROUX, COLLIN THOMAS, AND ASHA RAJAGOPAL
Institute for Cellular and Molecular Biology, Univ. of Texas at Austin, Austin, TX 78713 USA

1. INTRODUCTION

For those studying signal transduction, familiarity with the enzyme apyrase is likely to be restricted to the context of its use for scavenging unwanted ATP from a solution. Apyrases hydrolyze both the γ and β phosphate on ATP and the terminal phosphate on ADP, and they do so very efficiently. Assays that require the removal of ATP from solutions typically utilize apyrase to achieve this goal.

Apyrases first began to be discussed in relation to signal transduction when it was learned that extracellular ATP (xATP) is a signaling molecule in animals, and that there are purinergic receptors that transduce the xATP signal into calcium cascades and other changes in metabolism inside of cells (Zhang *et al* 1991). Apyrases would be relevant to the signaling role of xATP because most apyrases are ectoapyrases,: i.e., they are anchored in the plasma membrane with their active site facing out into the extracellular matrix (ECM), consequently they would be critical for the regulation of [xATP]. It is in this context that ectoapyrases have been most intensely studied: they are the best characterized enzymes for hyrolyzing xATP and thus turning off the effects of this inducer. The vast majority of functional studies on ectoapyrases have been done in animals systems, but, interestingly, the first apyrase gene was characterized in plants (Hsieh *et al* 1996). Is it possible that xATP and ectoapyrases have functions in plants? The purpose of this review is to summarize the initial studies on this question.

1.1 EXTRACELLULAR ATP: OCCURRENCE AND FUNCTION

The idea that ATP might exist outside of cells and might be physiologically functional is not a new one. The discovery that free ATP in the ocean is used for growth by marine organisms occurred over twenty years ago (Azam and Hodson 1977). Single-celled organisms actively efflux ATP and adenylate nucleotides, and *E. coli* secretes more than 99% of its cAMP, a net 9% of the total adenylate energy charge (Matin and Matin 1980). Yeast secrete significant amounts of ATP into the extracellular space in a glucose- and cAMP-dependent fashion, an efflux thought to be related to extracellular signaling or possibly to cellular detoxification (Boyum and Guidotti 1997).

How does ATP get out of cells? It is, of course, released when cells undergo autolysis, and it is also released when secretory vesicles fuse with the plasma membrane and empty their contents into the extracellular matrix (ECM) of cells. A third method by which it is moved out of cells is through specialized transport proteins. Work in animal systems has shown that MDR1, the primary ABC protein involved in the multi-drug resistance response of tumor cells, promotes the electrogenic efflux of ATP out of transformed cells at a rate of four million molecules per second (Abraham *et al* 1993). A similar efflux of ATP has also been reported through the CFTR channel (Reisin *et al* 1994), although there is contention over whether this is a direct or indirect function of the channel (Schweibert 1999). The release of ATP to the extracellular matrix in a regulated manner, mediated directly or indirectly by ABC transporters, is thought to be a way of regulating the intracellular adenylate pool involved in signaling. Homologs of the human MDR1 gene have been found in plants (Dudler and Hertig 1992), and have been shown to be involved in light-regulated hypocotyl elongation (Sidler *et al* 1998).

Given the multiple routes for ATP to be delivered to the outside of cells, it is not surprising that plants encounter ATP in the soil environment (Fig. 1). Thomas *et al* (1999) documented that soils inoculated with a stock soil flora derived from a single field sample showed a significant accumulation of ATP after 7 d, whereas those that were not inoculated accumulated far less. In both cases the amount of extracellular ATP measured in the soil was correlated positively with the number of colony-forming units.

2. MDR1 PROMOTES ATP EFFLUX IN YEAST AND IN PLANTS

Most of the soil ATP measured by Thomas *et al* (1999) (Fig. 1) was probably derived from soil microbes. However, plants themselves have

MDR1 transporters, and there is evidence that these transporters could also contribute to the release of ATP to the ECM of plant cells. Recent results from our laboratory indicate that the *Arabidopsis* homolog of MDR1 may serve as a plasma membrane ATP channel in yeast and plants. Wild type yeast cells do efflux ATP naturally, but the resultant xATP fails to accumulate due to the presence of periplasmic acid phophatases represented by the yeast *pho3* and *pho5* gene products. A yeast strain mutant in both *pho3* and *pho5* genes, strain YMR4, accumulates at least 10 times as much xATP as wild type yeast. When the MDR1 gene is expressed in a YMR4 strain, the xATP accumulation is 100 times that of an untransformed non-mutant strain and nearly 10 times that of the YMR4 cells line transformed with vector alone (Thomas *et al* 2000).

The same MDR1 gene that promoted elevated xATP in transformed yeast did so also when overexpressed in *Arabidopsis*. The assay used for these results measured [ATP] on foliar surfaces. It took advantage of the fact that *Arabidopsis* can be grown in sterile culture under high humidity conditions that suppress cuticle development. On such plants a droplet of buffer placed on and covering a leaf surface will establish liquid continuity with ECM fluid. When the droplet is removed some ECM fluid is removed with it, and the removed droplet has levels of ATP easily measurable by the luciferin-luciferase assay. The assay is quantitative, sensitive and reproducible. After assaying the plants in this way, we found that MDR1 expression correlated with higher xATP levels. Plants over-expressing MDR1 had as much as two to three times the level of xATP on the leaf surfaces as wild type (Thomas *et al* 2000). Additional tests of the validity of this assay are planned, but our initial results using it are consistent with the interpretation that MDR1 serves as an ATP conduit in plants.

3. HOW HIGH DOES [xATP] REACH IN THE ECM? HOW LOW IS THE K_M OF APYRASE FOR ATP?

The nM [xATP] measured in bulk soil (Fig. 1), whether contributed by soil microbes or by plants themselves, would presumably be a small fraction of the concentration on the root surface, where the density of microbes would be highest and where ATP secreted by plant roots would be at its highest concentration. Given that cellular [ATP] reaches above 5 mM (Di Virgilio 1998), it would not be surprising to see [xATP] in microenvironments of the ECM of plant cells reach many micromolar, at least transiently, although these measurements have never been made.

The above discussion raises the question, Is the K_m of ectoapyrase low enough to hydrolyze a significant fraction of xATP? The K_m of *Arabidopsis* ectoapyrase for ATP is 30 μM (Steinebrunner and Roux unpublished). This measurement was done on soluble apyrase; the K_m value for apyrase in its more native environment of the plasma membrane could plausibly be much lower. Indeed, if the ectoapyrase is to function in real soil environments, where the [ATP] in the bulk soil water is nM or below, then either its K_m in membranes must approach nM or the [xATP] in the microenvironment immediately surrounding the ECM apyrases must be well into the µM range. Studies in progress will resolve how much lower, if at all, does the K_m of ectoapyrase for ATP fall when this parameter is measured while apyrase is in its native membrane. The other side of the question, How high do [xATP] levels reach in the ECM?, remains unresolved for both animal and plant physiologists. Clearly this unknown has not slowed the advance of xATP studies in animal systems. Even today animal physiologists have to resort to theoretical arguments when they try to account for the fact that a purinergic receptor may require supra-µM xATP to be activated, because they do not have accurate measurements of the actual [xATP] in the microenvironments surrounding these receptors (Di Virgilio 1998).

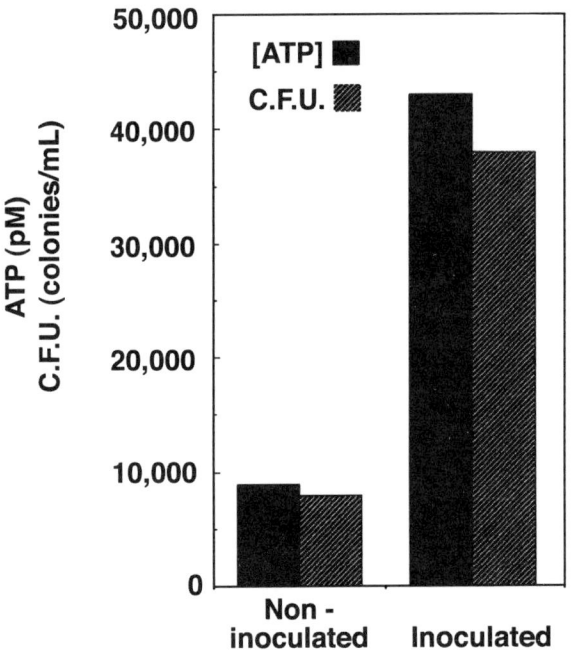

Figure 1. The [ATP] in bulk soil water is correlated positively with the microbial count (C.F.U. = colony-forming units).

4. ECTOAPYRASE IN ANIMALS: DAMPING THE ATP SIGNAL

In animal cells, any xATP generated would need to be hydrolyzed in part because xATP itself is a powerful elicitor of intracellular calcium signaling (Zheng *et al* 1991, Suko *et al* 1997) through receptor mediated G-protein signal transduction. By binding to purinergic receptors, extracellular ATP (xATP) triggers calcium cascades that lead to a variety of physiological changes, including programmed cell death (Di Virgilio 1998).

The need for rapid and controlled xATP hydrolysis is particularly important in the nervous system. There, xATP is involved in the chemical conduction across the synaptic cleft, where it binds to and activates one of two classes of receptors in the receiving membrane: the P2X or P2Y receptor. The former operates as a ligand-gated channel and the latter acts as a heterotrimeric G protein receptor. If xATP remains in the synapse, it can continue potentiation; therefore, like most secreted neurotransmitters, ATP is inactivated in the refractory period by a class of ATPases called apyrases (Todorov *et al* 1997).

5. ECTOAPYRASE IN PLANTS: NOVEL FUNCTION IN PHOSPHATE TRANSPORT

Most of our knowledge regarding xATP in eukaryotes focuses on how ATP gets outside the cell and what its targets are in the ECM. As indicated above, the extracellular enzymes best documented to catalyze the damping of the ATP signal are ectoapyrases. Although these enzymes have been extensively studied in animals, little is known about the function of plant ectoapyrases, or, indeed, even about the presence of ATP in the ECM of plants. Initial studies on the possible roles of ectoapyrases in plants have revealed unexpected and novel functions. The first apyrase gene characterized in plants was one in pea. Originally referred to an NTPase, this gene is up-regulated by light (Hsieh *et al* 1996), and the enzyme it encodes is stimulated by calcium and calmodulin (Chen *et al* 1987). In etiolated plants the enzyme has a dual localization, both in the nucleus (Tong *et al* 1993) and in the plasma membrane (Thomas *et al* 1999), where it is a typical ectoapyrase, with its active site facing into the ECM. However, in light-grown plants the apyrase is almost exclusively an ectoapyrase (Thomas and Roux unpublished).

To investigate the function of the pea ectoapyrase, the cDNA for the gene was overexpressed in light-grown *Arabidopsis* plants. Immunoassays showed that the resultant transgenic plants had much higher levels of ectoapyrase protein than wild-type plants. The first test was to compare how the transgenic and wild-type plants responded to treatment with xATP (Fig. 2). The transgenic plants showed more robust growth in the presence of ATP and preferentially transported the gamma phosphate from xATP into the plant (Thomas *et al* 1999). Unexpectedly, the apyrase also facilitated the uptake of inorganic phosphate in the transgenic lines, and it was able to complement a yeast mutant deficient in phosphate transport (Fig. 3). These observations raise the question of whether the apyrase, which is strongly localized in root, could itself be a phosphate transporter. The mammalian homolog of the pea apyrase, CD39, is known to form tetramers in the membrane and has a quaternary structure reminiscent of inwardly rectified potassium channels and P2X receptors (Wang *et al* 1998). However, it is uncertain whether apyrase itself is the pore or whether it channels phosphate to a separate transport system.

6. ECTOAPYRASE AND MDR1 CO-FUNCTION IN TOXIN RESISTANCE

An obvious conclusion from the results described above is that one function of apyrase would be to scavenge phosphate from xATP. However, when one considers that some of the xATP being delivered to the ECM is from the plant itself through its MDR1 channels, this raises the perplexing question of why would plants (or any other organism) secrete ATP in the first place? The idea that MDR proteins might efflux ATP into the ECM to be hydrolyzed without performing biological work is counterintuitive. What could be the benefit to cells effluxing ATP through MDR channels? One possible answer to this question comes from our recent finding that the drug resistance phenotype attributed to organisms strongly expressing MDR1 requires the efflux of ATP with the xenobiotic compound, and that the efficiency of the toxin efflux is dependent in part on the steepness of the ATP gradient between the inside and outside of cells (Thomas *et al* 2000). Our data seem to validate the model put forward by Abraham *et al* (1993), which suggests that the efflux of drugs through the MDR1 transport protein is coupled to the efflux of ATP.

According to this model the concentration of ATP is several orders of magnitude greater inside of the cell (mM) than in the bulk medium outside the cell (pM-nM). In the region near the cell surface, the ATP concentration is elevated due to the ATP channel activity of MDR1, but this ATP is rapidly hydrolyzed by the action of membrane bound ecto-ATPases. The functional interaction between the ATP efflux channel and the activity of the ecto-ATPases results in an actively maintained ATP gradient across the plasma membrane. This gradient can be used for work as it has a chemical and electrical potential. The symport of drugs with ATP can only continue insofar as the ATP gradient is maintained, hence the necessity of ecto-ATPase activity. This model predicts that disruption of either aspect of the two activities (ATP channel activity and/or ecto-ATPase activity) will impact the ability of MDR cells to resist drugs.

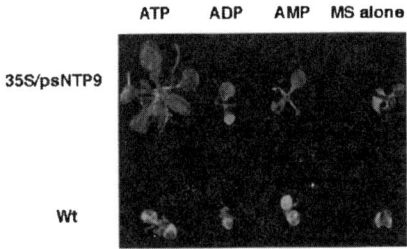

Figure 2. Relative growth of wild-type (Wt) versus transgenic plants expressing the pea apyrase in the presence and absence of 0,3 mM nucleoside phosphates.

We have obtained experimental evidence that strongly supports predictions of the model (Thomas *et al* 2000). Yeast cells lines deficient in ectophophatase activity (such as YMR4) are more sensitive to xenobiotic substances such as nigericin and cycloheximide, but expression of the MDR1 gene diminishes this toxin sensitivity. In plants, over-expression of MDR1 or ectoapyrase confers resistance to cycloheximide (Fig. 4). In *Arabidopsis*, overexpression of ectoapyrase or MDR1 can also impart resistance to herbicidal levels of the cytokinin, 2IP. We can abolish the resistance conferred by over-expression of MDR1, simply by growing the yeast or the plants in the presence of an ecto-ATPase inhibitor. These data hint at the primacy of ecto-ATPase activity in xenobiotic resistance, as disruption of steady-state extracellular ATP levels abolish any resistance offered by MDR1. We are presently testing crosses of MDR1 and ectoapyrase overexpressing lines to determine whether resistance is synergistic or additive.

Further evidence that members of the ABC-transporter family participate in resistance mechanisms comes from Urban *et al* (1999). They showed that the fungal plant pathogen *Magnaporthe grisea* up-regulates its ABC1 gene in response to a rice phytoalexin. A mutation in the promoter region of the ABC1 gene dramatically reduces transcript induction in the mutant, and it, unlike the wild-type pathogen, typically dies shortly after penetrating rice epidermal cells. These data show that the ABC1 gene helps confer in *M. grisea* a resistance to the defense mechanisms of plants it infects.

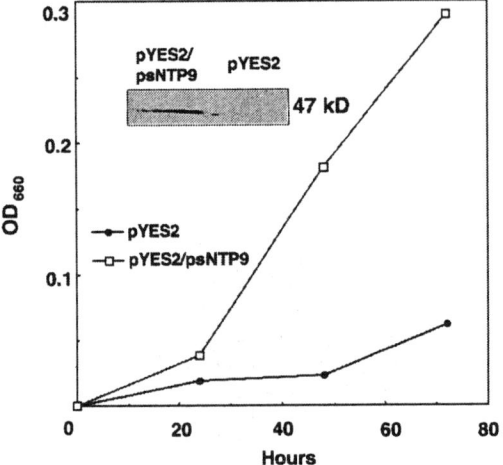

Figure 3. The pea apyrase gene (psNTP9) complements a yeast mutant defective in Pi-uptake. Shown is growth of yeast mutant (> OD_{660}) with the pYES2 vector alone and with pYES2/psNTP9. Inset shows immunoblot of the psNTP9 transformant and a control (vector alone). (Taken from Thomas *et al* 1999).

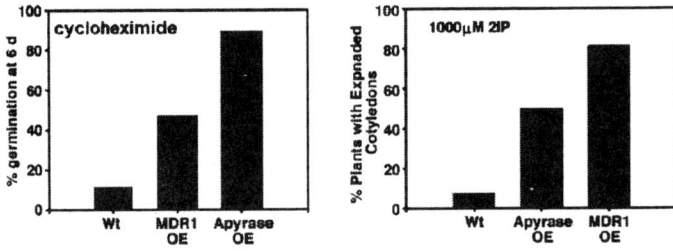

Figure 4. MDR1 and apyrase overexpression (OE) render *Arabidopsis* seeds more resistant to the inhibitory effects of cycloheximide on germination (A), and reverse the inhibitory effects of 1.0 mM 2-IP on seedling growth (B)

Although the MDR and related ABC transporters are well known for their function in exporting toxins from cells, it is plausible that they may also serve as a transporter for other growth regulating substances. The fact that MDR overexpression alters light-regulated hypocotyl growth (Sidler *et al* 1998) would also support this notion. We are currently testing this hypothesis as a way of more critically examining the unexpected connections between apyrase, ABC transporters, phosphate nutrition and light.

7. CAN ECTOAPYRASES PLAY AN ATP SIGNAL-DAMPING ROLE ALSO IN PLANTS?

The best characterized role for ectoapyrases in animals is to suppress the signal-inducing effects of xATP by destroying it. For it to play a similar role in plants there would have to be receptors responsive to ATP, ADP, adenosine, or some readily-generated product of ATP hydrolysis. In animals the best known receptors of ATP/ADP are the purinergic receptors P2X and P2Y. While members of the purinergic family of receptors continue to be cloned from mammalian systems, and there are hints of similar channels in protozoa, there were no plant homologs in any of the sequence databases identified as of this writing (Feb. 2000). On the other hand, the primary structure of these receptor proteins is not strongly conserved, so alignment approaches to finding homologs in plants would not be expected to be very successful.

An indication that purinergic receptors might be worth looking for in plants is that 6-benzylaminopurine, a plant-derived cytokinin, activates P2-purinoceptors in rat heart muscle cells. Pre-treatment of the cells with the cytokinin blocks any subsequent induction of muscle contraction by ATP, the normal agonist for the receptor, and antagonists of P2-purinoceptors block the contraction effects of the cytokinin. This suggests the possibility that cytokinin receptors in plants may be structurally related to purinergic receptors and may respond to some agonists of purinergic receptors, including ATP. If so, plant ectoapyrases could serve the role of silencing inductive effects of ATP in plants as it does in animals.

REFERENCES

Abraham, E., Prat, A., Gerweck, L., Seneveratne, T., Arceci, R., Kramer, R., Guidotti, G., and Cantiello, H., 1993, The multidrug resistance (mdr1) gene product functions as an ATP channel. *Proc. Natl. Acad. Sci. USA* **90**: 312-316.

Azam, F., and Hodson, R., 1977, Dissolved ATP in the sea and its utilization by marine bacteria. *Nature* **267**: 696-697.

Boyum, R., and Guidotti, G., 1997, Glucose dependent, cAMP- mediated ATP efflux from *Saccharomyces cerevisiae*. *Microbiology* **143**: 1901-1908.

Chen, Y., and Roux, S., 1986, Characterization of nucleoside triphosphatase activity in isolated pea nuclei and its photoreversible regulation by light. *Plant Physiol.* **81**: 609-613.

Dudler, R., and Hertig, C., 1992, Structure of an *mdr*-like gene from *Arabidopsis thaliana. J. Biol. Chem.* **267**: 5882-5888.

Froldi, G., Gallo, U., Ragazzi, E., and Caparrotta, L., 1999, 6-Benzylaminopurine: A plant derived cytokinin inducing positive inotropism by P2-purinoceptors. *Planta Medica* **65**: 245-249.

Hsieh, H., Tong, C., Thomas, C., and Roux, S., 1996, Light-modulated abundance of an mRNA encoding a calmodulin-regulated, chromatin-associated NTPase in pea. *Plant Mol. Biol.* **30**: 135-147.

Matin, A., and Matin, M., 1982, Cellular levels, excretion, and synthesis rates of cyclic AMP in *Escherichia coli* grown in continuous culture. *J. Bacteriol.* **149**: 801-807.

Reisin, I., Prat, A., Abraham, E., Amara, J., Gregory, R., Ausiello, D., and Cantiello, H., 1994, The cystic fibrosis transmembrane conductance regulator is a dual ATP and chloride channel. *J. Biol. Chem.* **269**: 20584-20591.

Schwiebert, E., 1999, ABC transporter-facilitated ATP conductive transport. *Am. J. Physiol.* **276**: C1-8

Sidler, M., Hassa, P., Hasan, S., Ringli, C., and Dudler, R., 1998, Involvement of an ABC transporter in developmental pathway regulating hypocotyl cell elongation in the light. *Plant Cell* **10**: 1623-1636.

Suko, Y., Kawahara, K., Fukuda ,Y., and Masuda, Y., 1997, Nuclear and cytosolic calcium signaling induced by extracellular ATP in rat kidney inner medullary collecting duct cells. *Biochem. Biophys. Res. Commun.* **234**: 224-229.

Thomas, C., Sun, Y., Naus, K., Lloyd, A., and Roux, S., 1999, Apyrase functions in plant phosphate nutrition and mobilizes phosphate from extracellular ATP. *Plant Physiol.* **99**: 543-551.

Todorov, L., Mihaylova-Todorova, S., Westfall, T., Sneddon, P., Kennedy, C., Bjur, R., and Westfall, D., 1997, Neuronal release of soluble nucleotidases and their role in neurotransmitter inactivation. *Nature* **387**: 76-79.

Tong, C., Dauwalder, M., Clawson, G., Hatem, C., and Roux, S., 1993, The major nucleoside triphosphatase in Pea (*Pisum sati*vum L.) nuclei and rat liver share common epitopes also present on nuclear lamins. *Plant Physiol.* **101**: 1005-10011.

Urban M., Bhargava, T., and Hamer, J., 1999, An ATP-driven efflux pump is a novel pathogenicity factor in rice blast disease. *EMBO J.* **18**: 512-521.

Wang, T-F., Ou, Y., and Guidotti, G., 1998, The transmembrane domains of ectoapyrase (CD39) affect its enzymatic activity and quaternary structure. *J. Biol. Chem.* **273**: 24814-24821

Zheng, L., Zychlinsky, A., Liu, C., Ojcius, D., and Young, J., 1991, Extracellular ATP as a trigger for apoptosis or programmed cell death. *J. Cell Biol.* **112**: 279-288.

Auxin- and GTP-binding Proteins and Protein Kinases from the Protonema of the Moss *Funaria hygrometrica*

[1]M. M. JOHRI, [2]KISHORE C. PANIGRAHI, [1]J. S. D'SOUZA, AND [1]D. MITRA

[1]*Department of Biological Sciences, Tata Institute of Fundamental Research, Homi Bhabha Road, Mumbai 400 005, India;* [2]*Pflanzenbiotechnologie, Universität Freiburg, Sonnestrasse 5, D-79104 Freiburg i.Br., Germany*

1. INTRODUCTION

The protonema of mosses such as *Funaria hygrometrica* Hedw., *Physcomitrella patens* (Hedw.) B.S.G. and *Ceratodon purpureus* (Hedw.) Brid. have been widely employed as paradigms to study the responses to plant hormones, light, gravity and abiotic stresses because the responses are rapid, discernible in single cells and thus cell autonomous. In many respects the protonema system is proving an excellent system to analyze signal transduction events. All phases of protonema development are regulated basically by light, gravity, temperature and the two phytohormones - auxin and cytokinin. The third hormone - abscisic acid (ABA), inhibits cell division and is believed to mediate responses to abiotic stresses (Bopp and Werner 1993). It could also be involved in the adaptation of mosses to abiotic stresses. Auxin and cytokinin regulate the protonemal development by directing or specifying the developmental potential of progeny cells to a predictable terminal fate. Although the major groups of phytohormones are ubiquitously distributed in plants, it is only starting from some of the bryophytes that well-defined responses to auxin, cytokinin and ABA have been found and it is conceivable that the hormone perception and action mechanisms evolved at the level of bryophytes. These hormones and ethylene occur naturally in several liverworts and mosses (reviewed in Bopp 1990, Johri 1990). There is no evidence for the occurrence and specific pharmacological effects of gibberellins in bryophytes as yet. The long-term experience with the cell line J-2 of the moss *F. hygrometrica* and the results of our studies to understand the possible role of auxin-binding proteins (ABPs), protein kinases (PKs) and heterotrimeric G-proteins (GPs) will be presented here.

1.1 Protonema Development in Cultures

Conventionally the protonema has been cultured on solid medium where its development involves the chloronemal and the caulonemal stages. The chloronema and caulonema filaments comprising these stages differ morphologically; the septae are straight in chloronema but oblique in caulonema. The chloroplasts are small and spindle-shaped in caulonema cells but large and round in chloronema cells. The chloronema filaments are also positively phototropic. Developmentally, the chloronema represents the default state of differentiation and any part of the protonema or gametophore (moss plant) can be used as an explant and the protonema regenerated. The protonema can also be grown on sterile cellophane discs overlaid on the solidified culture medium as described by Bopp *et al* (1964). As several discs with protonema can be accommodated simultaneously using a 15 or 20 cm diameter Petri dish, each protonemal disc can be transferred to a particular treatment. This method has proved very useful to investigate the effects of phytohormones and abiotic stresses such as salinity, osmoticum, low and high temperature.

The differentiation of caulonema has been found to be regulated by auxins (Johri and Desai 1973). Auxin has two effects, the inhibition of secondary chloronema formation and the stimulation of secondary caulonema production. The inhibitory effect of auxins is specifically antagonized by exogenous 3',5'-cAMP in *Funaria* cultures (Handa and Johri 1976). A far higher level of intracellular cAMP in chloronema than in caulonema and the stimulation of chloronema formation by the inhibitors of cyclic nucleotide phosphodiesterase (PDE) is consistent with the role of cAMP (Handa and Johri 1979). Cyclic AMP-specific PDEs and adenyl cyclase have been reported in the protonema (Hintermann and Parish 1979, Sharma and Johri 1982). With the recent isolation of a cDNA for a functional cyclic nucleotide-gated cation channel from *Arabidopsis* (Leng *et al* 1999), the role of cAMP and cGMP as second messengers needs to be reexamined. The overall regulation of cell differentiation is rather complex and regulated by several interacting factors.

In the protonema of *P. patens* besides auxin, a cytokinin is also involved in secondary chloronema inhibition (Ashton *et al* 1979) but a similar evidence for *F. hygrometrica* is not there. Cytokinin and auxin seem to be required for the formation of buds and the development of gametophores in several mosses (reviewed in Cove and Ashton 1984). The differentiation of a cell type seems to be regulated by several factors and it is the limiting one which becomes the demonstrable factor for the differentiation of a particular cell type. The side branch initial on a caulonema filament exhibits multiple developmental potentialities and depending on the phytohormone applied it can differentiate either into a chloronema cell, a caulonema cell or a bud initial. The role of a particular hormone can thus be visualized in directing or channelling the terminal fate of a progeny cell to a specific developmental fate by inhibiting the others.

In liquid cultures, whereas the chloronema cells proliferate profusely but typical caulonema are not formed. This makes it possible to grow chloronema cells on a preparative scale. Using chloronema cells of the cell line J-2 as inoculum and culturing them under completely defined conditions in suspension cultures, we have been able to obtain preparative amounts of predominantly caulonema filaments also (Johri 1974).

1.2 Axenic Cell Line J-2 and its Long-term Developmental Potential

An important technological breakthrough has been the development of the cell suspension culture system in the moss *F. hygrometrica* (Johri 1974). The cell line J-2 isolated from a single spore has now been maintained for over 29 years by repeated transfer to a low-calcium medium (LCM, contains 0.21 mM calcium). During this period, this cell line has neither lost the potential to differentiate nor has become polyploid. As the filaments are fragile, the protonema is grown in stationary suspension cultures in light at 25±1°C. When subcultured at high inoculum cell densities (>1 mg·ml^{-1}) in LCM, chloronemal cells multiply exponentially till a cell density of 18 to 20 mg·ml^{-1} (wet weight) is attained at the onset of stationary phase. The limiting level of calcium seems to retard the plastid multiplication and large number of tmema cells are formed. The tmema cells are known to be produced as a result of unequal cell division especially under unfavourable growth conditions (Bopp *et al* 1991). When a culture at the onset of stationary phase is shaken by swirling the flask gently, the protonemal filaments tend to break into single cells or 2-4 celled filaments each of which is capable of regeneration and normal differentiation upon transfer to fresh medium. In order to study the effect of various hormones, a known amount of LCM-grown inoculum (usually 0.05 mg·ml^{-1} wet weight of cells) is transferred to minimal medium with glucose (MMG which contains 4.2 mM calcium). Cells cultured in MMG multiply and differentiate normally (Johri and Desai 1973, Johri 1974) but are not pipettable and are not used as inoculum unless passaged several times using the LCM. Thus the experiments are essentially carried out using the MMG while the inoculum is maintained by repeated transfer to LCM. The chloronema cells can also be readily grown in liquid medium with a reduced level of other nutrients such as nitrate or phosphate but the pipettability is far better when calcium levels are modulated. The loss of pipettability is due to the production of protonemal rhizoids.

1.3 Regulation of Cell Differentiation in Moss Suspension Cultures

In suspension cultures auxin, cytokinin and ABA regulate respectively the differentiation of caulonema, shoot bud initials and brood cells (which are resting cells functionally similar to gemmae). The differentiation of caulonema marks a major developmental switch or pathway which is regulated by biologically active auxins (Johri and Desai 1973). Cells also respond to α-NAA and ethyl ester of IAA but not to 2,4-dichlorophenoxyacetic acid or indoleacetic acid (Johri and D'Souza 1990).

The initiation of cell differentiation and the sensitivity of chloronema cells to respond to auxin depends on the inoculum size and nutritional status. Below a cell density of 0.1 $mg \cdot ml^{-1}$, 1-10% caulonema filaments, exclusively primary ones, are formed. Applied IAA evokes two responses; at low levels (apparent Km 0.1 μM) it inhibits the formation of secondary chloronema and at a higher level (apparent Km 0.4 μm) it increases the production of secondary caulonema. By manipulating the level of exogenous auxin and the inoculum size, cultures containing predominantly caulonema (65-70%) are readily obtained (Johri 1974). As the secondary caulonema are formed at IAA concentrations higher than those inhibiting chloronema, both responses occur together in liquid cultures. During the 6-day period when caulonema are formed in response to IAA, the pH of the medium changes from initial value of about 5 to 6.5. This pH range during caulonema differentiation is similar to that reported for other auxin-induced responses. As IAA seems to be taken up through passive diffusion of undissociated molecules, the response to auxin can be optimized by adjusting the medium pH close to the pK of IAA.

The above strategy ultimately made it possible to obtain caulonema differentiation in auxin-free liquid medium also (Johri and D'Souza 1990). The caulonema were observed to differentiate without exogenous auxin after a lag of 6±1 days if the medium was buffered at pH 5.0 or if the nitrate levels were reduced. As the cultures age and nutrients become limited, the caulonema filaments formed look more like rhizoids (Johri and D'Souza 1990). Interestingly, auxin has earlier been reported to stimulate rhizoid production in several mosses (Bopp 1953, Gorton and Eakin 1957, Spiess *et al* 1976). Caulonema and rhizoids thus belong to the same class of cell type which is really represented by a spectrum of morphologies ranging between typical caulonema and typical rhizoids. The protonemal rhizoids formed in liquid cultures are multicellular with oblique septae and very few chloroplasts and turn brown as the cultures become old.

2. AUXIN-BINDING PROTEINS IN MOSS PROTONEMA

As already stated, caulonema can also differentiate without exogenous auxin after a lag of 6±1 days in a medium buffered at pH 5.0 or with a reduced levels of nutrients. During this period, the responsiveness of cells to auxin seems to undergo modulation (Johri and D'Souza 1990). The lag is prolonged beyond 6 days by the auxin antagonist *p*-chlorophenoxyisobutyric acid (PCIB) which reduces polar, basipetal transport in *Funaria* rhizoids (Rose and Bopp 1983). PCIB is also known to compete with IAA for auxin-binding sites in plants (Dohrmann *et al* 1978, Jacobs and Hertel 1978). Thus both basipetal transport and IAA-binding sites seem to be involved during caulonema differentiation. In the PCIB-treated cultures, there is no inhibition of growth and in fact there is a profuse stimulation of secondary chloronema growth. This result is consistent with the involvement of auxin in chloronema inhibition.

The prolongation of this lag with PCIB indicates that binding of IAA to membranes is involved. The moss microsomal membranes showed saturable and displaceable IAA binding of about 3.5 - 5 pmol/mg protein (Panigrahi 1998). In the microsomal fraction, three polypeptides of 28, 31 and 44 kDa cross-reacted with the polyclonal antibodies against the corn 22 kDa ABP1. However, none of these polypeptides cross-reacted with the anti-KDEL and anti-HDEL monoclonal antibodies and thus seem unlikely to be localized in the endoplasmic reticulum unless some other retention sequence is used in moss. Photoaffinity labelling with 5-azido[7^3–H]-indole-3-acetic acid using the acetone extracted microsomal membrane preparations showed the presence of several labelled polypeptides. The labelling of 28, 31, 33 and 40 kDa polypeptides was reduced in the presence of non-radioactive IAA. The 28 kDa polypeptide which is detected both by immunological and affinity labelling methods, could be an auxin-binding protein in moss protonema; it was observed to be present in the subcellular fraction containing predominantly ribosomes and membranes. The 28 kDa polypeptide in the moss thus shares epitopes with corn ABP1, but could be associated with ribosome or membrane fraction (Panigrahi 1998). Its role in auxin perception and basipetal polar auxin transport is yet to be understood.

3. CALCIUM-DEPENDENT PROTEIN KINASES FROM CHLORONEMA CELLS

The studies in several laboratories have established the role of calcium ions as a possible intracellular messenger in cytokinin-induced bud formation (Saunders and Hepler 1982, see Schumaker and Dietrich 1998). We have therefore been characterizing the calcium-sensing elements in the protonema and have been focusing on the calcium-dependent PKs. Using the conventional techniques of protein purification, a 70 kDa PK was first purified to homogeneity; it preferred lysine-rich histone as substrate and was fully active in the presence of 50 μM free calcium (Johri et al 1997). In the presence of sub-optimal levels of free calcium (for instance 23 μM), the histone phosphorylation was strictly dependent on the presence of 5-1000 nM moss calmodulin. At calmodulin levels of 100-1000 nM (optimum being around 400 nM), there was an increase in the autophosphorylation of the 70 kDa enzyme also. As judged on the basis of increase in histone phosphorylation, the phosphorylated PK 70 was far more active than the purified enzyme. Based on these properties, the PK 70 seems to be a calcium-dependent calmodulin-stimulated PK.

While the above studies were in progress, the in-gel kinase assays showed that the moss protonema contains multiple calcium-dependent protein kinases PKs of Mr 44, 48, 63 and 70 kDa (D'Souza and Johri 1999). All four require free calcium both for autophosphorylation and casein (substrate) phosphorylation. The phosphorylation of 44 kDa PK increased with an increase in cell density. A similar trend is also observed in the cells grown either in auxin-containing medium or in a medium buffered at pH 5, or when starved of nitrate. This PK shares epitopes with moss calmodulin and seems to be a PK with calmodulin-like domain (D'Souza and Johri 1999). No changes in the phosphorylation status of 48 and 63 PKs in relation to growth or hormone treatment have yet been observed.

The calcium-dependent PKs are a group of serine-threonine kinases which have a N-terminal catalytic and a C-terminal regulatory domain linked by an autoinhibitory region. The regulatory region has four Ca^{2+} binding motifs, the so-called EF-hands. From the chloronema cells, we have cloned a CDPK using the PCR based approach and PCR primers were designed against the aminoacid sequences GVMHRDLKPEN (sub domain VIb) and DKDGSGYIT (third EF hand). The degenerate primer PCR with genomic DNA yielded two fragments of sizes 900 and 800 bp. The 900 bp fragment showed a high degree of sequence conservation with the known CDPK sequences. A genomic library of *Funaria* constructed in λGEM-11 was screened with 900 bp fragment to obtain the full-length gene. Genomic clones have been identified, subcloned and are being sequenced. Northern blot analysis shows that the transcript of about 2.6 kb is upregulated by nutritional deprivation which can be brought by increase in cell density or by reducing the levels of nitrate, sulphate or phosphate in the liquid medium. In these media, both caulonema and rhizoids are formed profusely.

There is thus overwhelming evidence for the presence and utilization of the calcium messenger pathway in the moss system. We are trying to identify the specific stimuli which utilize these protein kinases in the signalling cascade. Since PK-44 responds to the same physiological conditions with culminate in caulonema differentiation, it could play a role in the cell differentiation process. The role of various calcium dependent PKs is currently under investigation.

3.1 Activation of a 38 kDa MAP Kinase by ABA

Using the protonema, we are trying to understand the mode of ABA action and its role in the acquisition of tolerance against stresses such as low temperature (8°C), salinity (200 mM NaCl), osmoticum (750 mM mannitol) and slow desiccation. Within minutes of treatment with 10 µM ABA, a 38 kDa MAP kinase was found to be activated in chloronema cells (D'Souza and Johri 2000). This activation is transient, specific to ABA and involves phosphorylation of tyrosine residues. A similar response is also observed upon treatment with NaCl but not with other abiotic stresses. Since the effects of ABA and NaCl on the activation are additive, both seem to act independently and finally the signals seem to converge at the level of MAP kinase. We wish to suggest that PK-38 is involved early in ABA action and possibly plays a role in the acquisition of tolerance to salinity.

3.2 GTP-binding Proteins in Chloronema and Caulonema

We had been developing improved methods for the detection of GTP-binding protein in plants. Based on the cholera or pertussis toxin enhanced ADP-ribosylation and immunological methods, we have demonstrated the presence of two classes of GPs in moss protonema and coleoptiles of corn and sorghum (Panigrahi and Johri 1998). These can be categorized into α-subunit of heterotrimeric GPs of 37-54 kDa and small Mr G-proteins of 18-32 kDa. The three plants investigated also showed yet another polypeptide of 92 kDa which shared epitopes with Gα-subunit. Similar GPs termed as extra-large ones have been described in the mammalian cell lines (Kehlenbach *et al* 1994). A 39 kDa β-subunit of heterotrimeric G-proteins was also detected by immunological method in moss.

Using sub-type specific antibodies directed against Gs, Gi1 and Gq α-subunits respectively, at least 12-15 polypeptides were recognized by different antibodies. The chloronema and caulonema cells showed a cell type specific pattern of Gα-subunits (Panigrahi 1998). The anti-Gs antibody detected a 100 kDa polypeptide specific to caulonema and three polypeptides of 68, 41 and 23 kDa specific to chloronema cells. The Gi1 antibody detected 56 and 50 kDa polypeptides specific respectively for caulonema and chloronema cell types. Similarly a cell type specific pattern was in the polypeptides recognized by anti-Gq antibody. It is interesting that the diversity of sub-type specific GPs exists not only at the species level but also at the level of cell types in a single species e.g., chloronema and caulonema cell types of moss *F. hygrometrica*. These studies thus show that there is a far greater diversity of G-proteins in plants that has been recognized so far and that several polypeptides sharing epitopes with Gs, Gi1 and Gq seem to be present. The role of GPs in the moss is yet to be fully understood. A GP-mediated regulation of calcium uptake has been suggested in *P. patens* where the nonhydrolyzable GTP analog has been found to stimulate the binding of dihydropyridine to calcium channel receptors in the plasma membrane preparations (Schumaker and Gizinski 1996).

3.3   ABA- and Stress- induced Polypeptides

The labelling experiments with ^{35}S-methionine show that after 30 min or so, ABA (10 µM) induces the synthesis of at least 10-15 new heat-stable polypeptides which range from 22-200 kDa (Ainapure 1998). After 12 to 18 hours, the proportion of heat-stable protein fraction in ABA-treated protonema is two times that of the control protonema (10-12% in ABA-treated and 6-7% in control). The polypeptides of 16, 17, 20, 97 and 102 kDa cross reacted with one of the four anti-LEA antibodies [anti-LEA-1, -LEA-2, -LEA-3 and -LEA-4]. The protonema subjected to desiccation, osmotic, salinity or low temperature stress treatment also showed the induction of LEA homologues which can be classified in two sets (Ainapure 1998). One set was common for all the stresses while the second one consisted of polypeptides specific to each stress. Thus the epitopes recognized by anti-LEA antibodies seem to be conserved in moss and the common polypeptides could be involved in general mechanism of stress tolerance, mediated by ABA. Similar to higher plants, in moss protonema also there is evidence for the involvement of both ABA-dependent and -independent mechanisms during response to various stresses.

ABA has been found to confer desiccation tolerance in mosses such as *Funaria* and *Physcomitrella* (Werner *et al* 1991, Goode *et al* 1992). Environmental stress and ABA induce the synthesis of several new polypeptides in *Funaria* protonema (Bopp and Werner 1993). In order to determine if the stress- and ABA-induced heat-stable proteins also included the heterotrimeric GPs (Panigrahi 1998), we have investigated for the presence of putative stimulatory G-protein α-subunits (HSGsαs). As a control, the anti-Gsα antibodies pre-incubated with the immunogenic Gsα peptide were used to probe the blots; polypeptides sharing epitopes with Gsα should be competed out while the non-specific ones will not be eliminated or competed. A detailed analysis is currently in progress but the following main results have emerged (unpublished data of Panigrahi, Ainapure and Johri).

The cells treated with ABA showed a total of six cross-reactive polypeptides which corresponded to 26, 29, 36, 39, 40 and 43 kDa. The HSGsαs were observed to be temporally regulated in response to ABA and their qualitative pattern of induction was observed to depend on the concentration of exogenous ABA. The protonema subjected to various abiotic stresses (slow desiccation, osmotic stress, salt stress and cold stress), showed only four or five HSGsαs but the qualitative as well as quantitative patterns were specific to each stress. The accumulation of HSGsαs was stress-specific and highest with salt stress. We wish to suggest that the signalling mechanisms involving these novel GPs play a role in the acquisition of tolerance to environmental stresses and ABA is involved in this response.

These results strongly suggest that once the protonema has survived the sub-lethal stress, a new set of some of the signalling molecules is utilized during growth in the stress-adapted state. In the studies described above the appearance of new HSGsαs was observed 60 min following ABA treatment. The overall mechanisms of stress-adaptation seems rather complex and other sub-classes of HSGαs such as Gqα and Giα were also observed.

4. CONCLUSIONS

The moss protonema is proving to be an excellent developmental system to study hormonal action mechanisms and the molecular basis of responses and adaption to abiotic stresses. The specific responses to hormones seem to have evolved early at the level of bryophytes but even in these cryptogams the level of complexity is comparable to the higher plants. It is interesting, that so far the signalling molecules such as G-proteins are believed to be constitutive but the present studies strongly suggest that these are inducible. Our results also provide a different insight towards understanding the phenomenon of acclimation in response to ABA in plants. Although we are far from understanding the detailed signalling mechanism in the moss protonema or even in other plants, the highly conserved signalling mechanisms in plants seem to have evolved in relation to survival strategies. There is a remarkable degree of similarity between the ABA responses in mosses and cereals (Knight *et al* 1995); the ABA- and osmotic stress-inducible promoter elements from the wheat Em gene are fully functional in *P. patens*. The partial sequence of *Funaria* calcium-dependent PK gene shows extensive homology with CDPK genes isolated from higher plants. These results support the conclusions of Reski *et al* (1997) that there is a high degree of conservation between moss and seed plant sequences and the codon usage in moss is very similar to that in dicotyledonous plants. We need to learn more about the identity of other signalling molecules in plants. There are several reports indicating the presence of other classes of GTP-binding proteins such as developmentally regulated and the extra large ones in addition to heterotrimeric and small Mr G-proteins (Wang et al 1997, Lee and Assmann 1999, Devitt et al 1999). The function of large GPs is largely unknown. There seems to exist enormous diversity of CDPKs and various sub-classes of GPs in moss. Much remains to be learnt about the involvement of a particular type of GPs and calcium-dependent PKs in the signal transduction events. Although a lot of new information is emerging, we are still far from understanding the molecular basis of hormonal action or cell differentiation. With the demonstration of highly efficient homologous recombination in *P. patens* (Schaefer and Zryd 1997), a far more rapid progress can be expected in future.

ACKNOWLEDGEMENTS

We are thankful to Dr. M. Udaya Kumar for providing the anti-LEA antibodies, to Dr. Richard Napier for anti-ABP1, anti-KDEL and anti-HDEL antibodies and to Dr. Paul Millner for the subclass specific antibodies against GTP-binding proteins.

REFERENCES

Ainapure, S. D., 1998, M.Sc Thesis - Studies on the abscisic acid and stress-induced proteins in the protonema of the moss *Funaria hygrometrica*. University of Mumbai, Mumbai, India.

Ashton, N.W., Grimsley, N.H., and Dove, D.J., 1979, Analysis of gametophytic development in the moss *Physcomitrella patens*, using auxin and cytokinin resistant mutants. *Planta* **144**: 427-435.

Bopp, M., 1953, Die Wirkung von Heteroauxin auf Protonemawachstum und Knospenbildung von *Funaria hygrometrica*. *Z. Bot.* **33**: 1-16.

Bopp, M., 1990, Plant hormones in lower plants. In *Plant Growth Substances 1988* (R. P.Pharis and S. B. Rood, eds.), Springer Verlag, Berlin, pp.1-10.

Bopp, M., Jahn, H., and Klein, B., 1964, Eine einfache Methode, das Substrat während der Entwicklung von Moosprotonemen zu wechseln. *Rev. Bryol. Lichénol.* **33**: 219-223.

Bopp, M., Quader, H., Thoni, Sawidis, Th., and Schnepf, E., 1991, Filament disruption in *Funaria* protonemata. I. Formation and disintegration of tmema cells. *J. Plant Physiol.* **137**: 273-284.

Bopp, M., and Werner, O., 1993, Abscisic acid and desiccation tolerance in mosses. *Bot. Acta.* **106**: 103-106.

Cove, D. J., and Ashton, N.W., 1984, The Hormonal Regulation of Gametophytic Development in Bryophytes. In *The Experimental Biology of Bryophytes* (A.F. Dyer and J. G. Duckett, eds.), Academic Press, London, pp.177-201.

Devitt, M.L., Kass, K.J., and Stafstrom, J.P., 1999, Characterization of DRGs, developmentally regulated GTP-binding proteins, from pea and *Arabidopsis*. *Plant Mol. Biol.* **39**: 75-82.

D'Souza, J.S, and Johri, M.M., 1999, Ca^{2+}dPKs from the protonema of the moss *Funaria hygrometrica*. Effect of IAA and cultural parameters on the activity of a 44 kDa Ca^{2+}dPK. *Plant Sci.* **145**: 23-32.

D'Souza, J.S, and Johri, M.M., 2000, ABA and NaCl activate myelin basic protein kinase activity in the chloronema cells of the moss *Funaria hygrometrica*. Submitted.

Dohrmann, U., Hertel, R., and Kowalik, H., 1978, Properties of auxin binding sites in different subcellular fractions from maize coleoptiles. *Planta* **140**: 97-106.

Goode, J.A., Stead, A.D. and Duckett, J.G., 1992, Redifferentiation of moss protonemata - An experimental and immunofluorescence study of brood cell formation. *Can. J. Bot.* **71**: 1510-1519.

Gorton B.S., and Eakin, R.E., 1957, Development of the gametophyte in the moss *Tortella caespitosa*. *Bot. Gaz.* **119**: 31-38.

Handa, A.K., and Johri, M.M., 1976, Cell differentiation by 3',5'-cyclic AMP in a lower plant. *Nature* **259**: 480-482.

Handa, A.K., and Johri, M.M., 1979, Involvement of cyclic adenosine-3',5'-monophosphate in chloronema differentiation in protonema cultures of *Funaria hygrometrica*. *Planta* **144**: 317-324.

Hintermann, R., and Parish, R. W., 1979, Determination of adenylate cyclase activity in a variety of organisms: evidence against the occurrence of enzyme in higher plants. *Planta* **146**: 459-461.

Jacobs, M., and Hertel, R., 1978, Auxin binding to subcellular fractions from *Cucurbita* hypocotyls: *In vitro* evidence for an auxin transport carrier. *Planta* **142**: 1-10.

Johri, M.M., 1974, Differentiation of caulonema cells by auxins in suspension cultures of *Funaria hygrometrica*. In *Plant Growth Substances* Hirokawa Publishing Co., Tokyo, pp. 925-933.

Johri, M.M., 1990, Hormonal regulation of development and differentiation in lower plants. In *Proc. Intl. Congr. of Plant Physiology* (S.K. Sinha, P.V. Sane, S.C. Bhargava and P.K. Agrawal, eds), In Print Exclusives, New Delhi, pp. 760-775.

Johri, M.M., and Desai, S., 1973, Auxin regulation of caulonema formation in moss protonema. *Nature New Biology* **245**: 223-224.

Johri, M.M., and D'Souza, J.S., 1990, Auxin regulation of cell differentiation in moss protonema. In *Plant Growth Substances* (R.P. Pharis and S.Rood, eds.), Springer Verlag, Berlin, pp. 407-418.

Johri, M.M., Panigrahi, K.C., and D'Souza, J.S., 1997, G-proteins, auxin-binding proteins and Ca^{2+}/CAM-dependent protein kinases from the protonema of the moss *Funaria*. Abstract No. 783. 5th *Intl. Congr. Plant Mol. Biol.*, Singapore, Sept. 21-27, 1997

Kehlenbach, R.H., Matthey, J., and Huttner, W.B., 1994, XLαs is a new type of G protein. *Nature* **372**: 804-809.

Knight, C.D., Sehgal, A., Atwal, K., Wallace, J.C., Dove, D.J., Coates, D., Quatrano, R.S., Bahadur, S., Stockley, P., and Cuming, A.C., 1995, Molecular responses to abscisic acid and osmotic stress are conserved between mosses and cereals. *Plant Cell* **7**: 499-506.

Lee, Y-Ru J., and Assmann, S. M., 1999, *Arabidopsis thaliana* extra-large GTP-binding protein (AtXLG1): a new class of G-protein. *Plant Mol. Biol.* **40**: 55-64.

Leng, Q., Mercier, R.W., Yao, W., and Berkowitz, G.A., 1999, Cloning and first functional characterization of a plant cyclic nucleotide-gated cation channel. *Plant Physiol.* **121**: 753-761.

Panigrahi, K.C.S., 1998, PhD Thesis - *Studies on the Auxin binding and GTP-binding Proteins in Plants*. University of Mumbai, Mumbai, India.

Panigrahi, K.C.S., and Johri, M.M., 1998, Improved methods to detect GTP-binding proteins from plants. *J. of Bioscience* **23**: 193-200.

Reski, R., Reynolds, S., Wehe, M., Kleber-Janke, T., and Kruse, S., 1997, Moss (*Physcomitrella patens*) expressed sequence tags include several sequences which are novel for plants. *Bot. Acta* **111**: 1-7.

Rose, S., and Bopp, M., 1983, Uptake and polar transport of indoleacetic acid in moss rhizoids. *Physiol. Plant.* **58**: 57-61.

Saunders, M.J., and Hepler, P.K., 1982, Calcium ionophore A23187 stimulates cytokinin like mitosis in *Funaria. Science* **217**: 943-945.

Schaefer, D.G., and Zryd, J.-P., 1997, Efficient gene targeting in the moss *Physcomitrella patens. Plant J.* **11**: 1195-1206.

Schumaker, K.S., and Dietrich, M.A., 1998, Hormone-induced signaling during moss development. *Annu. Rev. Plant Physiol. Plant Mol. Biol.* **49**: 501-523.

Schumaker, K.S., and Gizinski, M.J., 1996, G-proteins regulate dihydropyridine binding to moss plasma membranes. *J. Biol. Chem.* **271**: 21292-21296.

Sharma, S. and Johri, M.M., 1982, Partial purification and characterization of cyclic AMP phosphodiesterases from *Funaria hygrometrica. Arch. Biochem. Biophys.* **217**: 87-97.

Spiess, L.D., Lippincott, B.B., and Lippincott, J.A., 1976, Comparative effects of growth substances and *Agrobacterium* on the moss protonema to gametophore phase change. *J. Hattori Bot. Lab.* **41**: 185-192.

Wang, H., Lockwood, S.K. Hoeltzel, M.F., and Schiefelbein, J.W., 1997, The root hair defective 3 gene encodes an evolutionarily conserved protein with GTP-binding motifs and is required for regulated cell enlargement in *Arabidopsis. Genes Dev.* **11**: 799-811.

Werner, O., Ros Espin, R.M. Bopp, M., and Atzorn, R., 1991, ABA induced drought tolerance in *Funaria hygrometrica* Hedw. *Planta* **186**: 99-103.

Myoinositol Phosphates as Implicated in Metabolic Signaling and Calcium Homeostasis in Plants

SHASHIPRABHA DASGUPTA, DIPAK DASGUPTA, SUSWETA BISWAS, AND BIRENDRA B. BISWAS
Department of Biophysics, Molecular Biology and Genetics, University of Calcutta, Calcutta-700 009, Biochemistry Department, Bose Institute, Calcutta-700 054 and Biophysics Division, Saha Institute of Nuclear Physics, Calcutta-700 037, India

1. INTRODUCTION

Myoinositol trisphosphate, particularly Ins(1,4,5)P_3 is the intracellular messenger that mediates the effects of many cell surface receptors on the intracellular Ca^{2+} stores (Berridge and Irvine 1989). Although specific assays have identified high affinity Ins(1,4,5)P_3-binding sites in many animal tissues (Supattapone et al 1988, Mignery and Sudhof 1990), these have not been convincingly shown in all cases to be the receptors that mediate Ca^{2+} mobilization, nor it is clear whether the binding sites are different from one tissue to another. Myoinositol trisphosphate receptor from plants has also been reported and characterized (Biswas et al 1995).

A body of evidence is now available to indicate that phosphoinositide pathway generating InsP_3 and DAG exists in plants (Drøbak 1992). Because InsP_6 is present in high amounts in plant systems, an alternative pathway for generation of different InsP_3 from InsP_6 might be a possibility. The question arises whether InsP_3 thus generated mediate any Ca^{2+} mobilization in plant cells. The sequence of hydrolysis of phosphate groups from InsP_6 by phytase(s) from different systems including plants and microorganisms suggested that the hydrolysis of InsP_6 occurs adjacent to the free hydroxyl group. Therefore, the initial position of hydrolysis is the major determinant of subsequent points of hydrolysis (Cosgrove 1980, Barrientos et al 1994). Of the twenty possible InsP_3 isomers, the following three have been identified as phytase products: $Ins(1,2,3)P_3$, $Ins(1,2,6)P_3$, and $Ins(2,4,5)P_3$. The first two are formed as intermediates in the acid phytase catalysed hydrolysis of $InsP_6$. $Ins(1,2,3)P_3$ has also been shown to be a product of alkaline phytase from lily pollen (Loewus *et al* 1990). $Ins(2,4,5)P_3$ as one of the products of phytase has been reported from mungbean (Maitra *et al* 1988). $Ins(2,4,5)P_3$ has also been found to be effective in stimulating release of Ca^{2+}, though to a lesser extent than $Ins(1,4,5)P_3$ from intacellular stores. It was also noted that addition of a mixture of $Ins(1,4,5)P_3$ or $Ins(2,4,5)P_3$ and phytase leads to enhanced Ca^{2+} release relative to that observed with free $InsP_3$ (Samanta *et al* 1993). What is the mechanism by which $InsP_3$-phytase complex can act as a better elicitor for Ca^{2+} efflux from the intracellular stores is the crucial point to be elucidated. Some of the experimental data reported earlier (Dasgupta *et al* 1996, Dasgupta *et al* 1997) substantiated our assumption that $InsP_3$-phytase complex can interact with the $InsP_3$-receptor forming a ternary complex and the final proof came from the experiment that $Ins(1,3,4)P_3$ when complexed with the phytase enzyme can elicit Ca^{2+} release from the microsomal fraction but $Ins(1,3,4)P_3$ *per se* cannot do so as it does not bind the receptor.

2. RESULTS AND DISCUSSION

A series of experiments on the kinetics of hydrolysis of $InsP_6$ by mungbean phytase indicated that the time needed for optimal production of $InsP_3$ is around 30 min under experimental conditions and at that point of time 10% of $InsP_6$ could be accounted for as $InsP_3$. One of the intermediary phytase products, i.e. $Ins(2,4,5)P_3$, was tested for Ca^{2+} release from the microsomal/vacuolar fraction from mungbean hypocotyls and a concomitant Ca^{2+} release is discernible (Samanta *et al* 1993). This indicates that $InsP_3$ a phytase product from $InsP_6$ can elicit Ca^{2+} from intracellular stores of the plant cell. $Ins(2,4,5)P_3$ and $Ins(1,4,5)P_3$ both elicited Ca^{2+} release from ATP dependent azide insensitive Ca^{2+}-preloaded microsomes. The amount of Ca^{2+} released by $Ins(1,4,5)P_3$ at a concentration of maximal Ca^{2+} efflux was higher than that induced by $Ins(2,4,5)P_3$ under identical conditions. Other InsPs were ineffective at releasing Ca^{2+} (Tab. 1). The question arises whether $InsP_3$ produced by the action of phytase interacts with the receptor after dissociation from the phytase or $InsP_3$-phytase complex recognizes the receptor.

2.1 Interaction of Myoinositol Phosphates with Phytase as Indicated from its Conformational Changes

We compared the accessibility of tryptophan residues in free and myoinositol phosphate bound phytase with a view to checking the ligand-induced change in the conformation of the phytase. A graphical representation of tryptophan accessibility as a function of the number of phosphate groups in the myoinositol phosphates is presented in Fig. 1 using the modified Stern-Volmer plot (Dasgupta *et al* 1996). The notable feature is that there is a progressive decrease in accessibility starting from $InsP_6$ to InsP with the exceptions of both $Ins(1,4,5)P_3$ and $Ins(2,4,5)P_3$. The change in accessibility could originate from a conformational change in the enzyme.

Table 1. Effect of $InsP_3$-phytase complex on the release of Ca^{2+} from microsomes/vacuoles at 25°C

Ligand (µM)	Ca^{2+} release (nM/mg of protein)
Ins $(1,4,5)P_3$ (1.1)	1080 ± 120
Phytase – Ins $(1,4,5)P_3$ (0.1) Complex	2580 ± 180
5 kDa Phytase Fragment-Ins $(1,4,5)P_3$ (0.1) Complex	2720 ± 270
Ins $(2,4,5)P_3$ (0.1)	785 ± 115
Phytase - Ins $(1,4,5)P_3$ (0.1) Complex	1415 ± 68
Ins $(1,3,4)P_3$ (0.1)	120 ± 35
Phytase - Ins $(1,3,4)P_3$ (0.1) Complex	648 ± 108

Phytase was isolated and purified from the cotyledons of 72 h germinated seeds of *Vigna radiata* (Mandal *et al* 1972, Dasgupta *et al* 1996).

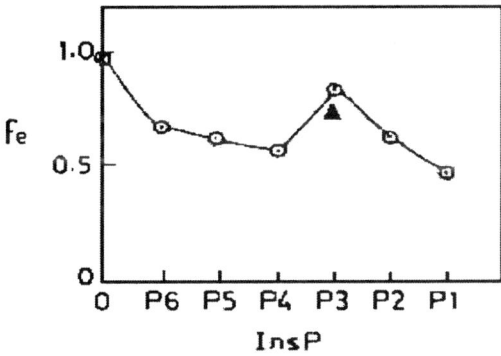

Figure 1. Plot of tryptophan accessibility of phytase against different myoinositol phosphates in phytase-substrate complex. The symbol (σ) denotes the value for $Ins(2,4,5)P_3$. In all cases acrylamide was added to an equilibrium mixture of phytase (0.3 µM) and myoinositol phosphates (25 µM) in 50 mM Tris-HCl buffer, pH 7.0 at 14°C.

2.2 Detection of High Affinity Binding Site in Phytase

Having established that the conformational change of phytase by $InsP_3$ does not follow the general trend shown for other inositol phosphates, we attempted to check the possibility that there was a second high affinity noncatalytic binding site in phytase for $Ins(1,4,5)P_3$ / $Ins(2,4,5)P_3$. In fact, a second high affinity site for $InsP_3$ in the enzyme was detected. The relevant binding isotherm for the high affinity and noncatalytic for high affinity and non-catalytic binding between phytase and $Ins(1,4,5)P_3$ as monitored from the binding assay using $[^3H]$-$Ins(1,4,5)P_3$ indicated the dissociation constant for $InsP_3$ as 75 ± 10 nM very similar to that obtained in the case of its receptor binding assay (Dasgupta *et al* 1996). The specificity of $InsP_3$ for this site has been demonstrated from the observation that a 50-fold excess of $InsP_6$ or other InsP could not displace labelled $Ins(1,4,5)P_3$ from phytase. However, $Ins(2,4,5)P_3$ or $Ins(1,4,5)P_3$ could do so. The binding of either of the two $InsP_3$ isomers to the noncatalytic site leads to significant change in the conformation of phytase. This was demonstrated from the change in the accessibility of tryptophan residues of the enzyme upon binding of $Ins(1,4,5)P_3$ / $Ins(2,4,5)P_3$ to the high affinity binding site of phytase. The accessibility of tryptophan residues to acrylamide for free phytase (f_e=0.98) changed in the presence of $Ins(2,4,5)P_3$ (f_e =0.55) and $Ins(1,3,4)P_3$ (f_e =0.65). A similar change in the accessibility took place when $Ins(1,4,5)P_3$ bound to the high affinity site in phytase (f_e =0.62). The accessibility (f_e) values were different when low affinity (Fig. 1) and high affinity (Fig. 2) binding sites of $InsP_3$ in phytase are saturated. The high affinity binding and its effects upon the conformation of phytase may account for the exceptional influence of $InsP_3$ in terms of the substrate induced conformational change in phytase (Fig. 1).

2.3 Formation of a Ternary Complex Involving Phytase $InsP_3$ and $InsP_3R$

Since the presence of a high affinity site in phytase for $Ins(2,4,5)P_3$ and $Ins(1,4,5)P_3$ is indicated, the possibility of the formation of a ternary complex was checked. The gel elution profile of the mixture in the case of $Ins(1,4,5)P_3$ when the three components were incubated is shown in Fig. 3a,b. The elution profile shows the presence of four peaks (Fig. 3b). Peak I corresponds to the ternary complex (mass = mass of phytase, 160 kDa + mass of $InsP_3R$, 400 kDa = 560 kDa). This was also verified from a standard calibration curve for molecular weight determination from elution volume (Fig. 3c). The presence of all three components in peak I was further confirmed by specific test for phytase and Ins P_3R and $InsP_3$ (Dasgupta *et al* 1996). Phytase binds to $InsP_3R$ in presence of $Ins(1,4,5)P_3$ or $Ins(2,4,5)P_3$ leading to the formation of a ternary complex. There is no interaction of phytase and $InsP_3R$ in absence of $InsP_3$ (Dasgupta *et al* 1996).

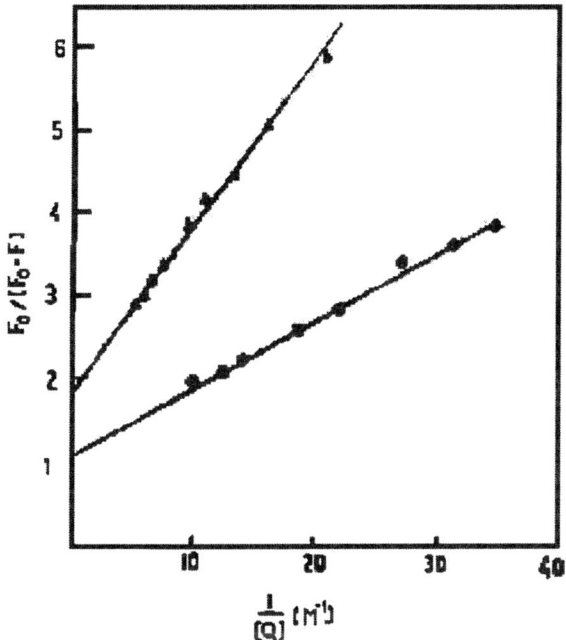

Figure 2. Comparison of the accessibilities of tryptophan residues in free and ligand bound (at high affinity site) phytase using acrylamide as quencher: Modified Stern-Volmer plots of $F_0/(F_0-F)$ against $1/Q$ for free phytase [(0.15µM) (●) and $Ins(2,4,5)P_3$ (0.3µM), phytase (0.15 µM) (Δ)] complex in 50 mM Tris-HCl buffer, pH 8.0, plus 25 mM NaCl at 5°C.

2.4 Differential Inhibition of High and Low Affinity Site for $InsP_3$ in Phytase

Differential effect of phenylglyoxal modification upon the two sites is demonstrated in Tab. 2. While treatment of phytase with the above modification agent for arginine side chain (Communi *et al* 1995) completely abolishes the catalytic activity, it has no influence upon the high affinity binding of phytase to $Ins(1,4,5)P_3$. It indicates that the two sites are different in terms of the amino acids present. This is further established from the limited proteolysis of the enzyme.

Table 2. Effect of phenyl glyoxal on activity of phytase and binding of Ins(1,4,5)P_3 to the high affinity site

Concentration of high phenyl glyoxal (mM)	% inhibition of phytase activity	% inhibition of affinity IP_3 binding
0	0	0
1	52.5	0
2	100	0

Phytase was incubated with phenyl glyoxal at 4°C for 1½ hr before checking for activity or binding of IP_3.

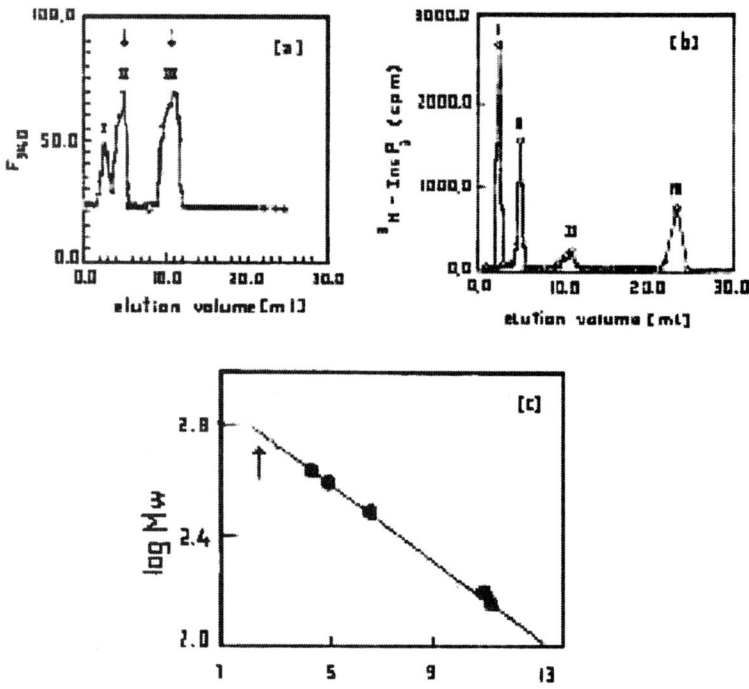

Figure 3. Formation of ternary complex, $InsP_3$-$InsP_3$R-phytase: Gel filtration (Sephadex G-200) of a mixture of $InsP_3$R (0.3 µM), phytase (0.3 µM) and β–ME, 3 mM EDTA and 25 mM NaCl, for 15 min at 5°C. Peaks I, II, III and IV refer to Ins(1,4,5)P_3- $InsP_3$R -phytase complex, Ins(1,4,5)P_3-$InsP_3$R complex, Ins(1,4,5)P_3-phytase complex and Ins(1,4,5)P_3, respectively. a) Plot of fluorescence (at 340 nm)

for each fraction against elution volume. Arrows indicate the positions of the markers, 400 kDa and 160 kDa. b) Plot of radioactivity for each fraction against elution volume. c) Molecular weight determination of ternary complex by the size exclusion chromatography on Sephadex G-200 (fractionation range: 5 kDa to 600 kDa). Peak I indicates the position of elution for the ternary complex corresponding to M_w = 560 kDa.

2.5 Separation of Phytase High and Low Affinity Site after Cleavage of Phytase by Trypsin

The result of the limited proteolysis of phytase by trypsin is shown in Fig. 4. The peptide fragments were fractionated on size-exclusion Sephadex G-100 column. Each fragment was tested for the following: i) $InsP_6$ hydrolysis, ii) high affinity $Ins(1,4,5)P_3$ binding and iii) Ca^{2+} release from microsomes/vacuoles in association with $Ins(1,4,5)P_3$. 62 kDa fragment is found to have the catalytic potential of phytase; on the other hand, 5 kDa fragment has the last two properties of high affinity $Ins(1,4,5)P_3$ in binding and Ca^{2+} release. It shows that high and low affiniity sites for $Ins(1,4,5)P_3$ are different in phytase.

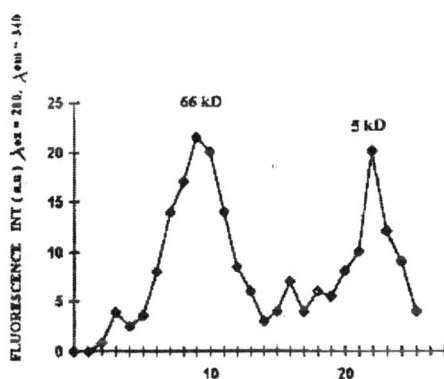

Figure 4. Gel filtration of trypsinized phytase through Sephadex G-100 column: The fractions were monitored fluorimetrically (λ_{ex} = 280 nm, λ_{em} = 340nm). The column was calibrated with lysozyme (14.6 kDa), ovalbumin (45 kDa), BSA (66 kDa) and T7RNAP (98 kDa). The molecular weights of the two major peptide fragments are indicated in the figure.

From electron microscopic studies, following the method of Hassan *et al* 1997, it is apparent that phytase is located along with the plasma membrane as well as in the cytosol (Fig. 5). This pattern of distribution of phytase is also discernible from fluorescence microscopic detection (Fig. 5). However, the prescence of InsP$_3$ along with the endoplasmic reticulum or vacuoles is apparent from immunoelectron microscopic scanning (data not given) as well as isolation of the receptor from microsomal pellet (Biswas *et al* 1995).

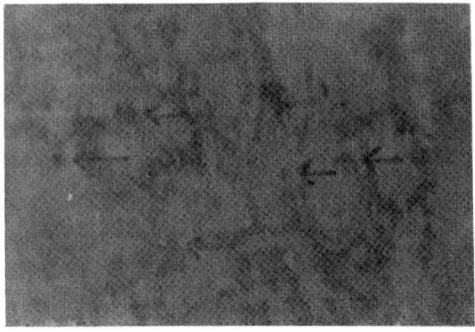

Figure 5. Membrane localization of phytase. Electron microscopy of cotyledon sections showing localization of phytase as processed by secondary antibody conjugated to gold particles (10 nm). 10,000x.

2.6 The Effect of the Ternary Complex InsP$_3$-Phytase-InsP$_3$R on Ca^{2+} Efflux from the Microsomes/Vacuoles

Since the InsP$_3$-phytase complex binds to InsP$_3$R and leads to the formation of a ternary complex, the question arises whether this has any relevance *in vivo*. The concentration response curves for free Ins(2,4,5)P$_3$ and Ins(2,4,5)P$_3$-phytase complex clearly showed that there was an increase in the level of Ca^{2+} release when the complex was added. A similar trend was noticed for Ins(1,4,5)P$_3$, though there was a difference in the relative extent of release of Ca^{2+} (Tab. 1). On the other hand, this could be ascribed to the fact that InsP$_3$-phytase complex binds to the endogenous microsomal / vacuolar receptor and elicits the release of Ca^{2+}. This is further confirmed that by the ability of Ins(1,3,4)P$_3$-phytase complex to release Ca^{2+} when free Ins(1,3,4)P$_3$ could elicit very insignificant amounts of Ca^{2+} release. This is also in conformity with the earlier observation that Ins(1,3,4)P$_3$ has a very low affinity for InsP$_3$R (Dasgupta *et al* 1996). The notable observation is that after trypsinization of phytase if 5 kDa fragment is complexed with InsP$_3$ and added to the microsomal fraction, more Ca^{2+} release is discernible than that when free Ins$_3$ is employed (Tab. 1).

2.7 Physiological Significance of the Alternative Pathway of InsP₃ Generation

An important role of the InsP$_3$-phytase-InsP$_3$R complex in the cell is to utilize InsP$_3$-generated either by phytase or other different pathways as an elicitor of intracellular Ca^{2+} efflux. It has further been substantiated (Fig. 6) that in presence of neomycin, which inhibits the phospholipase C activity responsible for generating Ins(1,4,5)P$_3$ from phosphoinositide-4,5 bisphosphate. The initial regeneration of mungbean plantlets from the cotyledonary explant is not effective (Pal *et al* 1991). Thus, in the light of the observations thus far recorded it is proposed that other inositol trisphosphates which *per se* are ineffective in Ca^{2+} mobilization in the cell may be made effective when complexed with the phytase (Biswas and Biswas 1996). The redundancy in the pathways of maintenance of Ca^{2+} concentration in the cell may be coupled to the noise induced stochastic opening of Ca^{2+} channels distinguishing particular cells(s) in a tissue to take a lead in the specific function(s) (Trewavas and Malho 1997). What is needed is to produce a knockout plant for PLC and to record the pattern of distribution of Ca^{2+} in a particular cell/tissue and to study the signaling mode the cells are entering into.

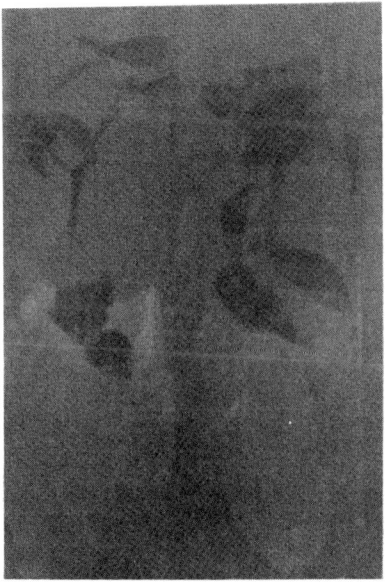

Figure 6. Regeneration from Mungbean cotelydonary explants in (a) the presence and (b) the absence of an inhibitor of phospholipase C.

3. CONCLUSION

An $InsP_3$-phytase complex has been found to interact with the $Ins(1,4,5)P_3$-receptor *in vitro* forming a ternary complex and for which only a nanomolar concentration of $InsP_3$ is required. In the nM range of concentrations, $InsP_3$ binds to a second site of phytase having 40-50 fold higher affinity than the normal catalytic site. The two sites are distinguishable by the proteolytic cleavage of the phytase. Limited proteolytic digestion of phytase by trypsin yielded four major peptide fragments. The fragment having a MW of 62 kDa has $InsP_6$ hydrolysing activity and a 5 kDa fragment has the high affinity site for $InsP_3$ binding. $InsP_3$, when bound to the non-catalytic site in phytase is not hydrolysed but induces a significant change in the conformation of phytase which in turn promotes $InsP_3$-receptor interaction forming $InsP_3$-phytase-receptor complex. Thus $InsP_3$ phytase complex is a better elicitor of Ca^{2+} efflux from microsomal/vacuolar fractions than the free $InsP_3$. This is further confirmed by the fact that whereas the $Ins(1,3,4)P_3$-phytase complex can elicit Ca^{2+} efflux from intracellular stores, $Ins(1,3,4)P_3$ *per se* is not effective. Intracellular localization of phytase and the receptor substantiates the effective role of $InsP_3$-phytase in Ca^{2+} homeostasis. Implicit is also the fact that it is the overall interaction leading to the requisite conformational change in the receptor that determines the potency of the $InsP_3$ isomers in their abilities of Ca^{2+} mobilization from the intracellular stores. This implies a novel pathway for signaling Ca^{2+} homeostasis mediated by $InsP_6$-phytase system in the plant cell. This is further substantiated by the observation that the regeneration of mungbean plantlet from the cotyledonary explant is not inhibited at the initial phase when neomycin was used to inhibit the phospholipase C to produce $Ins(1,4,5)P_3$ from phosphoinositide-4,5-bisphosphate.

ACKNOWLEDGEMENT

Financial support from the Department of Biotechnology and the Council of Scientific and Industrial Research, Govt. of India and technical assistance by Asim Poddar are thankfully acknowledged. Thanks are also due to American Chemical Society for permission of reproducing some of the figures from our paper published in *Biochemistry*.

REFERENCES

Barrientos, L., Scott, J. J., and Murthy, P.P.N., 1994, Specificity of hydrolysis of phytic acid by alkaline phytase from lily pollen. *Plant Physiol.* **106**: 1489-1495.

Berridge, M.J., and Irvine, R.F., 1989, Inositol triphosphate and diacylglycerol: two interactive second messengers, *Nature* **341**: 197-205.

Biswas, S., and Biswas, B.B., 1996, Metabolism of myoinositol phosphates and the alternative pathway in generation of myoinositol triphosphate involved in calcium mobilization in plants, *Sub. Cell Biochem.* **26**: 287-316.

Biswas, S., Dalal, B., Sen, M., and Biswas, B.B., 1995, Receptor for myoinositol triphosphate from the microsomal fraction of *Vigna radiata*, *Biochem. J.* **306**: 631-636.

Commun. D., Lecocq, R., Van Weyenberg, V., and Erneux, C., 1995, Active site labelling of Ins(1,4,5)P$_3$-kinase-A by phenylglyoxal, *Biochem. J.* **310**: 109-115.

Cosgrove, D.J., 1980, Intermediates in the dephosphorylation of inositolP$_6$ by phytase enzymes, in Inositol Phosphates, their Chemistry, Biochemistry and Physiology. Elsevier Amsterdam , pp 99-105.

Dasgupta, S., Dasgupta, D., Sen, M., Biswas, S., and Biswas, B.B., 1996, Interaction of myoinositol triphosphate – phytase complex with the receptor for intracellular Ca^{2+} mobilization in plants, *Biochemistr* **35**: 4994-5001.

Dasgupta, S., Dasgupta, D., Chatterjee, A., Biswas, S., and Biswas, B. B., 1997, Conformational changes in plant Ins(1,4,5)P$_3$ receptor on interaction with different myoinositol trisphosphates and its effect on Ca^{2+} release from microsomal fraction and liposomes, *Biochem. J.* **321**: 355-360.

Drøbak, B.K., 1992, The plant phosphoinositide system, *Biochem. J.* **288**: 697-712.

Hasson, T, Gillespie, D, Garcia, J. A., McDonald, R.B., Zhao, Yi-Dong, Yee, A.G., Mooseker, M.S., and Corey, D.P., 1977, Unconventional myosin in inner-ear sensory epithelia, *J. Cell Biol.* **137**: 1287-1307.

Loewus, F.A., Everard, J.D., and Young, K., 1990, Inositol metabolism: Precursor role and breakdown, in Inositol Metabolism in Plants (D.J. Movre, W.F. Boss and F.A. Loewus, eds), Wiley-Liss, New York, pp 21-45.

Maitra, R., Samanta, S., Mukherjee, M., Biswas, S., and Biswas, B.B., 1988, Isolation, characterization and biological function of myoinositol trisphosphate generated by phytase action on myoinositol hexaphosphate, *Ind. J. Biochem. Biophys.* **25**: 655-659.

Mandal, N.C., Burman, S., and Biswas, B. B., 1972, Isolation, purification and characterization of phytase from germinating mungbeans, *Phytochem.* **11**: 495-502.

Mignery, G.A., and Sudhof, T.C., 1990, The ligand binding site and transduction mechanism in the inositol-1,4,5- trisphosphate receptor, *EMBO J.* **9**: 3893-3898.

Pal, M., Ghosh, U., Chandra, M., Pal, A., and Biswas, B. B., 1991, Transformation and regeneration of mungbean (*Vigna radiata*), *Ind. J. Biochem. Biophys.* **28**: 449-455.

Samanta, S., Dalal, B., Biswas, S., and Biswas, B. B., 1993, Myoinositol trisphosphate phytase complex as an elicitor in calcium mobilization in plants, *Biochem. Biophys. Res. Commun.* **191**: 427-434.

Supattapone, S., Worley, P. F., Bazaban, J. M., and Snyder, S. H., 1988, Solubilization, purification and characterization of an inositol trisphosphate receptor, *J. Biochem.* **263**: 1530-1534.

Trewavas, A.J., and Malho, R., 1997, Signal Perception and Transduction: The origin of the phenotype. *Plant Cell* **9**: 1181-1195.

The Phosphoinositide (PI) Pathway and Signaling in Plants

IMARA Y. PERERA[1], INGO HEILMANN[2], AND WENDY F. BOSS[1]
[1]*Department of Botany, North Carolina State University, Raleigh, NC 27695 USA;* [2]*Institut für Pflanzenphysiologie und Mikrobiologie, Freie Universität Berlin, Germany*

1. INTRODUCTION

As sessile organisms, plants need to be able to perceive and respond to a wide range of environmental stimuli. Plants possess mechanisms to detect specific signals, transduce the information intra- and inter-cellularly, and initiate the appropriate responses. The phosphoinositide (PI) pathway is a major signal transduction pathway in plants involved in mediating the responses to various stresses (for reviews see Munnik *et al* 1998, Drøbak 1992).

In eukaryotes, the family of phosphoinositides are important metabolites and labile messengers involved in regulating cellular physiology. Key steps in the PI pathway are outlined in a simplified model in Fig. 1. The membrane associated phospholipid, phosphatidylinositol (PtdIns), is sequentially phosphorylated by specific lipid kinases (PI 4-kinase and PIP 5-kinase) (for a recent review on the plant lipid kinases see Drøbak *et al* 1999) to form phosphatidylinositol 4-phosphate (PtdInsP) and phosphatidylinositol 4,5-bisphosphate (PtdInsP$_2$), respectively. The stereospecificity of the six hydroxyls on the inositol ring, the multiple phosphorylated isomers, and the ability of the PIs to permeate both hydrophilic and hydrophobic environments allows for great functional diversity of the phosphoinositides. Both PtdInsP and PtdInsP$_2$ are regulators of cytoskeletal dynamics, vesicle trafficking and ion transport (Corvera *et al* 1999, Toker 1998, Janmey 1994). In response to a stimulus, PtdInsP$_2$ is hydrolyzed by phospholipase C (PLC), to produce the soluble second messenger inositol 1,4,5-trisphosphate (InsP$_3$) and diacylglycerol (DAG). It is well established that InsP$_3$ can trigger Ca^{2+} release from intracellular stores. In plants, the primary InsP$_3$-sensitive Ca^{2+}-store appears to be the vacuole (Canut *et al* 1993, Alexandre *et al* 1990, Schumaker and Sze 1987) although additional InsP$_3$-sensitive Ca^{2+} stores may exist (Sanders *et al* 1999, Muir and Sanders 1997).

1.1 Ca^{2+} and InsP$_3$ are Important Second Messengers in Plants

Transient increases in intracellular Ca^{2+} are an integral part of plant signaling and have been reported in plants after stimulation by cold (Knight *et al* 1996, Polisensky *et al* 1996), touch (Knight *et al* 1991), anoxia (Sedbrook *et al* 1996, Subbaiah *et al* 1994) and osmotic stress (Knight *et al* 1997, 1998). The high buffering capacity of the cytoplasm due in part to Ca^{2+} binding proteins such as calmodulin (CaM) or Ca^{2+} dependent protein kinases (CDPKs) limits the diffusion of Ca^{2+} within the cytoplasm and confines the Ca^{2+} changes to localized microdomains. However, a Ca^{2+} signal can be propagated through the cell as a wave or as oscillations by a regenerative process involving InsP$_3$-mediated Ca^{2+} release from intracellular stores and Ca^{2+} re-uptake *via* transporters. All plant PLCs characterized so far are of the delta type and Ca^{2+} sensitive. Therefore, a local increase in Ca^{2+} can induce the generation of InsP$_3$ which in turn can elicit further Ca^{2+} release. The spatial and temporal characteristics of the Ca^{2+} changes impart specificity to this universal second messenger and evoke the expression of specific genes involved in the appropriate stress response (Trewavas 1999, Sanders *et al* 1999, McAinsh and Hetherington 1998).

It has been shown that Ca^{2+} oscillations can be generated in plant cells by the microinjection of InsP$_3$ into stamen hair cells (Tucker and Boss 1996). Similarly, in pollen tubes, a Ca^{2+} wave can be propagated by the photo-release of caged InsP$_3$. The generation of the Ca^{2+} wave was blocked by heparin, which inhibits InsP$_3$ action (Franklin-Tong *et al* 1996). While the inter-dependency of InsP$_3$ and Ca^{2+} signals is beyond doubt, the challenge to plant scientists is to demonstrate changes in both InsP$_3$ and Ca^{2+} in the same plant system during the response to a physiological stimulus.

2. PHOSPHOINOSITIDE SIGNALING AND PLANT RESPONSES TO GRAVITY

Changes in InsP$_3$ and PI metabolism have been demonstrated in many different plant tissues in response to environmental stimuli and effectors including light (Kim *et al* 1996, Morse *et al* 1987), cold (Smolenska-Sym and Kacperska 1994, 1996), osmotic shock (Heilmann *et al* 1999, Pical *et al* 1999, Cho *et al* 1993, Srivastava and Jacoby 1989, Einsphar *et al* 1988), fungal elicitors (Kurosaki *et al* 1987), mastoparan (Franklin-Tong *et al* 1996, Cho *et al* 1995, Drøbak and Watkins 1994) and ABA (Lee *et al* 1996).

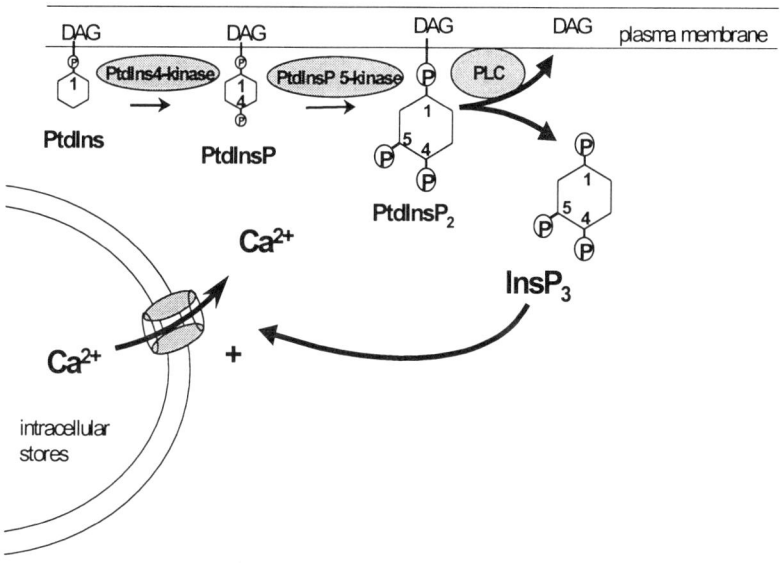

Figure 1. Model of the phosphoinositide pathway. Specific lipid kinases PtdIns 4-kinase and PtdInsP 5-kinase phosphorylate the membrane associated phospholipids, phosphatidylinositol (PtdIns) and phosphatidylinositol 4-phosphate (PtdInsP). Activation of phospholipase C (PLC) leads to the hydrolysis of phosphatidylinositol 4,5-bisphosphate (PtdInsP$_2$) yielding the soluble second messenger inositol 1,4,5-trisphosphate (InsP$_3$) and diacylglycerol (DAG). InsP$_3$ can mediate Ca^{2+} release from intracellular stores thereby triggering multiple downstream events.

Recently we reported that both short and long term increases in InsP$_3$ are associated with the gravitropic bending response in maize plants (Perera *et al* 1999). We investigated the involvement of the phosphoinositide pathway in gravisignaling using the internodal or stem pulvinus of maize plants as a model. The internodal pulvinus of maize plants is a disc-shaped tissue located at the base of the internode, immediately above the node. Cells containing starch (the putative gravity sensors) are confined solely to the pulvinus. Gravitropic curvature of the plant stem is due to differential cell elongation occurring exclusively on the lower side of the pulvinus. As the site of both graviperception and response, the pulvinus tissue is an ideal system to study the mechanisms for sensing changes in spatial orientation.

Within 10 s of gravistimulation, there was a transient 5-fold increase in $InsP_3$ in the lower half of the pulvinus. Subsequently, $InsP_3$ levels in the upper and lower halves oscillated asynchronously for at least 30 min. To determine whether the oscillations in $InsP_3$ affect the biosynthesis of $PtdInsP_2$, the precursor of $InsP_3$, we measured PIP 5-kinase activity in plasma membranes isolated from the upper and lower halves of gravistimulated pulvini. PIP 5-kinase activity in the lower pulvinus half increased transiently within 10 min of gravistimulation, and fluctuated between the upper and lower halves of the pulvinus over the first 2 hours of gravistimulation. Importantly, neither $InsP_3$ levels nor PIP 5-kinase activity changed in pulvini halves from vertical control plants.

Furthermore, we determined the presentation time (the duration of the gravistimulus necessary to induce a bending response) in maize to be between 2 and 4 hours of gravistimulation. To investigate whether changes in $InsP_3$ levels correlated with the establishment of differential growth, $InsP_3$ content was measured in upper and lower halves of maize pulvini over a period of several hours. Between 2 and 8 hours of gravistimulation, $InsP_3$ levels of the lower pulvinus half gradually increased up to 6–fold (Fig. 2). This increase was accompanied by an increase in PIP 5-kinase activity in the lower half of the pulvinus. This could reflect an increase in phosphoinositide turnover due to PLC activity in the lower pulvinus half. After 48 h, cells on the lower side of the pulvinus have completed the elongation response. Analysis of plasma membranes from maize pulvinus tissue at 48 h indicated that by this time the cells in the upper half had also shifted metabolism and exhibited 2- to 3-fold increased PIP 5-kinase activity (Perera, Heilmann and Boss unpublished results).

These data suggest that upon gravistimulation the cells in the pulvini undergo a change in the metabolic state reflected by increases in $InsP_3$ and $PtdInsP_2$ and by increased specific activity of PIP 5-kinase. As a diffusible second messenger, $InsP_3$ provides a mechanism to transmit and amplify the signal from the perceiving to the responding cells in the pulvinus. A sensing-mechanism combining both short term and long term changes in $InsP_3$ could enable a plant to distinguish between transient movements, caused by wind, and permanent lodging. The initial spike would serve as an initiation signal and the gradual increase would, in part, sustain the requisite biochemical processes that precede differential growth. Phosphoinositide metabolism is therefore not only involved in early signal transduction events following a change in spatial orientation, but possibly also in the establishment of tissue polarity and the coordination of differential growth during the gravitropic bending response of cereal grasses.

Short and long-term changes in PtdIns metabolism have also been reported in the response of plants to hyperosmotic stress. We detected an increase in InsP$_3$ within 90 s in *Galdieria sulphuraria*, when the osmolality of the culture medium was increased by as little as 2% (Heilmann *et al* 1999). The production of InsP$_3$ increased linearly up to 6-fold as the osmoticum was increased from 2 to 50%. Long-term changes in PtdIns metabolism and InsP$_3$ also have been documented in several plant species including *Dunaliella salina* (Einspahr *et al* 1988), red beet (Srivastava and Jacoby 1989) carrot cells (Cho *et al* 1993) and *Arabidopsis* (Pical *et al* 1999) within 5-10 min in response to hyperosmotic stress. Furthermore, key enzymes of the phosphoinositide pathway such as the PLC isoform PLC1 (Hirayama *et al* 1995) and a PtdInsP 5-kinase isoform (Mikami *et al* 1998) are upregulated in *Arabidopsis* in response to drought and salt stress. PtdInsP 5-kinase and PLC1 mRNA levels were shown to increase over the first 2-5 hours of stress.

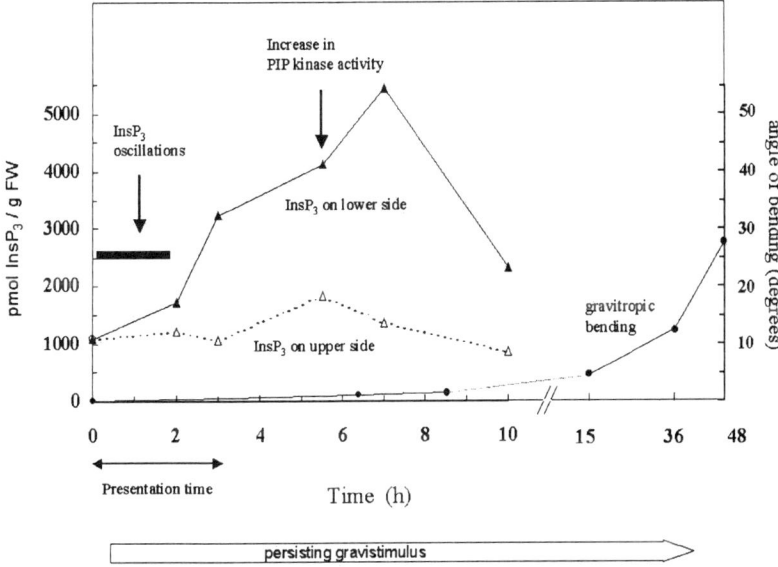

Figure 2. Transient and long term changes in InsP$_3$ in the gravistimulated maize pulvinus. Starting at 10 s and over the first 2 h of gravistimulation transient oscillatory changes in InsP$_3$ occur in both the upper and lower sides of the pulvinus (denoted by the solid bar). There is a gradual increase in InsP$_3$ only on the lower side (▲) compared with the upper (Δ) over the first 2-8 h of gravistimulation. Bending is first detectable at ~ 8h (•) and reaches a maximum at 48 h. The arrow indicates the timing of the increase in PIP 5-kinase activity on the lower side.

3. ADAPTATION TO THE ENVIRONMENT AND OSMOSIGNALING IN *GALDIERIA SULPHURARIA*

The intracellular distribution of $PtdInsP_2$ and of PIP 5-kinase was investigated in cultures of the red alga *G. sulphuraria* (Heilmann et al 1999). Cells grown heterotrophically in liquid culture undergo severe changes in metabolism as they enter the stationary phase of growth, possibly due to the depletion of the growth substrate. Upon the transition from logarithmic growth to the stationary phase of growth, the specific activity of PIP 5-kinase increased by 300% from a minimum value on day 7 after inoculation to a maximum on day 12 (Fig. 3A). Over this period, the amounts of $PtdInsP_2$ in the cells increased by 30%. In contrast, in plasma membranes the levels of $PtdInsP_2$ decreased by 70% between day 7 and day 12 (Fig. 3B). Furthermore, the specific activity of PIP 5-kinase in plasma membranes decreased by 65% during this time. Concurrently, the levels of PIP 5-kinase protein in plasma membranes from day 7 were higher than in plasma membranes from day 12 (Heilmann et al unpublished results).

Upon mild osmotic stimulation, $InsP_3$ increased transiently by up to 500% after 90 s in both 7-d-old and 12-d-old cells. The generation of $InsP_3$ at 90 s was preceded by a stimulus-induced transient increase in the specific activity of PIP 5-kinase at 60 s only in 12-d-old cells despite their already high total $PtdInsP_2$ content and high basal specific activity of PIP 5-kinase. We conclude that the distribution of $PtdInsP_2$ differs between 7-d-old and 12-d-old cells, and that the majority of $PtdInsP_2$ in 12-d-old cells is not involved in the production of $InsP_3$ signals, thereby requiring the activation of the PIP 5-kinase. These results imply the presence of distinct pools of $PtdInsP_2$ in the cells. Induced by the depletion of organic substrate for heterotrophic growth, *G. sulphuraria* cells undergo a shift in their physiological state at the transition from logarithmic growth to the stationary phase (between day 7 and day 12 after inoculation). This is reflected by changes in cellular phosphoinositide distribution and the reorganization of cellular pools of phosphoinositides.

4. CONCLUSIONS

The results presented here from two diverse plant systems highlight the changes in phosphoinositide metabolism that occur when cells adapt to a changing environment and initiate responses to extracellular cues (i. e. the shift in metabolism in the maize pulvinus during gravistimulation and the depletion of organic substrate for heterotrophic growth in *G. sulphuraria*). Most plant responses to stress involve a change in growth and therefore involve cell elongation and *de novo* membrane biogenesis, which could be regulated by PtdInsP$_2$ and InsP$_3$. The structural properties and multiple functions of phosphoinositides make them likely candidates for a central role in plant signal transduction, as well as in the orchestration of a coordinated cellular response to a variety of plant stresses.

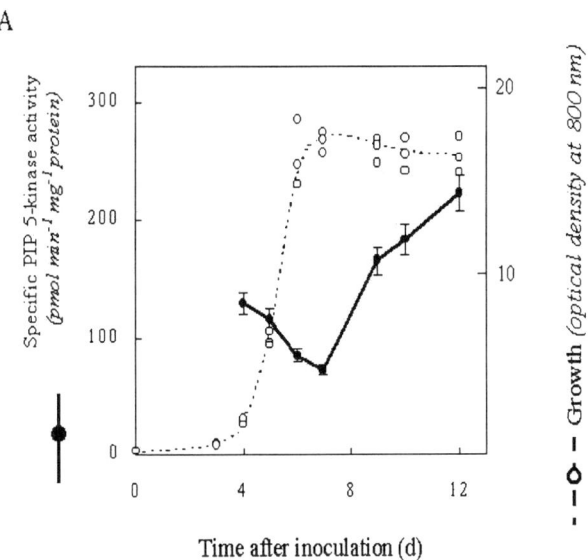

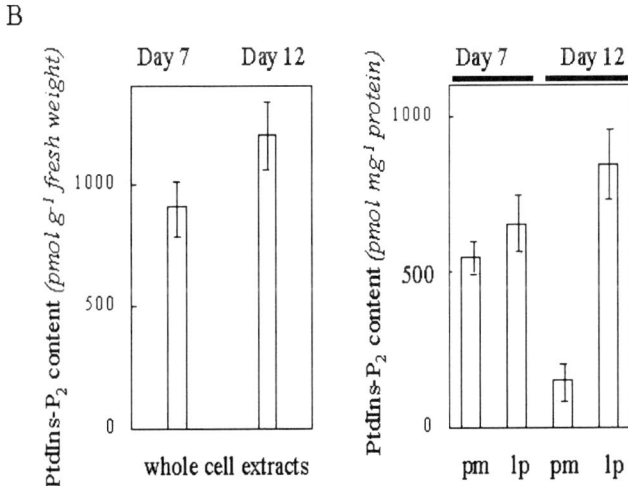

Figure 3. Changes in specific activity of PIP 5-kinase and levels of PtdIns-P$_2$ during the stationary phase of growth of the red alga *Galdieria sulphuraria*. A, Specific activity of PIP 5-kinase (solid line) in microsomes at various times of culture. Data are from one representative experiment, assayed in duplicate. The experiment was

repeated twice and the results were similar. The growth of the alga is indicated by the dashed line. B, Levels of PtdIns-P_2 in whole *G. sulphuraria* cells and in plasma membranes. Left panel, The content of PtdIns-P_2 was measured in whole 7-d-old and 12-d-old cells. Right panel, Plasma membranes were enriched by aqueous two phase partitioning on a 5.9% polymer system. PtdIns-P_2 content of plasma membrane (pm) and lower phase (lp) fractions from 7-d-old and 12-d-old cells. The data are the average of values from two experiments, assayed in duplicate.

ACKNOWLEDGEMENTS

This work was supported in part by NASA grant NAGW-4984, NSF grant MCB-9604285 and the NC Agricultural Research Service (WFB); and a DAAD fellowship HSPIII (IH).

REFERENCES

Alexandre, .J, Lassalles, J.P., and Kado, R.T., 1990, Opening of Ca^{2+} channels in isolated red beet root vacuole membrane by inositol 1,4,5-trisphosphate. *Nature* **343**: 567-569.

Allen, G.J, and Sanders, D., 1994, Osmotic stress enhances the competence of *Beta vulgaris* vacuoles to respond to inositol 1,4,5-trispohosphate. *Plant J.* **6**: 687-695.

Canut, H., Carrasco, A., Rossignol, M., and Ranjeva, R., 1993, Is vacuole the richest store of IP_3-mobilizable calcium in plant cells? *Plant Science* **90**: 135-143.

Cho, M.H., Tan, Z., Erneux, C., Shears, S., and Boss, W.F., 1995, The effects of mastoparan on the carrot cell plasma membrane polyphosphoinositide phospholipase C. *Plant Physiol.* **107**: 845-856.

Cho, M.H., Shears, S.B., and Boss, W.F., 1993, Changes in phosphatidylinositol metabolism in response to hyperosmotic stress in *Daucus carota* L. cells grown in suspension culture. *Plant Physiol.* **103**: 637-647.

Corvera, S., D'Arrigo, A., and Stenmark, H., 1999, Phosphoinositides in membrane traffic. *Curr. Opin. Cell Biol.* **11**: 460-465.

Drøbak, B.K., Dewey, R.E., and Boss, W.F., 1999, Phosphoinositide kinases and the synthesis of polyphosphoinositides in higher plant cells. In K.W. Jeon, ed., *International Review of Cytology*, Academic Press, **189**: 95-130.

Drøbak, B.K., and Watkins, P.A.C., 1994, Inositol (1,4,5) trisphosphate production in plant cells: stimulation by the venom peptides, melittin and mastoparan. *Biochem. Biophys. Res. Comm.* **205**: 739-745.

Drøbak, B.K., 1992, The plant phosphoinositide system. *Biochem. J.* **288**: 697-712.

Einspahr, K.J., Maeda, M., and Thompson, jr., G.A., 1988, Concurrent changes in *Dunaliella salina* ultrastructure and membrane phospholipid metabolism after hyperosmotic shock. *J. Cell Biol.* **107**: 529-538.

Franklin-Tong, V.E., DrØbak, B.K., Allan, A.C., Watkins, P.A.C., and Trewavas, A.J., 1996, Growth of pollen tubes of *Papaver rhoeas* is regulated by a slow-moving calcium wave propagated by inositol 1,4,5-trisphosphate. *Plant Cell* **8**: 1305-1321.

Heilmann, I., Perera. I.Y., Gross, W., and Boss, W.F., 1999, Changes in phosphoinositide metabolism with days in culture affect signal transduction pathways in *Galdieria sulphuraria*. *Plant Physiol.* **119**: 1331-1339.

Hirayama, T., Ohto, C., Mizoguchi, T., and Shinozaki, K., 1995, A gene encoding a phosphatidyl-inositol-specific phospholipase C is induced by dehydration and salt stress in *Arabidopsis thaliana*. *Proc. Natl. Acad. Sci. USA* **92**: 3903-3907.

Janmey, P.A., 1994,. Phosphoinositides and calcium as regulators of cellular actin assembly and disassembly. *Annu. Rev. Physiol.* **56**: 169-191.

Kim, H.Y., Coté, G.G., and Crain, R.C., 1996, Inositol 1,4,5-trisphosphate may mediate regulation of K^+ channels by light and darkness in *Samanea saman* motor cells. *Planta* **198**: 279-287.

Knight, H., Brandt, S., and Knight, M.R., 1998,. A history of stress alters drought calcium signaling pathways in *Arabidopsis*. *Plant J.* **16**(6): 681-687.

Knight, H., Trewavas, A.J., and Knight, M.R., 1997, Calcium signaling in *Arabidopsis thaliana* responding to drought and salinity. *Plant J.* **12**: 1067-1078.

Knight, H., Trewavas, A.J., and Knight, M.R., 1996, Cold calcium signaling in *Arabidopsis* involves two cellular pools and a change in calcium signature after acclimation. *Plant Cell* **8**: 489-503.

Knight, M.R., Campbel,l A.K., Smith, S.M., and Trewavas, A.J., 1991, Transgenic plant aequorin reports the effects of touch and cold-shock and elicitors on cytoplasmic calcium. *Nature* **352**: 524-526.

Kurosaki, F., Tsurusawa, Y., and Nishi, A., 1987, Breakdown of phosphatidylinositol during the elicitation of phytoalexin production in cultured carrot cells. *Plant Physiol.* **85**: 601-604.

Lee, Y., Choi, Y.B., Suh, S., Lee, J., Assmann, S.M., Joe, C.O., Kelleher, J.F., and Crain, R.C., 1996, Abscisic acid-induced phosphoinositide turnover in guard cell protoplats on *Vicia faba*. *Plant Physiol.* **110**: 987-996.

McAnish, M.R., and Hetherington, A. M., 1998, Encoding specificity in Ca^{2+} signalling systems. *Trends Plant Sci.* **3**: 32-36.

Mikami, K., Katagiri, T., Iuchi, S., Yamaguchi-Shinozaki, K., and Shinozaki, K., 1998, A gene encoding phosphatidylinositol-4-phosphate 5-kinase is induced by water stress and abscisic acid in *Arabidopsis thaliana*. *Plant J.* **15**: 563-568.

Morse, M.J., Crain, R.C., and Satter, R.L., 1987, Light-stimulated inositol phospholipid turnover in *Samanea saman*. *Proc. Natl. Acad. Sci. USA* **84**: 7075-7078.

Muir, S.R., and Sanders, D., 1997, Inositol 1,4,5-trisphosphate-sensitive Ca^{2+} release across nonvacuolar membranes in cauliflower. *Plant Physiol.* **114**: 1511-1521.

Munnik, T., Irvine, R.F., and Musgrave, A., 1998, Phospholipid signalling in plants. *Biochim. Biophys. Acta* **1389**: 222-272.

Perera, I.Y., Heilmann, I., and Boss, W.F., 1999, Transient and sustained increases in nositol 1,4,5-trisphosphate precede the differential growth response in gravistimulated maize pulvini. *Proc. Natl. Acad. Sci. USA* **96**: 5838-5834.

Pical, C., Westergren. T., Dove, S.K., Larsson, C., and Sommarin, M., 1999, Salinity and hyperosmotic stress induce rapid increases in phosphatidylinositol 4,5-bisphosphate, diaclyglycerol pyrophosphate, and phosphatidylcholine in *Arabidopsis thaliana* cells. *J. Biol. Chem.* **274**: 38232-38240.

Polisensky, D.H., and Braam, J., 1996, Cold-shock regulation of the *Arabidopsis* TCH genes and the effects of modulating intracellular calcium levels. *Plant Physiol.* **111**: 1271-1279.

Sanders, D., Brownlee, C., and Harper, J.F., 1999, Communicating with calcium. *Plant Cell* **11**: 691-706.

Schumaker, K.S., and Sze, H., 1998. Inositol 1,4,5-trisphosphate releases Ca^{2+} from vacuolar membrane vesicles of oat roots. *J. Biol. Chem.* **262**: 3944-3946.

Sedbrook, J.C., Kronebusch. P.J., Borisy, G.G., Trewavas, A.J., and Masson, P.H., 1996, Transgenic *AEQUORIN* reveals organ-specific cytosolic Ca^{2+} responses to anoxia in *Arabidopsis thaliana* seedlings. *Plant Physiol.* **111**: 243-257.

Smolenska-Sym, G., and Kacperska, A., 1996, Inositol 1,4,5-trisphosphate formation in leaves of winter oilseed rape plants in response to freezing, tissue water potential and abscisic acid. *Physiol. Plant.* **96**: 692-698.

Smolenska-Sym, G., and Kacperska, A., 1994, Phophatidylinositol metabolism in low temperature-affected winter oilseed rape leaves. *Physiol. Plant.* **91**: 1-8.

Srivastava, A., Pines, M., and Jacoby, B., 1989, Enhanced potassium uptake and phosphatidylinositol-phosphate turnover by hypertonic mannitol shock. *Physiol. Plant.* **77**: 320-325.

Subbaiah, C.C., Bush, D.S., and Sachs, M.M., 1994, Elevation of cytolsolic calcium precedes anoxic gene expression in maize suspension-cultured cells. *Plant Cell* **6**: 1747-1762.

Trewavas, A., 1999, Le calcium, c'est la vie: calcium makes waves. *Plant Physiol.* **120**: 1-6.

Toker, A., 1998, The synthesis and cellular roles of phosphatidylinositol 4,5-bisphosphate. *Curr. Opin. Cell Biol.* **10**: 254-261.

Tucker, E.B., and Boss, W.F., 1996, Mastoparan-induced intracellular Ca^{2+} fluxes may regulate cell-to-cell communication in plants. *Plant Physiol.* **111**: 459-467.

Phytochrome Signal Transduction in Cucumber Cotyledons

DOLORS VIDAL, CARMEN BERGARECHE, M. TERESA GIL, AND ESTHER SIMON
Department of Plant Biology, Faculty of Biology, University of Barcelona, Diagonal 645, 08028 Barcelona, Spain

1. INTRODUCTION

Nitrate reductase (NR) is the key enzyme in nitrate assimilation in plants. The specific calcium requirement by NR and its role in enzyme function are not well stablished. Previous experiments showed that Pfr (far-red light-absorbing form of phytochrome) promoted NRA in etiolated cotyledons of *Cucumis sativus* through a red-far-red reversible response, and an increase in cytoplasmatic calcium mimicked the Pfr response (Bergareche *et al* 1994). PI turnover is a major mechanism for transmembrane signalling in response to external stimuli (Lehle, 1990). Several lines of evidence support the presence of plant PI, the enzymes involved in their metabolism and a Ca^{2+}-regulated amplification system responsive to IP_3. It has been proposed that inositol phospholipids are hydrolyzed in the light and their products, IP_3 and diacylglycerol (DAG), increase cytosolic Ca^{2+} and activate kinases, respectively (Berridge 1987). Furthermore, Pfr affects the levels of PI in etiolated leaves of *Zea mays* (Guron *et al* 1992). PI turnover replaces the light requirement for the induction of NR in maize (Chandok and Sopory 1994).

On the other hand, phorbol myristate acetate, an activator of protein kinase C, increases NRA in the absence of light and a PKC-type kinase appears to be involved in the light mediated expression of NR (Raghuram and Sopory 1995). According to Chandok and Sopory (1996) phosphorylation *via* PKC-type kinase may be a key event in Pfr transduction and Pfr may initiate more than one signalling event in the regulation of NR.

However, PKC-type kinases do not appear to play a general role in Pfr transduction pathways, since tyrosine/kinase and serine/threonine kinase are involved in Pfr-mediated control of chalcone synthase and chlorophyll a/b-binding protein respectively (Schäfer et al 1997). Besides, Romero and Lam (1993) did not find any effect of protein kinase inhibitors on Pfr-control of the Lhcb gene. Thus, the role of protein phosphorylation and the type of protein kinase involved in Pfr-mediated responses are not fully established.

Here we examined whether the PI pathway and protein phosphorylation participated in Pfr-transduction in cucumber NRA response, by testing the effect of inhibitors and activators of this pathway and analyzing light-dependent protein phosphorylation.

2. MATERIALS AND METHODS

Cucumis sativus seeds were germinated on moist vermiculite at 25°C in the dark. Etiolated seedlings of 6 days old were used in all experiments. For NRA and IP_3 assays, etiolated cotyledons were incubated in 50 mM $Ca(NO_3)_2$ for 7 h in the dark and were then transferred for 2 h to Petri dishes containing the different chemicals assayed. Then, 15 min red light (R) irradiation (17,4 µE $m^{-2} \cdot s^{-1}$) was applied to cotyledons, followed by a dark incubation in $Ca(NO_3)_2$ for 15 h. At the end of the incubation period, samples were taken and analyzed. For protein phosphorylation assays, 15 min of R light was applied to etiolated seedlings. Then, cotyledons were excised, frozen and homogenized (Chandok and Sopory 1992) at 4°C. The homogenate was centrifuged at 17,000xg for 30 min; the pellet was discarded, and the supernatant was centrifuged again at 100,000xg for 1 hour and assayed for protein phosphorylation. The R light source was as previously described (Bergareche et al 1994). All extractions and assays were performed under green safe light.

NR assays were conducted *in vivo* using the method described elsewhere (Bergareche and Simón 1988). A IP_3 analysis based on the binding reaction of IP_3 to bovine adrenal preparations was used to measured endogenous IP_3. Frozen cotyledons were homogenized in the same medium as for NR assay, mixed with ice-cold 20% perchloric acid, kept on ice for 20 min and centrifuged at 10.000xg for 15 min at 4°C. Supernatants were titrated to pH 7.5 and tested by the ^{3}H-IP_3 system (Amersham). The bound IP_3 was then separated by centrifugation and the radioactivity was counted in a ß-scintillation counter.

Total protein phosphorylation was measured as decribed by Chandok and Sopory (1992) in a medium with histone H1 (as exogenous substrate) and 100 µM ATP containing about $5 \cdot 10^5$ cpm [γ ^{32}P]ATP per sample. Reactions were carried out at 30°C for 15 min. Radioactivity of precipitates was measured in a scintillation counter. Protein phosphorylation was also assayed in a reaction mixture containing 80 μM [γ ^{32}P] ATP (3,000 Ci·mmol^{-1}, Amersham). After incubation at 37°C for 10 min, the enzymatic reaction was stopped by adding SDS-sample buffer. Protein samples were subjected to SDS-PAGE (Laemmli 1970). Radiolabeled polypeptides were visualized by autoradiography.

3. RESULTS AND DISCUSSION

3.1 The Role of Inositol Phospholipides on NRA and IP$_3$ Levels

3.1.1 Effect of neomycin sulfate and lithium on NRA

Dark and R-induced cotyledons were incubated in neomycin sulfate. NRA levels in R-irradiated cotyledons diminished gradually as neomycin concentration increased (Fig. 1). 30-50 μM neomycin completely inhibited the R-induced increase in NRA. Non-irradiated cotyledons did not show any change in NRA when treated with neomycin sulfate.

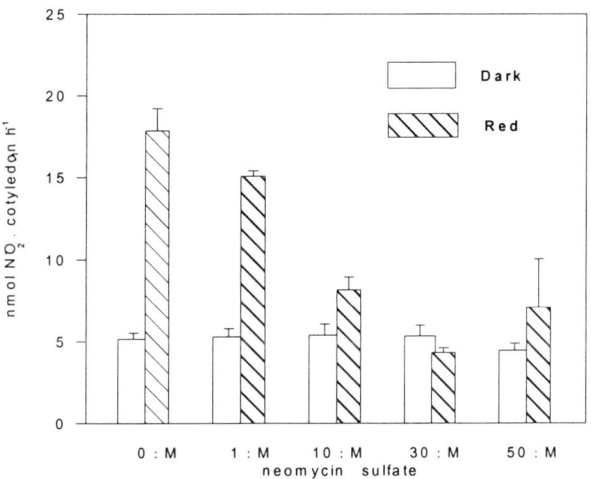

Figure 1. Effect of neomycin and red light on nitrate reductase activity of cucumber cotyledons. A 2 h pulse of neomycin sulfate was applied 7 h after the beginning of calcium nitrate incubation and a 15 min R light pulse was applied after removing the neomycin.

When etiolated cucumber cotyledons were incubated for 2 h in 10 μM LiCl before R pulse, then R irradiated for 15 min and then kept in darkness in a 50 mM Ca(NO$_3$)$_2$ solution for 15 h, NRA decreased 32% in relation to control (non treated cotyledons) (Fig. 2). The Li$^+$ concentration assayed was too low to revert the Pfr effect. 10 μM *myo*-inositol reverted the inhibitory effect of Li$^+$. No differences to the irradiated control were observed when *myo*-inositol was supplied alone.

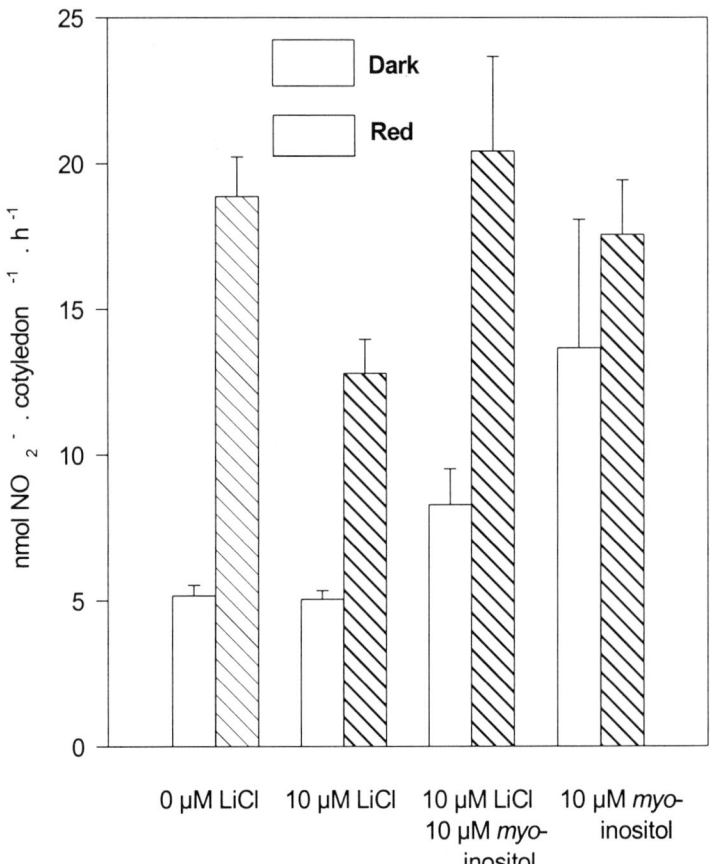

Figure 2. Effect of lithium, myo-inositol and red light on nitrate reductase activity of cucumber cotyledons. A 2 h pulse of LiCl was applied 7 h after the beginning of Ca(NO$_3$)$_2$ incubation and a 15 min R light pulse was applied after removing LiCl.

Lithium application had no effect on NRA in etiolated cotyledons, but the supply of *myo*-inositol with or without Li^+ increased NRA 31 % and 85 % respectively (Fig. 2). Addition of 10 μM LiCl also inhibited the R-induced protoplasts swelling and 1 mM *myo*-inositol prevented the Li^+ effect (Bossen *et al* 1990). Inhibition of Li^+ was also reported in connection with Pfr-induced hook opening (Glaber 1988). Chandok and Sopory (1998) reported that Li^+ inhibits R-light stimulated nitrate-induced NRA in etiolated maize leaves. However, they found that the effect of Li^+ was greater if it was given immediately after R light. In our experiments, the effect of Li^+ was clearest if it was added before R irradiation, as described by Hartmann and Pfaffmann (1990) in moss protonemata. Our data clearly indicated that R light effect on NRA was inhibited specifically by Li^+ and neomicyn. Thus, the results are consistent with the involvement of PI metabolism in Pfr control of NRA in cucumber cotyledons.

3.1.2 Effect of neomycin sulfate and lithium on IP_3

Neomycin sulfate application decreased IP_3 levels in R-treated cotyledons whereas the opposite effect was obtained for etiolated cotyledons (Tab. 1).

The effect of Li^+ was clearer if it was added 2 h before irradiation. Incubation in LiCl of etiolated cotyledons for 2 h increased IP_3 levels of irradiated cotyledons about 10-fold and the effect was almost completely reversed by *myo*-inositol. When *myo*-inositol was supplied alone, IP_3 levels increased about 5-fold. A very small increase in IP_3 was detected in etiolated cotyledons treated with lithium and/or *myo*-inositol (Fig. 3).

Table 1. IP_3 levels in R and etiolated cucumber cotyledons treated with neomycin sulfate expressed in percentage of the control.

Neomycin sulfate	IP_3 (% of control) R irradiated cotyledons	IP_3 (% of control) etiolated cotyledons
0 μM	100 %	100 %
1 μM	53.3 %	183.2 %
10 μM	79.9 %	139.1 %
30 μM	93.3 %	133.5 %

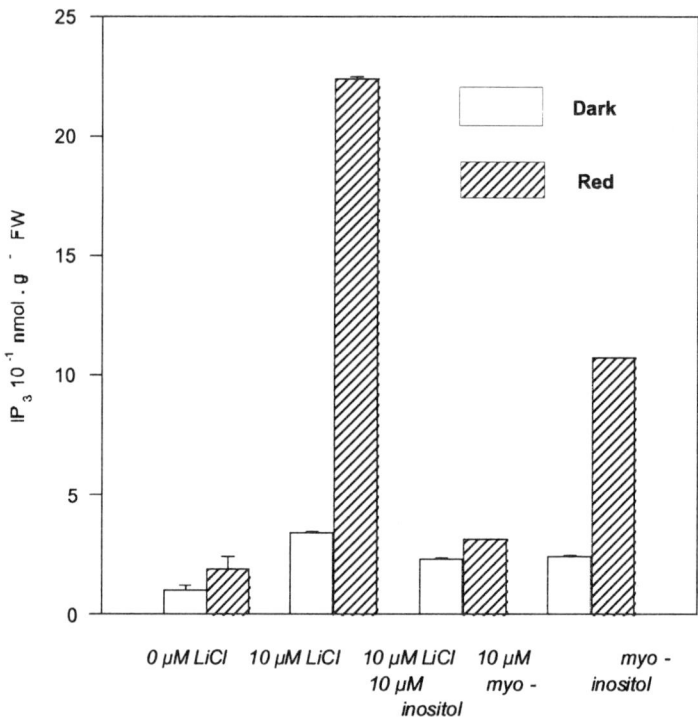

Figure 3. Effect of lithium, myo-inositol and R light on IP_3 concentration of cucumber cotyledons. A 2 h pulse of LiCl was applied 7 h after the beginning of $Ca(NO_3)_2$ incubation and a 15 min R light pulse was applied after removing LiCl and/or myo-inositol.

3.2 Protein Phosphorylation

A 15 min R light treatment increased the total ^{32}P incorporated into the proteins of the extract, as quantified by scintillation counting (Tab. 2). The addition of Ca^{2+} (1-5 mM) to the extracts mimicked the effect of R light and increased the total phosphorylated proteins.

Table 2. Effect of 15 min R light applied to etiolated cucumber seedlings on total protein phosphorylation, as quantified by liquid scintillation. Effect of the addition of Ca^{2+} to the etiolated and R-treated cucumber cotyledon extracts. Data are expressed as percentage of the control.

$CaCl_2$	R-irradiated cotyledons	Etiolated cotyledons
0 µM	145 %	100 %
5 µM	216 %	194 %

3.2.2 Phosphorylation of polypeptides depending on R light and Ca^{2+}

To examine how Ca^{2+} affects kinase activity, 1 and 5 mM Ca^{2+} were added to the extracts and kinase activity was measured after SDS-PAGE and autoradiography. The addition of Ca^{2+} increased the rate of protein phosphorylation, included the histone added to the extract, which increased 7% with 1 mM Ca^{2+} and 24% with 5 mM Ca^{2+}.

The 15 min of R irradiation on cucumber cotyledons changed the kinase activity of the extracts. Thus, R treatment increased histone phosphorylation and the phosphorylation of low molecular weight polypeptides, mainly bands at 30, 24-20 and 18 kDa (Fig. 4). Nevertheless, the rate of phosphorylation of high molecular weight (>45 kDa) proteins decreased. The addition of 10 mM $CaCl_2$ increased phosphorylation of 24-20 and 18 kDa bands (data non shown).

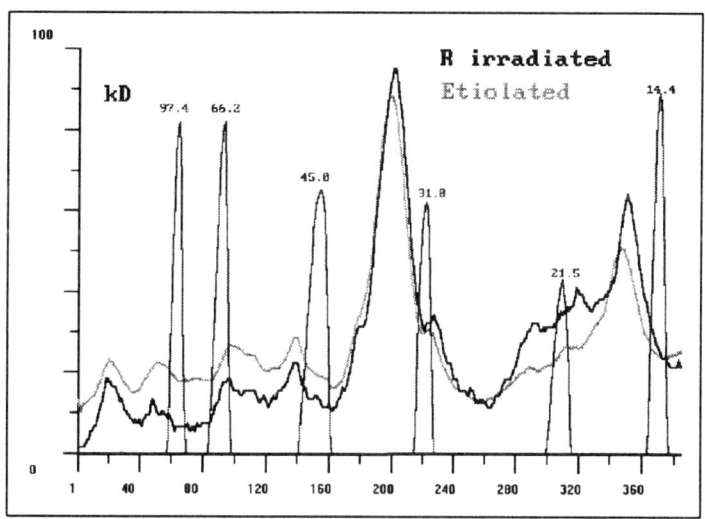

Figure 4. Effect of a 15 min R light pulse on protein phosphorylation of *Cucumis sativus* cotyledon extract. Autoradiography profiles corresponding to etiolated (grey line) and irradiated (black line) cotyledon extracts.

Phosphorylation of low molecular weight proteins stimulated by R light has been described in etiolated pea and rice seedlings (Ogura et al 1999, Hamada et al 1999).

In both cases, R-dependent phosphorylation was related to nucleoside diphosphate kinase. Park and Song (1990) also found a 16 kDa protein phosphorylated dependent on R light in oat seedling extracts. Characterization of the protein kinases involved in R light mediated phosphorylation in cucumber cotyledons are in progress in our laboratory.

ACKNOWLEDGMENTS

We gratefully acknowledge support by grants PB94-0890 and PB97-0928 from the Ministerio de Educación y Ciencia and we thank Robin Rycroft for correcting the English text.

REFERENCES

Bergareche, C., Ayuso, R., Masgrau, C. and Simón, E., 1994, Nitrate reductase in cotyledons of cucumber seedlings as affected by nitrate, phytochrome and calcium. *Physiol. Plant.* **91**: 257-262.

Bergareche, C., and Simón, E., 1988, Nitrate reductase and nitrate content under two forms and three levels of nitrogen nutrition in *Lolium perenne* L. *J. Plant Physiol.* **132**: 28-33.

Berridge, M. J., 1987, Inositol trisphosphate and diacylglycerol: Two interacting second messengers. *Annu. Rev. Biochem.* **56**: 159-193.

Bossen, M.E., Kendrick, R.E. and Vredenberg, W.J., 1990, The involvement of a G-protein in phytochrome-regulated, Ca^{2+}-dependent swelling of etiolated wheat protoplasts. *Physiol. Plant.* **80**: 55-62.

Chandok, M.R., and Sopory, S.K., 1992, Phorbol myristate acetate replaces phytochrome-mediated stimulation of nitrate reductase in maize. *Phytochemistry* **31**: 2255-2258.

Chandok, M.R., and Sopory, S.K., 1994, 5-Hydroxytryptamine affects the turnover of phosphoinositides in maize and stimulates nitrate reductase in absence of light. *FEBS Lett.* **356**: 39-42.

Chandok, M.R., and Sopory, S.K., 1998, Lithium inhibits phytochrome-stimulated nitrate reductase activity and transcript level in etiolated maize leaves. *Plant Science* **139**: 195-203.

Glaber, J., 1988, PhD Tesis *Lichtinduzierte Signalverarbeitung beim Hypokotylhaken von Phaseolus vulgaris*. University of Mainz, Germany.

Guron, K., Chandock, M.R., and Sopory, S. K., 1992, Phytochrome-mediated rapid changes in the level of phosphoinositides in etiolated leaves of *Zea mays*. *Photochem. Photobiol.* **56**: 691-695.

Hamada, T., Hasunuma, K., and Komatsu, S., 1999, Phosphorylation of proteins in the stem section of etiolated rice seedling irradiated with red light. *Biol. Pharm. Bull.* **22**: 122-126.

Hartmann, E., and Pfaffmann H., 1990, Phosphatidilinositol and phytochrome-mediated phototropism of moss protonemal tip cells. In *Inositol metabolism in Plants* (D.J. Morré, W.F. Boss and F.A. Loewus, eds.) Wiley-Liss, New York. pp. 259-275.

Laemmli, U.K., 1970, Cleavage of structural proteins during the assembly of the head of bacteriophage T4. *Nature* **227**: 680-685.

Lehle, L., 1990, Phosphatidyl inositol metabolism and its role in signal transduction in growing plants. *Plant Mol. Biol.* **15**: 647-658.

Ogura, T., Tanaka, N., Yabe, N., Komatsu, S., and Hasunuma, K., 1999, Characterization of protein complexes containing nucleoside diphosphate kinase with characteristics of light signal transduction through phytochrome in etiolated pea seedlings. *Photochem. Photobiol.* **69**: 397-403.

Park, H.J., and Song, P.S., 1990, The phytochrome-mediated phosphorylation in crude extract and nuclei from etiolated *Avena* seedlings. *Photochem. Photobiol.* **51S**: 8-95.

Raghuram, N., and Sopory, S.K., 1995, Evidence for some common signal transduction events for opposite regulation of nitrate reductase and phy-A gene expression by light. *Plant Mol. Biol.* **29**: 25-35.

Romero, L.C., and Lam, E., 1993, Guanine nucleotide binding protein involvement in early steps of phytochrome-regulated gene expression. *Proc. Natl. Acad. Sci. USA* **90**: 1465-1469.

Schäfer, E., Kunkel, T., and Frohnmeyer, H., 1997, Signal transduction in the photocontrol of chalcone synthase gene expression. *Plant Cell Environ.* **20**: 722-727.

Calcium, Calmodulin and Phosphoinositides in Leaflet Movements Mediated by Phytochrome
Nyctinastic and rhythmic movements

LUISA MOYSSET, EVA FERNANDEZ, LUIS A. GOMEZ, AND ESTHER SIMON
Department of Plant Biology, Faculty of Biology, University of Barcelona, Avd. Diagonal, 645, 08028, Barcelona, Spain

1. INTRODUCTION

Calcium transduction pathways are involved in plant signaling, including light and hormonal responses (Sanders *et al* 1999). Calmodulin and phosphoinositide metabolism may also be other components of the transduction pathway (Yang 1996). Cytoplasmic Ca^{2+} has been related with some phytochrome-mediated responses such as protoplast swelling, germination of fern spores and seed germination (Tretyn *et al* 1991). Microinjection experiments reported by Neuhaus *et al* (1993) strongly support this involvement. Thus, a pre-eminent role of Ca^{2+} in phytochrome transduction pathways has been proposed (Roux 1994). Protein phosphorylation and calcium-binding proteins such as calmodulin, could be downstream in the phytochrome transduction pathway (Schäfer *et al* 1997).

Leaflets of various legumes show rhythmic and nyctinastic movements brought about by the pulvinus, a small organ at the base of the leaf blade. Reversible turgor changes in pulvini motor cells are considered responsible for leaflet movement. In turn, they are driven by water and ion fluxes, mainly of K^+ and Cl^- (Satter and Galston 1981). *Albizzia lophantha* and *Robinia pseudoacacia* show nyctinastic and rhythmic movements. Both nyctinastic and rhythmic movements are controlled by an endogenous clock and its interaction with phytochrome photoconversion. Pfr (far-red light-absorbing form of phytochrome), through a low fluence response, enhances closure in nyctinastic movement (Moysset and Simón 1989) and rephases the circadian rhythmic movement of *A. lophantha* leaflets (Moysset *et al* 1994), as it does in *Samanea saman* (Simón *et al* 1979) and *Robinia pseudoacacia* (Gómez and Simón 1995).

Several data support the involvement of Ca^{2+} in phytochrome-mediated nyctinastic movements of *A. lophantha* leaflets. Thus, calcium ionophore A23187 emulated the effect of red light while TMB-8, an intracellular calcium antagonist, inhibited the action of Pfr (Moysset and Simón 1989). Besides, the reversible inhibition of nyctinastic closure of *A. lophantha* leaflets caused by EGTA, a calcium chelating agent, and by lanthanum, a calcium-channel blocker, have also been found (Moysset and Simón 1990, Moysset *et al* 1994).

On the other hand, it has been suggested that intracellular Ca^{2+} is involved in circadian clock functions. Thus, Lakin-Thomas (1985) proposed a model based on oscillations in intracellular Ca^{2+} compartimentation for *Neurospora* rhythm, whereas Kippert (1986) proposed that the gradient of Ca^{2+} between cytoplasm and mitochondrial matrix is the circadian oscillator. Goto *et al* (1985) considered the Ca^{2+}-calmodulin complex a key element in the control loop for the circadian oscillator of cell division rhythm in *Euglena*.

Here we examine whether calcium, calmodulin and phosphatidylinositol pathways participate in the events connecting phytochrome photoconversion and nyctinastic- and rhythmic-movements of *A. lophantha* and *R. pseudoacacia* by testing the effect of some calmodulin antagonists, LiCl and *myo*-inositol on leaflet movement.

2. MATERIAL AND METHODS

Plant growth conditions, light sources, light and chemical treatments and measurement of leaflet movements were as previously described (Gómez and Simón 1995, Moysset and Simón 1989). Experiments testing the effect of lithium and calmodulin antagonists on phytochrome-controlled nyctinastic closure were carried out as described in Moysset and Simón (1989). Experiments conducted to study rhythmic movements were performed by applying calmodulin antagonist pulses of 2 h, which perturb the free-running movement in continuous darkness (DD). Pairs of leaflets were excised at the end of the light period, floated on 50 mM sucrose and kept in DD. Sucrose is required to maintain oscillations for several cycles (Moysset *et al* 1994). Calmodulin antagonist pulses were applied between 8 h and 32 h of darkness. Leaflets were removed from sucrose solution, transferred to calmodulin antagonist solutions for 2 h and returned to fresh sucrose solutions. Measurements were performed at 2 h intervals, starting at 32 h DD. The time series were obtained from means of 10 values of leaflet angles. They were fitted to a five-harmonic discrete Fourier transform with the program PCNONLIN (Statistical Consultants, Inc.), and phase-shifts were calculated according Gómez and Simón (1995).

3. RESULTS AND DISCUSSION

3.1 Effect of Lithium on Nyctinastic Closure

Lithium ions are known to inhibit inositol-1-phosphatase and could therefore reduce the availability of phosphatidylinositol-4,5-bisphosphate (PIP_2) (Halcher and Sherman 1980). LiCl inhibited the nyctinastic closure both in R and FR irradiated *Albizzia* leaflets (Fig. 1). When R light irradiation increased the Pfr level the effect of Li^+ depended on the concentration, and increased as the concentration rose from 2 to 20 mM. On the FR irradiated leaflets the effect was not concentration dependent.

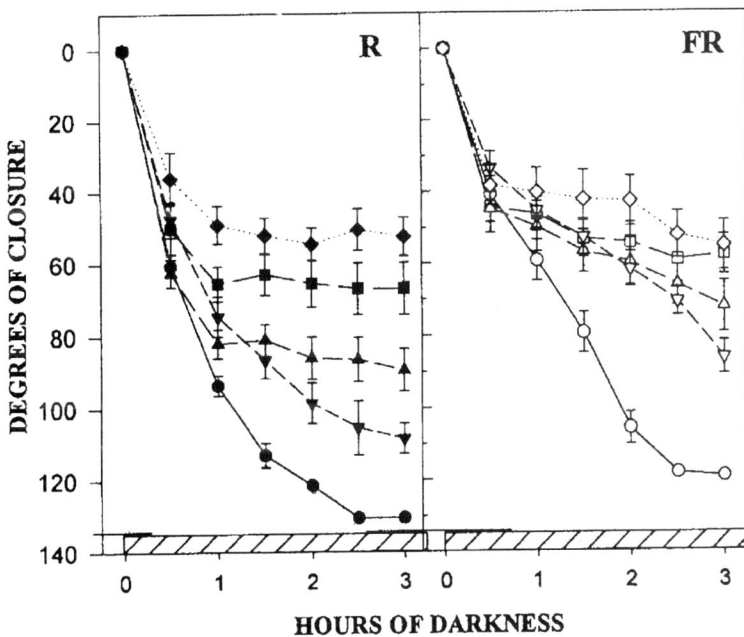

Figure 1. Effect of LiCl on phytochrome controlled nyctinastic closure of *Albizzia lophantha* at 8 h of photoperiod. Leaflet pairs floated on distilled water (•), 2 (▼), 5 (▲), 10 (■) and 20 mM (♦). Each point represents the mean value of three experiments. Vertical bars represent SEM. R, red irradiated leaflets. FR, far-red irradiated leaflets.

The addition of *myo*-inositol after 2 h of Li^+ treatment and immediately before R and FR irradiations partially prevent the inhibitory effect of Li^+ (Fig. 2). On R-irradiated leaflets this inhibitory effect depended on the concentration, and increased as it rose from 50 to 100 mM. On the FR-irradiated leaflets the effect was not concentration dependent.

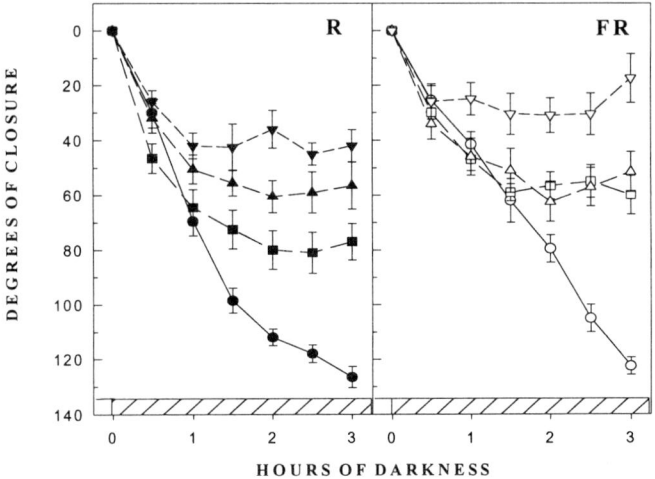

Figure 2. Effect of myo-inositol on LiCl inhibition of phytochrome controlled nyctinastic closure of *Albizzia lophantha* at 8 h of photoperiod. Leaflet pairs were incubated in LiCl 5 mM (▼) for 2 h in white light, after which they were transferred to myo-inositol 50 (▲) or 100 (■) mM, irradiated with R or FR light and transferred to darkness. Control leaflet floated on distilled water (●). For more details see legend to Fig. 1.

These results suggest that inositol metabolism is a component of the Pfr transduction pathway, as was also found in R-induced swelling of etiolated wheat protoplast (Bossen *et al* 1990) and phytochrome-regulated gravitropism of maize roots (Perdue *et al* 1988). The fact that *myo*-inositol only partially prevented the Li^+ effect could mean that phytochrome-controlled nyctinastic closure is not only dependent on the hydrolysis of PIP_2. This corroborates previous results showing that external Ca^{2+} also contributes to the phytochrome effect mediating nyctinastic closure (Moysset and Simón 1990).

3.2 Effect of Calmodulin Antagonists on Nyctinastic Closure

Calmodulin antagonists (ANTs-CAM) like trifluoperazine (TFP), chlorpromazine (CPZ), W7, calmidazolium and 48/80 compound inhibited nyctinastic closure in a concentration dependent manner. TFP (10, 50 and 100 µM) inhibited leaflet closure of *A. lophantha* (data non shown). This effect was more pronounced in R irradiated leaflets than in FR. TFP inhibited the closure of R irradiated leaflets at 10 µM whereas an equivalent effect was obtained in FR irradiated leaflets at 10 times higher concentrations. The inhibitory effect was conspicuous 30 min after R irradiation and was maintained until the end of the experiment.

Similar results were observed for CPZ, but at concentrations 5-10 times higher than TFP. These data suggest that TPF is a more effective calmodulin antagonist than CPZ. This order of effectiveness is paralleled by the inhibition of CAM-induced activation of phosphodiesterase (Hidaka *et al* 1988).

An external supply of 48/80 to *A. lophantha* leaflets also reduced the nyctinastic closure on R irradiated leaflets and did not significantly inhibit the closure in FR (Fig. 3).

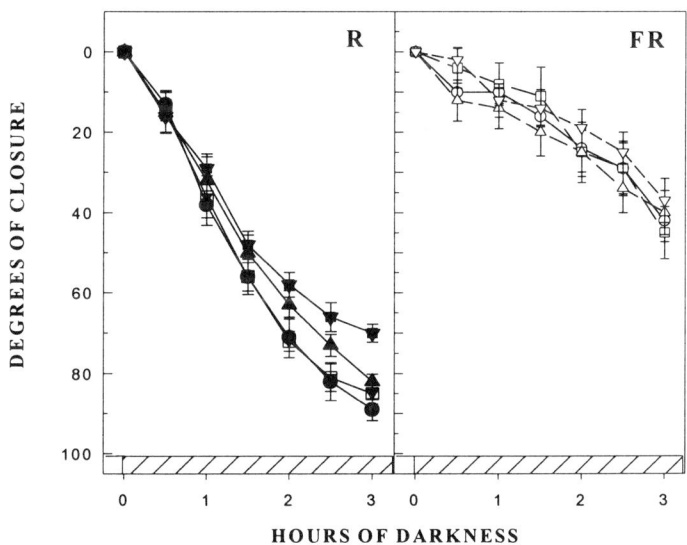

Figure 3. Effect of 48/80 on phytochrome controlled nyctinastic closure of *Albizzia lophantha* at 8 h of photoperiod. Leaflet pairs were floated on distilled water (●), 0.5 (▼), 1 (▲) and 5 (■) µg/l 48/80 compound. For more details see legend to Fig. 1.

3.3 Effect of Calmodulin Antagonists on Rhythmic Movement

ANTs-CAM also modified the rhythmic movement of *A. lophantha* leaflets. Two-hour pulses of calmidazolium, W-7 and TFP caused phase shifting depending on the circadian time (CT) at which the chemicals were applied (Tab. 1). Two hour pulses of W-5, the inactive analog of W-7, had no effect on the rhythm.

Table 1. Effect of calmodulin antagonists on phase-shifting of rhythmic movement of *A. lophantha* leaflets

		CT 10	CT 20
Calmidazolium	1 µM	-0.5 ± 0.3	0.2 ± 0.1
	10 µM	-0.5 ± 0.1	1.1 ± 0.2
	50 µM	-1.5 ± 0.3	1.8 ± 0.2
W-7	50 µM	0.2 ± 0.3	0.9 ± 0.1
	100 µM	-1.7 ± 0.3	1.0 ± 0.2
	250 µM	-2.4 ± 0.4	2.2 ± 0.7
Trifluoperazine	10 µM	-1.8 ± 0.4	0.6 ± 0.2
	50 µM	-2.1 ± 0.7	2.1 ± 1.0

TFP was the most effective ANTs-CAM tested since at 10 µM it induced phase shifts similar to those induced by calmidazolium 50 µM and W-7 100 µM (Tab. 1). Similar results were found in *R. pseudoacacia,* although in this case all ANTs-CAM tested generated greater phase shifts than in *A. lophantha* (Gómez et al 1999). This is not surprising since the circadian clock of *Albizzia* is more persistent than the clock of other related legumes such as *R. pseudoacacia* or *Samanea*. Thus, 15 min of R light cause 10-12 h advances or delays in *Robina* (Gómez and Simón 1995) whereas the same stimulus only caused up to 3 h of phase shift in *A. lophanta* (Fig. 4B). As shown in Fig. 4B 2 h pulses of 10 mM $CaCl_2$ also generated phase shifts of *Albizzia* leaflets rhythm similar to those induced by R light.

Similar phase-response curves (PRC) were obtained with the different ANTs-CAM tested in *A. lophantha* with maximum delays and advances at CT 12-CT 14 (Fig. 4B). As seen in Fig 4B and 4A the PRC shape for TFP in *Albizzia* differs from that found in *Robinia*. In *Albizzia,* TFP caused delays of the phase from CT 4 to CT13 and maximum advances of phase at the end of the subjective day (CT 14). In *Robinia* maximum advances were found in the middle of the subjective day and the shift from delays to advances occurred on transition from subjective night to subjective day. These differences could be due to differences in the photoperiodic regime under which plants were entrained prior to free running conditions. Thus, *A.*

lophantha received 16 h light and 8 h dark daily while *R. pseudoacacia* was entrained in 12 h light/12 h dark cycle.

The shape of PRC for TFP in *Albizzia* is similar to those found in *Neurospora* (Nakashima 1986) except that the crossover point in *Neurospora* (CT 6-CT 10) occurs almost 4 h earlier than in *Albizzia* (CT 12-CT 14). In both cases, Ca^{2+}, calcium ionophore (A23187) and ANTs-CAM caused delays of phase during the subjective day and maximum advances of phase induced by ANTs-CAM were evident 4-8 h before maximum advances caused by Ca^{2+} and calcium ionophore (A23187).

Taken together, our data suggest that the Ca^{2+}/CAM complex may play a dual role in the control of leaflet movements. It could be involved in the phytochrome-regulated signal transduction cascade coordinating the activity of the endogenous oscillator with environmental light. It might also be a feature of the *A. lophantha* circadian oscillator, as postulated in *Robinia* (Gómez *et al* 1999), in a similar way to that proposed by Goto *et al* (1985) in *Euglena* and *Lemna* rhythms, in which the Ca^{2+}/CAM complex controls phosphorylation /dephosphorylation during half the circadian cycle.

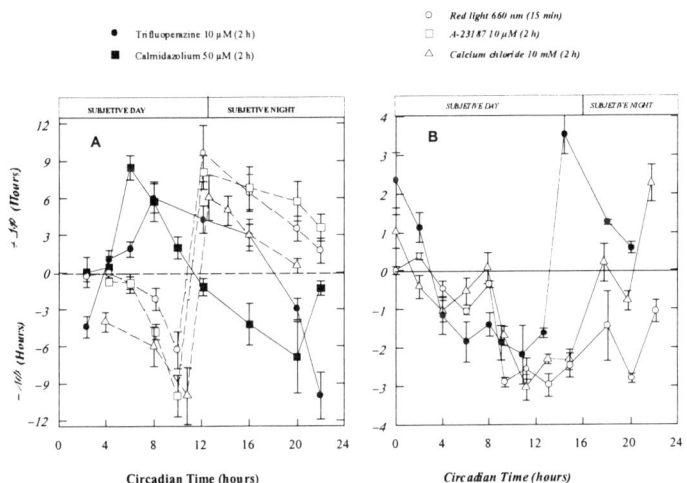

Figure 4. Phase-response curve for the effect of calcium, calmodulin inhibitors, calcium and R light on the free-running rhythm of *R. pseudoacacia* (A) and *Albizzia lophantha* (B) leaflets movements in DD. (A), redrawn in part from Gómez and Simón (1995, 1999).

4. CONCLUSIONS

The inhibitory effect of LiCl and ANTs-CAM on phytochrome-mediated nyctinastic closure of *A. lophantha* indicates that inositol metabolism and Ca^{2+}/CAM are involved in the Pfr transduction pathways. Calcium and calmodulin also appear to be involved in rhythmic movements, and they may connect the Pfr signal with the circadian clock, or indeed form part of the endogenous oscillator itself.

ACKNOWLEDGMENTS

We gratefully acknowledge support by grants PB94-0890 and PB97-0928 from the Ministerio de Educación y Ciencia and we thank Robin Rycroft for correcting the English text.

REFERENCES

Gómez, L.A., and Simón, E., 1995, Circadian rhythm of *Robinia pseudoacacia* leaflet movements: role of calcium and phytochrome. *Photochem. Photobiol.* **61**: 210-215.

Gómez, L.A., Moysset, L. and Simon, E., 1999, Effects of calmodulin inhibitors and blue light on rhythmic movement of *Robinia pseudoacacia* leaflets. *Photochem. Photobiol.* **69**: 722-727.

Goto, K., Laval-Martin, D.L., and Edmunds, Jr.L.N., 1985, Biochemical modeling of an autonomously oscillatory circadian clock in *Euglena. Science* **228**: 1284-1288.

Halcher,L.M., and Sherman, W.R., 1980, The effects of lithium ion and other agents on the activity of *myo*-inositol-1-phosphatase from bovine brain. *J. Biol. Chem.* **255**: 10896-10901.

Hidaka, H., Inagaki, M., Nishikawa, M., and Tanaka, T., 1988, Selective inhibitors of calmodulin-dependent phosphodiesterase and other enzymes. *Meth. Enzymol.* **159**: 652-660.

Kipper, F., 1986, Endocytobiotic coordination, intracellular calcium signalling and the origen of endogenous rhythms. *Annu. NY Acad. Sci.* **503**: 476-495.

Lakin-Thomas, P.L., 1985, Biochemical genetics of circadian rhythm in *Neurospora crassa*, studies on the cell strain. Ph.D. thesis. University of California. San Diego.

Moysset, L., Gómez, L.A., and Simón, E., 1994, Effects of lanthanum on rhythmic and nyctinastic leaflet movements in *Albizzia lophantha, J. Expt. Bot.* **45**: 85-93

Moysset, L., and Simón, E., 1989, Role of calcium in phytochrome-controlled nyctinastic movements of *Albizzia lophantha, Plant Physiol.* **90**: 1108-1114.

Moysset, L., and Simón, E., 1990, Effect of EGTA on nyctinastic closure mediated by phytochrome in *Albizzia lophantha, Plant Cell Physiol.* **31**: 187-193.

Nakashima, H., 1986, Phase shifting of the circadian conidiation rhythm in *Neurospora crassa* by calmodulin antagonists, *J. Biol. Rhythms* **9**: 27-41.

Neuhaus, G., Bowler, C., Kern, R., and Chua, N. H., 1993, Calcium/calmodulin-dependent and -independent phytochrome signal transduction pathways, *Cell* **73**: 937-952.

Perdue, D.O., LaFavre, A.K. and Leopold, A.C., 1988, Calcium in the regulation of gravitropism by light, *Plant Physiol.* **86**: 1276-1280.

Roux, S.J., 1994, Signal transduction in phytochrome responses, In Photomorphogenesis in Plants, Kendrick R.E. & Kronenberg G.H.M., eds, Kluwer Academic Publishers, pp 187-209.

Satter R. L., and Galston, A. W., 1981, Mechanisms of control of leaf movements. *Annu. Rev. Plant Physiol.* **32**: 83-110.

Sanders, D., Browniee, C., and Harper, J. F., 1999, Communicating with calcium. *Plant Cell* **11**: 691-706.

Schäfer, E., Kunkel, T., and Frohnmeyer, H., 1997, Signal transduction in the photocontrol of chalcone synthase gene expression. *Plant Cell Environ.* **20**: 722-727.

Simón, E., Satter, R.L., and Galston, A.W., 1976, Circadian rhythmicity in excised *Samanea* pulvini. II. Resetting the clock by phytochrome photoconversion. *Plant Physiol.* **58**: 421-425.

Tretyn, A., Kendrick, R. E., and Wagner, G., 1991, The role(s) of calcium ions in phytochrome action. *Photochem. Photobiol.* **54**: 1135-1155.

Yang, Z., 1996, Signal transducing proteins in plants: an overview, In Signal transduction in plant growth and development, Verma D.P.S., ed, Springer, pp 1-37.

Calmodulin and Plant Responses to the Environment
Modulation of Plant Tolerance to Toxic Metals by a Plasma Membrane Calcium/Calmodulin Binding Channel

[1]RAMANJULU SUNKAR, [1]TZAHI ARAZI, [1]BOAZ KAPLAN, [1]DVORA DOLEV, AND [1,2]HILLEL FROMM

[1]*Department of Plant Sciences, The Weizmann Institute of Science, Rehovot 76100, Israel;* [2]*Centre for Plant Sciences, Leeds Institute for Biotechnology and Agriculture, The University of Leeds, Leeds LS2 9JT, UK*

1. INTRODUCTION

Calcium ions (Ca^{2+}) function as a widespread messenger in mediating responses of eukaryotes to environmental signals, often through Ca^{2+}-modulated proteins like calmodulin (CaM) (for review see Snedden and Fromm 1998, Zielinski 1998). To understand how plants respond to environmental changes we focus on the identification and characterisation of the cellular targets of Ca^{2+}/CaM. These proteins are believed to play a key role in regulating cellular and biochemical processes associated with plant adaption to the environment (e.g. metabolic regulation, transcriptional regulation, ion transport and cell structure modifications).

Heavy metals are present in soils as natural components and as a result of human activities. Natural minerals deposits containing particularly large quantities of heavy metals are present in many regions of the globe. Metal-rich mine tailings, metal smelting, gas exhausts, burning of fossil fuels, down wash from power lines, indiscriminate use of fertilizers in agriculture and sludge dumping are the most important human activities that accelerated dramatically due to world industrialization. In fact, the direct emission of trace metals into the air, water and soil is measured in millions of tons per annum (Nriagu and Pacyna 1988). Moreover, the levels of most toxic metals emitted into the atmosphere because of human activities are several times higher than those released by natural sources. For example, the anthropogenic emission of Pb into the atmosphere was estimated to be more than 300-fold that of its natural emission (Ayres 1992).

Animals tend to deal with environmental hazards by simply avoiding them. Plants, more limited in mobility, have developed unique protective strategies. By understanding these strategies, we may improve plant tolerance to environmental hazards, as well as implement biotechnologies aimed at cleaning the environment. Genes encoding proteins that are involved in the transport of metal ions are crucial in this area of research. Specifically, these form potential targets for improving plant tolerance to toxic metals or increasing the accumulation of metal ions for phytoremediation (Raskin 1996). However, only a few genes of this kind have been cloned (e.g. Schachtman *et al* 1997) and very little is known about their mode of action or subcellular membrane localization.

2. CLONING, SEQUENCE AND FUNCTIONAL ANALYSIS OF A CALMODULIN-BINDING CHANNEL PROTEIN FROM TOBACCO

By screeining plant cDNA expression libraries with ^{35}S-radiolabeled recombinant CaM, we identified a novel protein designated NtCBP4 (*Nicotiana tabacum* Calmodulin-Binding channel-like protein). Subsequently, using similar techniques as well as using RT-PCR we isolated three other isoforms of this protein family from tobacco and members of this protein family from tomato, patato and alfalfa. Recently, a similar protein was found in barley (Schuurink *et al* 1998) and six members in *Arabidopsis thaliana* (Kohler *et al* 1999). Thus, multigene families encoding different isoforms of these channel proteins exist in several, perhaps in all higher plant species. NtCBP4 and other isoforms share sequence similarities with known K^+-selective channel from animals and plants, as well as with animal cyclic nucleotide-gated non selective cation channels. These similarities cover a large region of NtCBP4 from their first transmembrane segment to the end of their predicted cyclic nucleotide-binding domain (cNBD). Hydropathic analysis of NtCBP4 predicted the occurrence of a short hydrophylic N-terminal, a hydrophobic region containing six transmembrane spanning domains, S1-S6, a pore region located between S5 and S6, and a hydrophilic C-terminal. For further considerations of sequence analysis see Arazi *et al* (1999, 2000).

Pore-lining regions define the ion selectivity filter of ion channels, and hence are conserved among proteins with the same ion selectivity (Sater *et al* 1994). The NtCBP4 predicted pore-lining sequences is not identical to any of the known consensus sequences that define ion selectivity (Fig. 1) (Sater *et al* 1994). However, it does share similar features with the K^+ selective channels. Nevertheless, NtCBP4 lacks residues found to be essential for K⁺ selectivity, such as the complete GYGD motif (Nakamura *et al* 1997, Doyle *et al* 1998). Instead, NtCBP4 contains GQNL in its place (Fig. 1). The pore region of NtCBP4 is also similar to the pore of cyclic-nucleotide gated (CNG) channels from animals, which are permeable to Ca^{2+} and other cations. Taken together, the NtCBP4 pore might function as a nonselective cation filter.

```
                       PORE REGION                    |       S6
NtCBP4     .FFYCFWWGLQNLSSLGQNLQTSTFIWEMCFAVFISIAGLVLFAF....LIGNMQTCLQSST
KcsA        TYPRALWWSVETATTVGYGDLYPVTLWGRLVAVVVMVAGITSFGL....VTAALATWFVGRE
Dmshaker    SIPDAFWWAVVTMTTVGYGDMTPVGFWGKIVGSLCVVAGVLTIALPVP.VIVSNFNYFY...
AtAKT1      RYVTSMYWSITTLTTVGYGDLHPVNTKEMIFDIFYMLFNLGLTAY....LIGNMTNLVVHGT
RCNG2       EYIYCLYWSTLTLTTIG.ETPPPVKDEEYLFVIFDFLIGVLIFAT....IVGNVGSMISNMN
```

Figure 1. Amino acid sequence of P and S6 regions of NtCBP4 compared with that of K+-selective channels from *Streptomyces lividans* (Kcsa), *Drosophila* (Dmshaker), *Arabidopsis* (AtAKT1), and a non-selective cation channel from rat (RCNG2) (accession numbers Z37969, X07131, X93022 and X55519, respectively). Structural elements of the KcsA channel are indicated above the sequence. Residues of the ion-selectivity filter, and lining of the cavity and inner pore are presented in bold and shadowed letters, respectively. Conserved amino acid residues are in grey background.

In addition to their hydrophobic core, NtCBP4 and related proteins contain a conserved cNBD that resides at the C-terminus. This domain belongs to a family of structurally similar cNBD that include the *E. coli* CAP, the cAMP and cGMP activated protein kinases and the CNG channels (Shabb and Corbin 1992). To date, experimental evidence is still lacking for the binding of cyclic nucleotide to a plant protein. However, studies of the *Arabidopsis* channels KAT1 and AKT1 indicate a cGMP-dependent shift in their activation potentials towards more negative voltages (Gaymard *et al* 1996). The amino acid sequence of the NtCBP4 cyclic nucleotide-binding domain fits almost perfectly to its known signature pattern, and all the conserved residues that are important for cNMP binding (Shabb and Corbin 1992) are present, suggesting that under certain physiological conditions this domain would bind cNMP. Interestingly, the NtCBP4 cNBD sequence suggests a higher affinity for cGMP than for cAMP (based on Shabb and Corbin 1992). This is in accordance with the fact that cGMP is now well established as a second messenger in plants and is involved in light, gibberellic acid and nitric oxide signal transduction (Bowler and Chua 1994, Penson *et al* 1996, Durner *et al* 1998).

Recently, studies of a NtCBP4-related protein from *Arabidopsis* (AtCNGC2) revealed its ability to function as a CNG cation channel in heterologous systems (Leng *et al* 1999) supporting the presumed functional role of the cNBD of this protein family. These studies are also consistent with earlier studies by Kurosaki *et al* (1994) investigating carrot cells. These authors reported on the presence of plasma membrane ion channels that are activated by cAMP, are permeable to Ca^{2+}, and are negatively regulated by CaM.

2.1 NtCBP4 is Associated with the Tobacco Plasma Membrane

Before investigating the function of NtCBP4, we addressed questions concerning its expression and association with plant membranes in wild type and its transgenic plants. The 81-kDa NtCBP4 protein was immunodetected in microsomes from tobacco roots and shoots, indicating its presence in both organs. Tobacco membrane solubilization studies revealed that NtCBP4 is an integral membrane protein (Arazi *et al* 1999), which is in agreement with sequence analysis that predicted six transmembrane domains. We further determined the subcellular localization of NtCBP4 by separation of membranes on sucrose gradients and by aqueous two-phase partitioning. Each of these membrane fractionation procedures revealed, independently, that NtCBP4 co-fractionates with the plasma membrane, both for the endogenous tobacco NtCBP4 in WT plants and with the protein overexpressed in transgenic plants (Arazi *et al* 1999). In conclusion, this data support the postulated function of NtCBP4 as a component of a plasma membrane ion channel.

2.2 NtCBP4 Overexpression Alters Plant Sensitivity to Specific Metals

Transgenic techniques have become a powerful tool to approach important biological problems in multicellular organisms. Therefore, we established several transgenic tobacco lines that express the NtCBP4 RNA in the sense orientation under the transcriptional control of the promoter of the CaMV 35S gene, in order to elucidate the function of NtCBP4 *in planta*. Several independently derived transgenic lines were obtained, differing in their level of NtCBP4. A few lines had about double the amount of NtCBP4 protein compared to the WT plants. Other lines had intermediate levels of NtCBP4 and certain lines exhibited levels of NtCBP4 similar to WT. A transgenic line with an antisense construct had about the same levels of

NtCBP4 as the WT. These transgenic lines did not show any developmental aberrations under normal green house conditions. Because NtCBP4 was assumed to function as a plasma membrane non-selective cation channel, wild type and transgenic plants were exposed to different toxic metals including Na^+, Ba^{2+}, Cd^{2+}, Co^{2+}, Ni^{2+}, La^{3+} and Mn^{2+} as chloride salts, Pb^{2+} as $Pb(NO_3)_2$ and Zn^{2+} as $ZnSO_4$. We found that transgenic plants with higher levels of NtCBP4 are relatively tolerant to Ni^{2+} (Fig. 2) but hypersensitive to Pb^{2+} (Fig. 3).

2.3 Ni^{2+} Tolerance is Associated with Reduced Ni^{2+} Accumulation in Transgenic Plants

Nickel and a few other heavy metals are essential for plant growth and development as micronutrients. In order to obtain these metals, plant use a variety of metal uptake systems at the plasma membrane. However, an excess of microelements in the tissue, or entry of other non-essential toxic metals may pose a serious threat to plants. Because of the dual role of Ni^{2+} as an essential microelement on the one hand and as a toxic environmental factor on the other, complete exclusion is not possible. Therefore, genetically engineering of plants with reduced metal uptake can contribute to improving their tolerance to Ni^{2+}. Exclusion of toxic metal ions by the plasma membrane is conceptually the best mechanism for preventing damage to cellular functions under metal stress conditions.

The tolerance of transgenic lines overexpressing NtCBP4 to Ni^{2+} was apparent in young seedlings (Fig. 2a) and in plants grown to four weeks as well as in transgenic calli (Arazi *et al* 1999). Thus, resistance to Ni^{2+} is not a function of a specific developmental stage of the plant and seems to be cell-autonomous. The concentration-dependent Ni^{2+} accumulation in shoots of wild type and transgenic plants has revealed that the overexpression restrict the accumulation of Ni^{2+} (Fig. 2b).

2.4 Pb^{2+} Hypersensitivity is Associated with Enhanced Pb^{2+} Accumulation in Transgenic Plants

The transgenic lines overexpressing NtCBP4 were found to be hypersensitive to Pb^{2+} compared to wild type plants (Fig. 3a). Because Pb^{2+} is a non-essential element in plants and is extremely toxic, plant cells are not likely to possess specific Pb^{2+} transporters. However, certain metal uptake transporters in plants are relatively non-selective, such that both metal nutrients and non-essential toxic metals are taken up, which leads to metal toxicity in plants. On the other hand, plants that hyperaccumulate heavy metals should be useful for phytoremediation, a strategy of using plants for removing toxic metals (Raskin 1996). In this study, we have shown that NtCBP4 can enhance the accumulation of Pb^{2+} in tobacco plants (Fig. 3b). In animals, voltage-gated Ca^{2+} channels are known to be permeable to Pb^{2+} (Simons and Pocock 1987). To date, plant transporters mediating Pb^{2+} uptake into plants have not been characterizied at the molecular level. However, it has been proposed that Pb may permeate through plasma membrane Ca^{2+} channels in plants (Huang and Cunningham 1996). Pb^{2+} hypersensitivity of transgenic plants likely results from enhanced Pb^{2+} entry through NtCBP4.

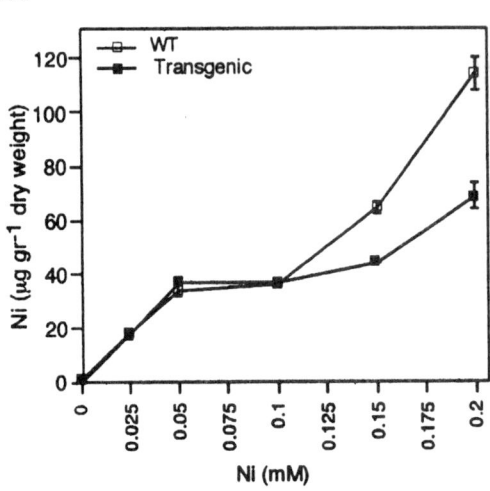

Figure 2. Ni^{2+} tolerance and accumulation in transgenic plants.
(a). WT and transgenic seedlings were grown with or without 0,2 mM $NiCl_2$ for 12 days.
(b). Concentration-dependent Ni^{2+} accumulation in 4-week-old WT and transgenic plants after 1 day exposure.

a.

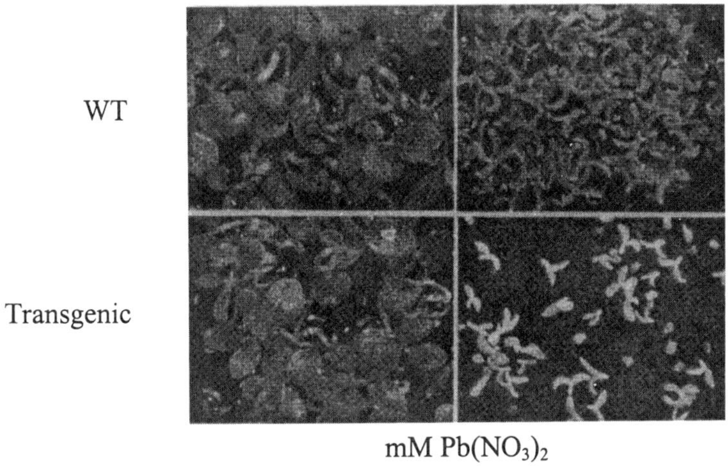

WT

Transgenic

mM Pb(NO$_3$)$_2$

b.

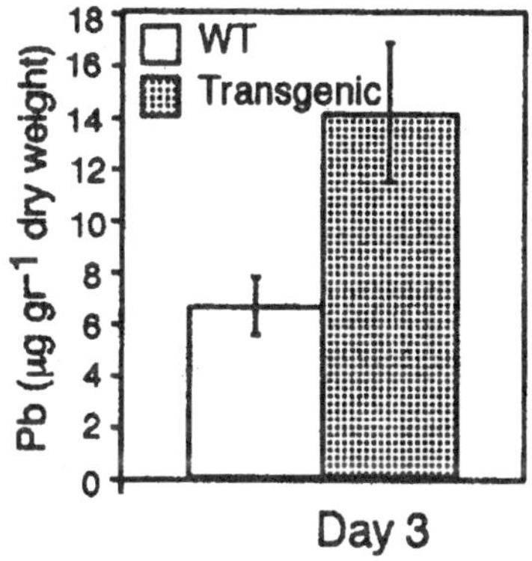

Figure 3. Pb^{2+} hypersensitivity and accumulation in transgenic plants.

(a) WT and transgenic seedlings grown with or without 1 mM Pb(NO$_3$)$_2$ for 12 days
(b) Pb accumulation in shoots of WT and transgenic plants. The plants were grown for four weeks before exposing them to 0,1 mM Pb(NO$_3$)$_2$ for 3 days and Pb^{2+} content was determined.

3. A MODEL FOR THE FUNCTION AND REGULATION OF NtCBP4

In spite of the remarkable effects of NtCBP4 on tolerance and accumulation of heavy metals, its physiological role is still unknown. Taken together the data presented so far, a model for the structure, function and regulation of NtCBP4 in plants is proposed in Fig. 4. In this model, NtCBP4 functions as a plasma membrane CNG non-selective cation channel. This and related plants channels, might be involved in mediating plant responses to biotic and abiotic stimuli from the environment or signals originating from within the plant. Binding of cNMP (preferentially cGMP) to NtCBP4 will likely take place when the cytoplasmic concentration of cGMP rises, as a result of an external stimulus. Certain stimuli are known to induce a rise in cGMP in plants, including gibberellic acid in barley aleurone cells (Penson *et al* 1996), and NO in spruce needles (Pfeiffer *et al* 1994) and tobacco plants (Durner *et al* 1998). cGMP binding to NtCBP4 will result in an influx of cations (including Ca^{2+}) into the cytoplasm, causing a rise in the intracellular concentration. A rise in cytoplasmic Ca^{2+} will activate CaM, resulting in binding to NtCBP4. This will impair the binding of cGMP to NtCBP4 and therefore inhibit channel opening. Ion channels that are regulated by cyclic nucleotide and Ca^{2+}/CaM might be involved in the reciprocally repressive cross-talk between cGMP and Ca^{2+}/CaM-mediated pathways in plants (e.g. in phototransduction; Bowler and Chua 1994).

However, one should consider the posssibility that the gene family of CNG channels in plants encodes isoforms that may respond to Ca^{2+}/CaM and cyclic nucleotides in various different ways. These isoforms may differ in their ability to bind CaM-related proteins at different concentrations of Ca^{2+}. In addition, one cannot exclude the possibility that some of these channels, which likely function as multimeric complexes (homomeric or heteromeric) are activated by Ca^{2+}/CaM and/or might be inactivated by cyclic nucleotides. A major challenge is to determine whether these features can be attributed to NtCBP4 and the related plant proteins. The powerful combination of molecular-genetic approaches, Ca^{2+} imaging and electrophysiological techniques should allow us to address these questions in the near future.

4. CONCLUSIONS

Using radiolabeled recombinant CaM as a probe to screen a tobacco cDNA library, we isolated NtCBP4, which shares certain similarities with vertebrate and invertebrate K^+ - and non-selective cation channels. Using transgenic plants we found that NtCBP4 can modulate plant tolerance to heavy metals. To our knowledge, this is the first report of a specific plant protein that affects plant tolerance to, or accumulation of Ni^{2+} and Pb^{2+}. This gene may prove useful for implementing selective ion tolerance in crops and improving phytoremediation strategies.

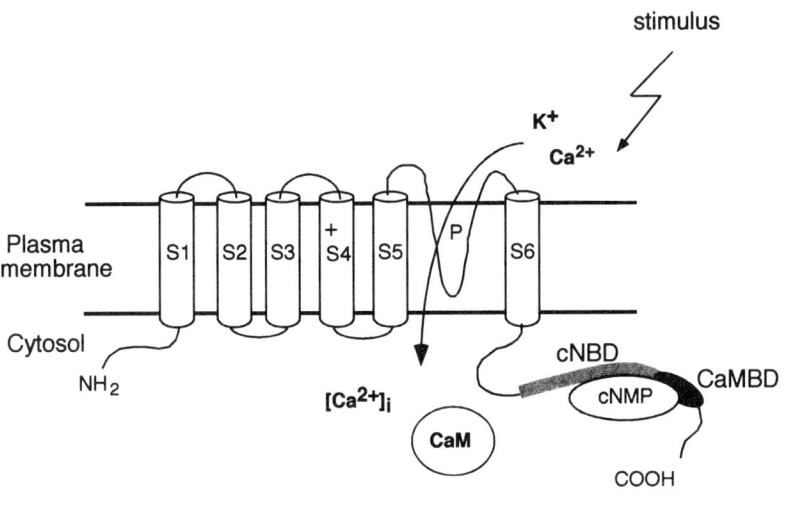

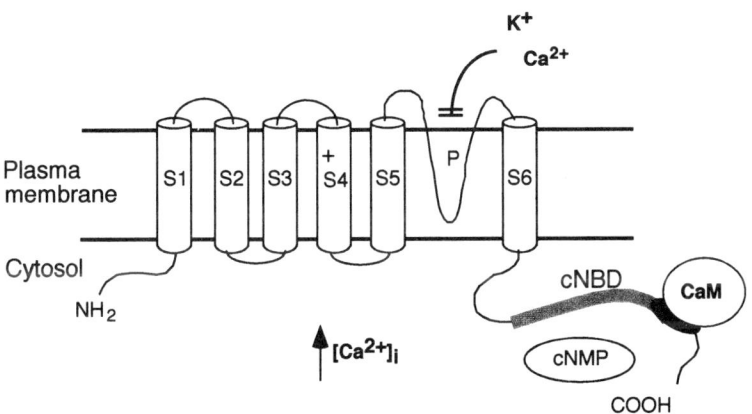

Figure 4. A working model of CBPs function and regulation. The presumptive membrane-spanning regions and pore are shown as S1-S6 and P respectively. Putative cyclic nucleotide-binding domain and identified CaM-binding site are shown.

ACKNOWLEDGEMENTS

R.S. holds a Sir Charles Clore postdoctoral fellowship and T.A. is a recipient of a graduate fellowship from the Sir Charles Clore Foundation. This work was supported by a grant from the Israel Science Foundation to H.F.

REFERENCES

Arazi, T., Sunkar, R., Kaplan, B., and Fromm, H., 1999, A tobacco plasma membrane calmodulin-binding transporter confers Ni^{2+} tolerance and Pb^{2+} hypersensitivity in transgenic plants. *Plant J.* **20**: 171-182.

Arazi, T., Kaplan, B., and Fromm, H., 2000, A high-affinity calmodulin-binding site in a tobacco plasma membrane channel protein coincides with a characteristic element of cyclic nucleotide-binding domains. *Plant Mol. Biol.* (in press).

Ayres, R.U., 1992, Toxic heavy metals: Materials cycle optimization. *Proc. Natl. Acad. Sci. USA* **89**: 815-820.

Bowler, C., and Chua, N-H., 1994, Emerging themes on plant signal transduction. *Plant Cell* **6**: 1529-1541.

Doyle, D.A., Morais, Cabral, J., Pfuetzner, R.A., Kuo, A., Gulbis, J.M., Cohen, S.L., Chait, B.T., and MacKinnon, R., 1998, The structure of a potassium channel: molecular basis of K^+ conduction and selectivity. *Science* **280**: 69-77.

Durner, J., Wendehenne, D., and Klessig, D.F., 1998, Defence gene induction in tobacco by nitric oxide, cyclic GMP, and cyclic ADP-ribose. *Proc. Natl. Acad. Sci. USA* **95**: 10328-10333.

Gaymard, F., Cerutti, M., Horeau, C., Lemaillet, G., Urbach, S., Ravallec, M., Devauchelle, G., Sentenac, H., and Thibaud, J.B., 1996, The Baculovirus/insect cell system as an alternative to *Xenopus oocytes*. First characterization of the AKTI K^+ channel from *Arabidopsis thaliana*. *J. Biol. Chem.* **271**: 22863-22870.

Huang, J.W., and Cunningham, S.D., 1996, Lead phytoextraction species variation in lead uptake and translocation. *New Phytol.* **134**: 75-84.

Kohler, C., Merkle, T., and Neuhaus, G., 1999, Characterization of a novel gene family of putative cyclic nucleotide- and calmodulin-regulated ion channels in *Arabidopsis thaliana*. *Plant J.* **18**: 97-104.

Kurosaki, F., Kaburakai, H., and Nishi, A., 1994, Involvement of plasma membrane-located calmodulin in the response decay of cyclic nucleotide-gated cation channel of cultured carrot cells. *FEBS Lett.* **340**: 193-196.

Leng, Q., Mercier, R.W., Yao, W., and Berkowitz, G.A., 1999, Cloning and first functional characterization of a plant cyclic nucleotide-gated cation channel. *Plant Physiol.* **121**: 753-761.

Nakamura, R.L., Anderson, J.A., and Gaber, R.F., 1997, Determination of key structural requirements of a K^+ channel pore. *J. Biol. Chem.* **272**: 1011-1018.

Nriagu, J.O., and Pacyna, J.M., 1988, Quantitative assessment of worldwide contamination of air, water and soils by trace metals. *Nature* **333**: 134-139.

Penson, S.P., Schurink, R.C., Fath, A., Gubler, F., Jacobsen, J.V., and Jones, R.L., 1996, cGMP is required for gibberellic acid-induced gene expression in barley aleurone. *Plant Cell* **8**: 2325-2333.

Pfeiffer, S., Janistyn, B., Jessner, G., Pichorner, H., and Ebermann, R., 1994, Gaseous nitric oxide stimulates guanosine-3'-5'-cyclic monophosphate (cGMP) formation in spruce needles. *Phytochem.* **36**: 259-262.

Raskin, I., 1996, Plant genetic engineering may help with environmental cleanup. *Proc. Natl. Acad. Sci. USA* **93**: 3164-3166.

Sater, W.A., Yang, J., and Tsien, R.W., 1994, Structural basis of ion channel permeation and selectivity. *Curr. Opin. Neuro.* **4**: 313-323.

Schachtman, D.P., Kumar, R., Schroeder, J.I., and Marsh, E.L., 1997, Molecular and functional characterization of a novel low-affinity cation transporter (LCT1) in higher plants. *Proc. Natl. Acad. Sci. USA* **94**: 11079-11084.

Schuurink, R.C., Shartzer, S.F., Fath, A., and Jones, R.L., 1998, Characterization of a calmodulin-binding transporter from the plasma membrane of barley aleurone. *Proc. Natl. Acad. Sci. USA* **95**: 1944-1949.

Shabb, J.B., and Corbin, J.D., 1992, Cyclic nucleotide-binding domains in proteins having diverse functions. *J. Biol. Chem.* **267**: 5723-5726.

Simons, T.J.B., and Pocock, G., 1987, Lead enters bovine medullary cells through calcium channels. *J. Neurochem.* **48**: 383-389.

Snedden, W.A., and Fromm, H., 1998, Calmodulin, calmodulin-related proteins and plant responses to the environment. *Trends Plant Sci.* **3**: 299-304.

Zielinski, R.E., 1998, Calmodulin and calmodulin-binding proteins in plants. *Annu. Rev. Plant Physiol. Plant Mol. Biol.* **49**: 697-725.

Calcium Signaling: Downstream Components in Plants

[1,3]GIRDAR K. PANEY, [1,3]VEENA, [2]RENU DESWAL, [3]SONA PANDEY,
[3]S. B. TEWARI, [1]W. TYAGI, [1]VANGA SIVA REDDY, [3]ALOK
BHATTACHARYA, AND [1]SUDHIR K. SOPORY
[1]*International Center for Genetic Engineering and Biotechnology Aruna Asaf Ali Road, New Delhi 110067, India,* [2]*National Center for Plant Genome Research, Jawaharlal Nehru University Campus, New Delhi, and* [3]*School of Life Sciences, Jawaharlal Nehru University, New Delhi, India*

1. INTRODUCTION

Calcium ions are involved in coupling a large number of extra-cellular stimuli to a variety of physiological processes in eukaryotes. At the cellular level Ca^{2+} enables the cells to communicate with the exterior milieu by acting as a major intracellular second messenger.

The information encoded in the calcium signals can be decoded by the downstream components, which includes calmodulin (CaM), a ubiquitous calcium receptor, and calcium dependent protein kinases. Both these pathways are inter-linked and also interact with a number of other factors, which are downstream in the signal transduction pathway. Thus, all these components, like, CaM, CaM-binding proteins and calcium-dependent kinases, cross talk with each other at different levels and form a very complex network. In fact this may be required for generating specific responses in view of multifunctional role played by calcium in plant signaling.

Calcium is known to regulate three different families of protein kinases in plants viz. calcium-dependent protein kinases (CDPKs), the Ca^{2+}/phospholipid-dependent kinases (PKCs) and Ca^{2+}/CaM-dependent kinases (CaM-kinases). In plants the main transducer of calcium signals known till date are kinases from CDPK family which require only calcium for their activation (Harper *et al* 1991). Ca^{2+}/CaM-dependent kinases and Ca^{2+}/phospholipid-dependent kinases are well known in animal systems. In

Signal Transduction in Plants: Current Advances.
Edited by Sopory *et al.,* Kluwer Academic/Plenum Publishers, 2001.

plants, existence of PKC has been reported in a few systems (Chandok and Sopory 1998).

Existence of CaM in plants and its induction/regulation in response to various stimuli (Jena *et al* 1989, Braam *et al* 1992) is well known. Several evidences for the presence of calmodulin-dependent kinases (Watillon *et al* 1992, 1993, 1995, Lu *et al* 1996, Takezawa *et al* 1996, Pandey and Sopory 1998) suggest the possibility of involvement of CaM kinases in a signaling pathway parallel to CDPK-mediated signaling. In this paper we show that the level of a pea homologue of Ca^{2+}/CaM kinase is regulated by salt and cold. In addition to CaM, there are several other Ca^{2+}-binding proteins in plants some of which are involved in buffering or maintaining Ca^{2+} homeostasis (Ikura 1996). The list of calcium-binding proteins, CaM-binding proteins and kinases is increasing (Celio *et al* 1996, see Zielinski 1998, see Sopory and Munshi 1998, Kudla *et al* 1999, Shi *et al* 1999). It is expected that plants will have a number of calcium-binding proteins performing various functions. We have found the presence of homologue(s) of an *E. histolytica* calcium binding protein (EhCaBP) and EhCaBP stimulated kinase(s) in plants, which may be involved in alternate calcium signaling pathway regulating development and adaptation under unfavorable environmental conditions.

2. PEA HOMOLOGUE OF ZmCCaMK: REGULATION BY COLD AND SALT

The presence of a novel Ca^{2+}/CaM kinases have been reported in plants (see Sopory and Munshi 1998). Earlier we have purified and characterized a CaM kinase (ZmCCaMK) from maize and polyclonal antibidies were raised against it (Pandey and Sopory 1998). By using ZmCCaMK antibodies we have identified an immuno-homologue of this kinase in pea with a molecular mass of 72 kDa. The pea immuno-homologue (pCCaMK) was present in crude extract and also in purified nuclear extract indicating nuclear localization of the CaM kinase homologue. Separation of proteins from various plants parts like, leaves, tendrils, root, shoot by SDS-PAGE followed by western blotting with ZmCCaMK antibodies showed that CaM kinase level is more in shoots as compared to roots and seed. As the expression level and the activity of a number of kinases is affected under different stress conditions, the effect of various stresses was checked on the expression levels of CaM kinase. When the protein level was compared between roots and shoots, it was seen that it was stimulated only in roots in response to stress. Different concentrations of NaCl (50 mM, 100 mM, 150 mM, 200 mM) and kinetic studies (24 hours, 48 hours and 92 hours) using 5 days old pea roots revealed that 24 hours of salt treatment induced higher protein level when compared to the control. The maximum level was attained at 24 hours that remained constant up to 48 hours. The protein level decreased slightly up to 72 hours, yet remained more than the untreated plants. There was no effect of mannitol indicating that salt induced up-regulation of the kinase is specific and not a general stress or drought response. To check the effect of temperature stress on the expression level of pCCaMK, 5 days old pea plants were given cold shock (4°C) and heat shock (37°C) for 24 hours. On comparison with 6 days old plants taken as control (grown at 25°C), the protein level was induced only in case of cold shock. All these results suggest that it is not a general stress regulated kinase but may have a specific role in cold and salt stress mediated signaling pathways. Convergence of signaling pathways related to cold and salt stress has been shown previously (Monroy and Dhindsa 1995). It was further found that in absence of any stress, exogenous Ca^{2+} could regulate pCCaMK. However only cold and not the salt mediated up regulation was inhibited by EGTA and W7 thus indicating that cold and salt signaling may be operative *via* different downstream components (see Fig. 1).

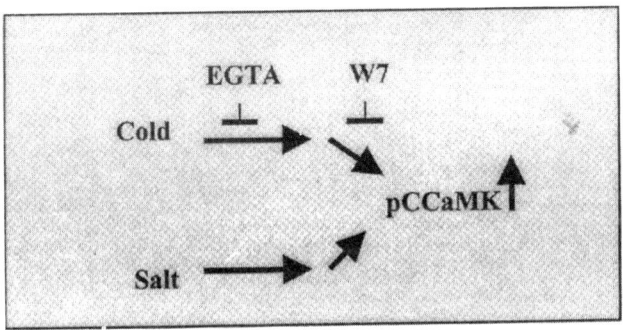

Figure 1. A model depicting the possible pathways for the regulation of pCCaMkinase. In absence of any stress, exogenous Ca^{2+} could up regulate pCCaMkK. However only cold and not the salt mediated up regulation was inhibited by EGTA and W7 thus indicating that cold and salt signaling may be operative *via* different downstream components.

3. PRESENCE OF EhCaBP HOMOLOGUE(S) IN PLANTS

EhCaBP is a calcium binding protein isolated from *E. histolytica*. It has 4 calcium binding sites (EF-hand motif) like calmodulin (CaM). However it can not complement CaM functions (Prasad *et al* 1992, 1993, Yadava *et al*

1997). By using *EhCaBP* cDNA as probe, Southern hybridization of *Brassica juncea, Nicotiana tabaccum and Pisum sativum* genomic DNA digested with different restriction enzymes was done. Multiple bands were observed in Southern hybridization. To rule out the possibility that these are not *CaM* homologues, we have done Southern hybridization by using *Arabidopsis thaliana CaM-3* gene as probe (Pandey 1999) and found that the number and banding pattern is different, which strengthen the fact that *EhCaBP* homologous sequences are different than *CaM*.

By northern hybridization also we have detected *EhCaBP* homologous transcript. The size of transcript detected in all the above plants was ~1.4 kb as compared to ~800 bp CaM transcript (data not given). This clearly suggests that the *EhCaBP* homologous transcript is different than CaM.

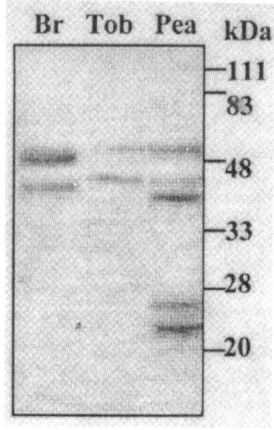

Figure 2. Detection of EhCaBP immuno-homologue(s) in higher plants by western blot analysis. Protein (40 µg) from light grown leaf was resolved on 15% SDS-PAGE, transferred onto nitrocellulose membrane and probed with 1:2000 diluted EhCaBP polyclonal affinity purified antibodies. Br stands for *Brassica* and Tob is tobacco.

Using western blot analysis multiple immuno-homologues of EhCaBP were detected in *Brassica juncea, Nicotiana tabacum* and *Pisum sativum*. In the leaf extract of *Brassica* two polypeptides of ~55 kDa and ~40 KDa, in tobacco also two polypeptides of ~55 kDa and ~43 kDa whereas in pea leaf extract 5 polypeptides of ~55 kDa, ~43 kDa, ~37 kDa, ~26 kDa and ~22 kDa were detected (Fig. 2).

4. EhCaBP STIMULATED KINASE IN *Brassica juncea*

In addition to the currently known class of kinases, we have identified a different class of protein kinase activity in plants. We have purified and partially characterized a protein kinase activity from *Brassica juncea* that is stimulated by a structural, but not functional homologue of CaM purified from *E. histolytica* (EhCaBP). This Ca^{2+}/EhCaBP stimulated protein kinase (BjCCaBPK) was purified to homogeneity from 8-days-old etiolated *B. juncea* seedlings using ammonium sulphate precipitation, ion exchange chromatography on DEAE-Sepharose, affinity chromatography on EhCaBP-Sepharose followed by gel elution. The protein showed a single polypeptide of 43 kDa on SDS-PAGE. The optimum pH for the kinase activity was 8.0. It was stimulated over 6 fold by EhCaBP (10.5 nM) but not by calmodulin when used at equimolar concentration. Moreover the kinase also did not bind CaM-Sepharose. There was no inhibition of the kinase activity in presence of W-7 (a CaM antagonist), KN-62 (a specific calcium/calmodulin kinase inhibitor) and anti-CaM antibodies. Even staurosporine (a protein kinase C inhibitor) had no effect on the BjCCaBPK activity. Furthermore a CaM-kinase specific substrate, Syntide-2, proved to be a poor substrate for the BjCCaBPK compared with histone III-S. The phosphorylation of histone III-S involved serine residues. In our laboratory, we have biochemically characterized two other protein kinases, a ZmCCaMK which belong to Ca^{2+}/CaM kinase (Pandey and Sopory 1998) and a PKC type kinase, cPKC from maize (Chandok and Sopory 1998). In Tab. 1, we have compared the properties of these kinases with BjCCaBP kinase and based on the differences in the properties we are suggesting that BjCCaBPK may be a novel protein kinase which does not fall in the CDPKs, CaM kinases, and PKCs categories and has an affinity towards a calcium binding protein like EhCaBP (Deswal *et al* 2000).

4.1 DETECTION OF EhCaBP STIMULATED KINASE ACTIVITY IN PEA

To find out if EhCaBP stimulated kinase activity is restricted to *B. juncea* or present in other higher plants, we also checked for this activity in pea. *In vitro* phosphorylation of histone type III S was done using partially purified protein extract from pea. Partial purification was achieved using ammonium sulfate precipitation followed by pI (iso-electric point) precipitation. When *in vitro* phosphorylation of histone type III S was done by using pea partially purified extract in the presence and absence of Ca^{2+} (100 µM Ca^{2+}), EhCaBP (10.5 nM), EhCaBP antibodies (1: 50 diluted) and pre-immune sera, several other polypeptides besides histone (~38 kDa, 55 kDa and 62 kDa) were also phosphorylated (Fig. 3).

Table 1. Comparison of biochemical properties of BjCCaBPK (EhCaBP stimulated kinase from *Brassica juncea*), ZmCCaMK (a Ca^{2+}/CaM kinase from maize) and ZmcPKC (a PKC type kinase from maize).

Properties	BjCCaBPK	ZmCCaMK	ZmcPKC
Optimum requirement			
PH	7.8	7.5	7.5
Ca^{2+}	2.11×10^{-9}	1×10^{-6}	1.48×10^{-8}
EhCaBP	+	-	-
CaM	-	+	-
Lipids	-	-	+
Ca^{2+} binding	+	+	+
Inhibited by			
KN-62	-	+	?
W-7	-	+	-
Staurosporine	-	-	+
H-7	+	+	+
Substrate	Histone>syntide>BSA	syntide>Histone>Casein	Histone

When EhCaBP polyclonal antibody (1:50 dil.) was added the phosphorylation was inhibited and was equivalent to phosphorylation in the absence of Ca^{2+}. The phosphorylation of two other polypeptides at approx. 55 kDa and 68 kDa were also stimulated by addition of Ca^{2+} plus EhCaBP and brought down by EhCaBP antibodies, whereas, pre-immune sera could not block the EhCaBP stimulated kinase activity.

5. ROLE OF EhCaBP HOMOLOGUE(S) IN PLANTS

To elaborate the role of EhCaBP homologue(s) in plants, transgenic approach was followed. Transformation of tobacco with EhCaBP gene in sense and antisense orientation in pBI121 vector with respect to CaMV 35S promoter was done by *Agrobacterium* mediated leaf disc method. The transgenic plants were confirmed by high stringency southern, northern, western and immuno-precipitation assay. The EhCaBP-antisense transgenic plants showed altered morphological changes (Pandey 1999). The main characteristic features noticed were an increase in greenness and leaves that are more round, thick, small, and more hairy when compared to wild type and sense transgenic plants (Fig. 4). The whole plant looks compact and all above characters were maintained in T_1 generation. However in glass house growth condition only two characters were maintained like more greenness and thickness of the leaves.

The total chlorophyll content in such plants was 60-70% higher than wild type plants. Furthermore the levels of phytohormones (a few cytokinin and ethylene) and polyamines were altered in antisense plants compared to wild and sense plants (Pandey 1999).

The EhCaBP expressing plants did not show any change in the morphology, however the seeds of T_1 plants were found to grow under high salt concentration (200 mM NaCl). This suggested that the expression of EhCaBP leads to tolerance under higher salinity level. Whether similar calcium binding proteins are involved in stress tolerance mechanism in tolerant varieties needs to be looked into in future studies.

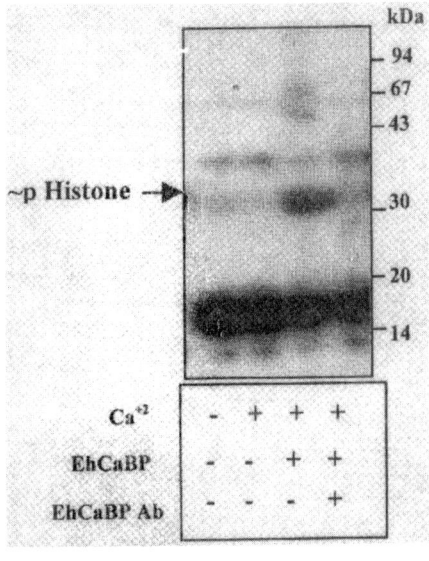

Figure 3. EhCaBP stimulated protein kinase activity in *P. sativum* extracts. The crude protein from pea leaf was partially purified by ammonium sulphate precipitation followed by pI (iso-electric point) precipitation. Kinase assay was performed by using 40 μg of the histone type III S as a substrate and 28 μg of the partially purified extract. After kinase assay, 4x protein sample buffer was added and loaded on 12% SDS-PAGE, dried and exposed to autoradiography. The arrow head shows the phosphorylated histone.

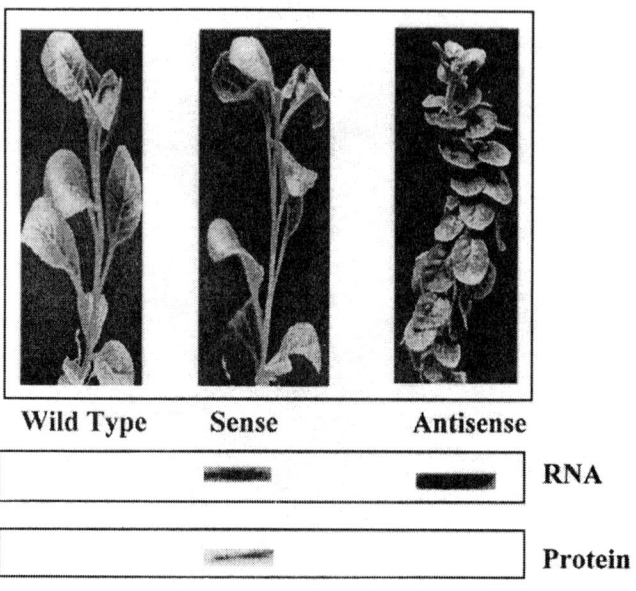

Figure 4. Variation in the morphology of *Nicotiana tabacum* cv Xanthi plants transformed with EhCaBP gene in sense and antisense orientation with respect to CaMV 35S promoter. EhCaBP transgenic (sense and antisense) and wild type plants (10 week old) on MS-basal medium containing 300 µg /ml kanamycin and wild type plant in minus kanamycin medium at (25°C±2, 16 hours of light (100 µE·m^{-2}·s^{-1}) and 8 hours of darkness.

6. ROLE OF GLYOXALASE I: A CALCIUM BINDING ENZYME

Besides protein kinases and phosphatases that are regulated by calcium and calcium binding protein, there are enzymes that are directly regulated by these factors. One such enzyme that we have identified is glyoxalase I which is involved in methylglyoxal detoxification and maintaining glutathione homeostasis. We have earlier purified glyoxalase I (Deswal *et al* 1993) and shown that it is a calcium binding enzyme and its activity is modulated by calcium/calmodulin (Deswal and Sopory 1999). To determine the functional significance of glyoxalase I in plants, we have cloned its gene (Veena *et al* 1999) and raised transgenic plants via *Agrobacterium* mediated transformation. The transgenic plants over-expressing glyoxalase I showed tolerance to salt (NaCl). In addition to salt, the transgenic tobacco plants over-expressing *Gly I* were also found to be tolerant to heavy metals. Leaf discs obtained from wild type and sense transgenic plants over-expressing high (NtBIS1-11) and medium (NtBIS1-15) levels of glyoxalase I were incubated with 20 mM of zinc chloride for 96 hours. Leaf discs of wild type plants showed an early bleaching as compared to leaf discs of sense transgenic plants which was further confirmed by estimation of chlorophyll (Chl) content in these leaf discs. Treatment of the leaf discs of NtBIS1-11 and NtBIS1-15 with 20 mM zinc chloride resulted in the loss of Chl by 26% and 58% respectively whereas loss of Chl in wild type was 91%. The tolerance level of transgenic plants to zinc was dependent on the expression level of glyoxalase I protein. The biochemical basis of this tolerance mechanism is not clear. However it could be that during stress, the levels of methyl glyoxal which is a toxic to compound increases and that leads to decreased cell division (Scaife 1969, Szent-Gyogi and Egyud 1966) and causes cell death (Kalapose *et al* 1991). A system producing more glyoxalase could therefore convert methyl glyoxal into non-toxic form, thus preventing the system from its cytotoxic effects during stress condition. Besides detoxification of methylglyoxal, the glyoxalase system could also play a role in providing tolerance under stress by recycling glutathione that would be trapped spontaneously by methyl glyoxal to form hemi thioacetal (Creighton *et al* 1988, Thornalley 1990), there by maintaining the glutathione homeostasis.

7. CONCLUSIONS

It is clear from the work of a number of laboratories and from different plant systems that calcium is an important universal signaling component involved in coupling of environmental and developmental signals to altered gene expression and physiological changes. The mechanism by which calcium increase in the cytoplasm leads to specific responses is not clear. It seems there are multiple downstream components, activated by calcium, that follow specific signal transduction pathways to elicit the desired response (Fig. 5). It has now been shown that in addition to CDPKs the plants have other calcium dependent kinases also viz. Ca^{2+}/CaM kinases (Pandey and Sopory 1998, Takezawa *et al* 1996), protein kinase C (Nanmori *et al* 1994, Chandok and Sopory, 1998) and CIPKs, AtCBL interacting kinase (Shi *et al* 1999) other novel kinases like EhCaBPk reported in this chapter. Besides the activation of kinases, calcium and calcium binding proteins, like CaM, can directly activate different enzymes.

A few such enzymes have been reported earlier (see Snedden and Fromm 1998) and we have now shown that glyoxalase I is also a calcium binding protein. Our work on the transgenic plants harboring EhCaBP and glyoxalase gene, in both sense and antisense orientation, and work from other laboratories (see Sopory and Munshi 1998, Pandey *et al* 1999) have shown that calcium binding proteins, kinases and phosphatases (Liu and Zhu 1998) play an important role in developmental responses and in conferring tolerance to abiotic stress in plants. It is expected that plants will have a large number of such proteins to transduce messages and regulate various processes. An indication of this also comes from the genome sequence of chromososme 2 and 4 of *Arabidopsis* (Lin *et al* 1999, Meyer *et al* 1999). Identification and functional significance of all such proteins will engage the workers in this field in the next decade.

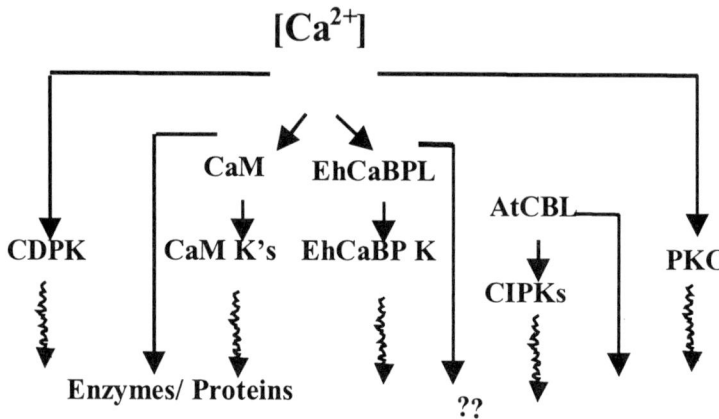

Figure 5 Downstream calcium signal transduction pathways in plants. Besides calmodulin (CaM), in plants calcineurin B-like proteins (AtCBL) exist which act as a calcium sensor and interact with novel CIPKs type of kinases. There could be a possibility of an alternate pathways like those mediated *via* homologue(s) of EhCaBP (EhCaBP like protein, EhCaBPL) and EhCaBP-stimulated kinase(s) in addition to already known kinases in plants like CaM-kinases, calcium dependent protein kinase (CDPKs), protein kinase C (PKCs).

ACKNOWLEDGMENT

The research was partially funded from the internal grants of ICGEB, New Delhi and a research grant from CSIR (India). The authors are thankful to Dr. Autar K. Mattoo (USDA, USA), Dr. V. R. Sashidhar (Bangalore), Dr. M. V. Rajam (Delhi) for their help in the measurement of ethylene, cytokinin and polyamines respectively.

REFERENCES

Braam, J., Sistrunk, M. L., Polisensky, D. H., Xu, W., Purugganan, M. M., Antosiewicz, D. M., Campbell, P., and Johnson, K.A., 1997, Plant response to environmental stress: regulation and functions of the *Arabidopsis* TCH genes. *Planta* **203**: S35-S41.

Celio, M.R., Pauls, T., and Schwaller, B., 1996, *Guidebook to calcium binding proteins. (Celio ed.) AS.*

Chandok, M.R., and Sopory, S. K., 1998, ZmcPKC70, a protein kinase C-type enzymes from maize: biochemical characterization, regulation by PMA and its possible involvement in nitrate reductase gene expression. *J. Biol. Chem.* **273**: 19235-19242.

Creighton, D. J., Migliorini, M., Pourmotabbed, T., and Guha, M. K., 1988, Optimization of efficiency in the glyoxalase pathway. *Biochemistry* **27**: 7376-7384.

Deswal, R., and Sopory, S.K., 1991, Purification and partial characterization of glyoxalase I from higher plant *Brassica juncea*. *FEBS Lett.* **282**: 277-280.

Deswal, R., Pandey, G. K., Chandok, M. R., Yadava, N., Bhattacharya, A., and Sopory, S. K., 2000, A novel protein kinase from *Brassica juncea* stimulated by a protozoan calcium binding protein: Purification and partial characterization. *Eur. J. Biochem.* **257**: 3181-3188.

Deswal, R., and Sopory, S. K., 1999, Glyoxalase 1 from *Brassica juncea* is a calmodulin stimulated protein. *Biochem. Biophys. Acta* **1450**: 460-467.

Harper, J.E., Sussman, M.R., Schaller, G.E., Putnam-Evans, C., Charboneau, H., and Harmon, A.C., 1991, A calcium dependent protein kinase with a regulatory domain similar to calmodulin. *Science* **252**: 951-954.

Ikura, M., 1996, Calcium binding and conformational response in EF-hand proteins. *Trends Biol. Sci.* **21**: 14-17.

Jena, P.K., Reddy, A.S.N., and Poovaiah, B. W., 1989, Molecular cloning and sequencing of a cDNA for plant calmodulin: signal induced changes in the expression of calmodulin. *Proc. Natl. Acad. Sci. USA* **86**: 3644-3648.

Kalapose, M. P., Schaff, Z. S., Garzo, T., Antonie, F., and Mndl, J., 1991, Accumulation of phenols in isolated hepatocytes after pretreatment with methylglyoxal. *Toxicol. Lett.* **58**: 181-191.

Kudla, J., Xu, Q., Harter, K., Gruissem, W., and Luan, S., 1999, Genes for calcineurin B-like protein in *Arabidopsis* are differentially regulated by stress signals. *Proc. Natl. Acad. Sci. USA* **96**: 4718-4723.

Lin, X. *et al* 1998, Sequence and analysis of chromosome 2 of the plant *Arabidopsis thaliana*. *Nature* **402**: 761-768.

Liu, J., and Zhu, J.-K., 1998, A calcium sensor homolog required for plant salt tolerance. *Science* **280**: 1943-1954.

Lu, Y.T., Hidaka, H., and Feldman, L.J., 1996, Characterization of a calcium/calmodulin protein kinase homologue from maize roots showing light-regulated gravitropism. *Planta* **199**: 18-24.

Mayer, K. *et al* 1998, Sequence and analysis of chromosome 4 of the plant *Arabidopsis thaliana*. *Nature* **402**: 769-777.

Monroy, A.F., and Dhindsa, R.S., 1995, Low temperature signal transduction: Induction of cold acclimation specific gene of alfalfa by calcium at 25°C. *Plant Cell* **7**: 321-331.

Nanmori, T., Taguchi, W., Kinugusa, M., Oje, Y., Sahara, S., Fukami, Y., and Kikkawa, W., 1994, Purification and characterization of protein kinase C from a higher plant, *Brassica compestris* L. *Biochem. Biophys. Res. Commun.* **203**: 311-318.

Pandey, G.K., 1999, *PhD Thesis* Presence and role of homologues of EhCaBP, a calcium binding protein of *E. histolytica* in higher plants and characterization of a novel kinase stimulated by EhCaBP from *Brassica juncea*. Jawaharlal Nehru University, New Delhi, India.

Pandey, S., and Sopory, S. K., 1998, Biochemical evidence for a calmodulin-stimulated calcium-dependent protein kinase in maize. *Eur. J. Biochem.* **255**: 718-726.

Pandey, S., Tewari, S.B., Upadhyaya, K.C., and Sopory, S.K., 2000, Calcium Signaling: Linking environmental signals to cellular functions. *Crit. Rev. Plant Sci.* **19**: 291-318.

Prasad, J., Bhattacharya, S., and Bhattacharya, A., 1992, Cloning and sequence analysis of a calcium binding protein gene from a pathogenic strain of *Entamoeba histolytica*. *Mol. Biochem. Parasitol.* **52**: 137.

Prasad, J., Bhattacharya, S,. and Bhattacharya, A., 1992, The calcium binding protein of *Entamoeba histolytica*: Expression in *Esherichia coli* and immunochemical charaterization. *Cell Mol. Biol. Res.* **39**: 167.

Scaife, J. F., 1969, Mitotic inhibition induced in human kidney cells by methylgloxal and kethoxal. *Experientia* **25**: 178-179.

Shi, J., Kim, K.-N., Ritz, O., Albrecht, V., Gupta, R., Harter, K., Luan, S., and Kudla, J., 1999, Novel protein kinases associated with calcineurin B-like calcium sensors in *Arabidopsis*. *Plant Cell* **11**: 2393-2405.

Snedden, W.A., and Fromm, H., 1998, Calmodulin, calmodulin-related proteins and plant responses to the environment. *Trends Biol. Sci.* **3**: 299-304.

Sopory, S.K., and Munshi, M., 1998, Protein kinases and phosphatases and their role in cellular signaling in plants. *Crit. Rev. Plant Sci.* **17**: 245-318.

Szent-Gyogi, A., and Egyud, L. G., 1966, On the regulation of cell division. *Proc. Natl. Acad. Sci. USA* **56**: 203-207.

Takezawa, D., Ramachandren, S., Paranjiape, V., and Poovaiah, B.W., 1996, Dual regulation of chimeric plant serine/threonine kinase by calcium and calcium/calmodulin. *J. Biol. Chem.* **271**: 8126-8132.

Thornalley, P.J, 1993, The glyoxalase system in health and disease. *Mol. Aspects Med.* **14**: 287-371.

Veena, Reddy, V. S., and Sopory, S. K., 1999, Glyoxalase1 from *Brassica juncea*: molecular cloning, regulation and its overexpression confer tolerance in transgenic tobacco under stress. *Plant J.* **17**: 385-395.

Watillion, B., Kettermann, R., Boxus, P., and Burney, A., 1993, A calcium/calmodulin-binding serine/threonine protein kinase homologoues to the mammalian type II calcium/calmodulin-dependent protein kinase is expressed in plant cells. *Plant Physiol.* **101**: 1381-1384.

Watillion, B., Kettermann, R., Boxus, P., and Burney, A., 1995, Structure of calmodulin-binding protein kinase gene from Apple. *Plant Physiol.* **108**: 847-848.

Wattilon, B., Kettermann, R., Boxus, P., and Burney, A., 1992, Cloning and characterization of an apple (*Malus domestica* L Borkn) cDNA encoding a calmodulin-binding region of type II mammalian Ca^{2+}/calmodulin-dependent protein kinase. *Plant Sci.* **81**: 227-235.

Yadava, N., Chandok, M.R., Prasad, J., Bhattacharya, S., Sopory, S.K., and Bhattacharya, A.. 1997, Characterization of EhCaBP a calcium-binding protein of *Entamoeba histolytica* and its binding proteins. *Mol. Biochem. Parasitol.* **84**: 69-82.

Zielinski, R.E., 1998, Calmodulin and calmodulin binding proteins in plants. *Annu. Rev. Plant Physiol. Plant Mol. Biol.* **49**: 697-725.

Zielinski, R.E., Ling, V,. and Perera, I., 1990, Plant protein phosphorylation, protein kinases, calcium and calmodulin. In Current topics in plant biochemistry and physiology Vol. 9, Columbia, MO: *Interdisc. Plant Biochem. Physiol. Prog. Univ. Missouri*, pp: 141-152.

Apoplast Calmodulin: The Identification, Functions and Transmembrane Mechanism

DAYE SUN AND LIGENG MA
Institute of Molecular Cell Biology, Hebei Normal University, Shijiazhuang, Hebei 050016 P. R. of China

1. INTRODUCTION

It is well known that calmodulin (CaM) is an important intracellular receptor protein for the Ca^{2+} messenger. Recently, it has been found that it is also present and functions at the extracellular area of both plant and animal cells. Our work indicates that CaM is not only present at the apoplast surface in the plant kingdom universally, but also has several important biological functions and is involved in transmembrane signaling mechanisms.

2. IDENTIFICATION, LOCALIZATION AND PURIFICATION OF APOPLAST CaM

It was in 1984 in S. Roux's laboratory, while studying the distribution of subcellular CaM, that the water soluble form of cell wall CaM was first detected from oat coleoptile cells by means of vacuum filtration-centrifugation and radioimmunoassay (Biro *et al* 1984). In the same year, MacNeil *et al* (Northern General Hospital, Sheffield, UK) also reported the presence of CaM activity at the extracellular area of B16 melanoma cells (MacNeil *et al* 1984). Though interesting, these results were not followed up in any great detail.

Since 1988, our group has re-examined this work and identified both water soluble and salt soluble forms of CaM from wheat coleoptile cell walls

(Ye *et al* 1988). Later, we also localized it in cell wall of corn root tip cells by immuno-electronmicroscopy (Li *et al* 1993). In fact, we have detected apoplast CaM from all plant species that we have checked so far by using several methods like radioimmunoassay, ELISA, enzymology, and so on. These species include oat, wheat, corn, cauliflower, carrot, tobacco, *Angelica dahurica etc*. We believe therefore that the presence of apoplast CaM is universal in the plant kingdom.

We have partially purified the apoplast CaM, and compared its properties with intracellular CaM and found that the main properties are almost the same, like molecular weight, hydrophobicity, Ca^{2+}-dependent ability to activate target enzyme (PDE), *etc* (Li *et al* 1993); but there are also some differences, as for example, Ca^{2+} binding affinity, intrinsic and sensitized Tb^{3+} fluorescence, and the amount of CaM required to achieve half maximal activity of NAD kinase (unpublished data).

3. THE BIOLOGICAL FUNCTIONS OF APOPLAST CaM

Though it is still not clear what is the biological significance of appoplast CaM, yet our studies have revealed its following biological functions:

- promotion of cell proliferation as seen in suspension culture cells (Sun *et al* 1994),
- acceleration of cell wall regeneration and first division of the cell from protoplasts (Sun *et al* 1995),
- regulation of pollen germination and tube growth (Ma *et al* 1997),
- stimulation of trans-plasma membrane redox reactions (Sun *et al* 1998),
- regulation of short term reaction of apoplast Al^{3+} (Ma *et al* 2000).

To study the above functions, we mainly used membrane-impermeable antagonists like agarose-W7, CaM antibody and purified CaM. In one experiment, adding 10^{-9} to 10^{-6} M pure CaM enhanced the rate of pollen germination and tube growth in the culture conditions. The optimal CaM concentration was about 10^{-7} M and concentrations higher than 10^{-6} M reduced the effects. The same amount of BSA had no effect, indicating that the CaM effect was specific and not due to any nutritional effect of proteins. In another experiment, adding the membrane–impermeable CaM macromolecule inhibitor, anti-CaM serum (20 µg/ml) and W7-agarose (0.5 mM) inhibited pollen germination and tube growth. Higher concentrations were completely inhibitory. This indicated that endogenous apoplast CaM may not only be involved in this process, but also functions as a major factor in the regulation of pollen germination and tube growth. The anti-CaM serum is specific to CaM and cannot penetrate the membrane so also the beads of agarose-W7 which are even larger than pollen cell and have no possibility to enter cells (Ma et al 1997). Lately, similar results were also obtained in our *in vivo* experiments by microinjecting CaM and anti-CaM antibody into style transmitting tissue. These experiments showed that the endogenous apoplast CaM regulates pollen germination and tube growth, whereas exogenous CaM can accelerate the above processes.

Recently, we found that the apoplast CaM may be involved in the regulation of light-independent *rbcS* gene expression and this seems to be one of the most important functions of apoplast CaM. *RbcS* gene encodes for the small subunits of rubisco. Expression of *rbcS* gene is controlled in both a light-dependent and -independent manner. A lot of work has been done for elucidating the mechanism of former; however, the mechanism of light-independent expression of *rbcs* gene is still unknown. Using a suspension-culture cell line of transgenic tobacco which contained the GUS reporter gene fused with the promoter region of rice *rbcS* gene, we studied the time course of GUS activity and apoplast CaM concentration dynamics under total dark conditions. It was found that the GUS activity increased to the optimal level on the 10th day and then decreased. The CaM concentration achieved its peak on 9-10th day, about one day earlier than GUS activity peak. So both phenomena may be closely related. The dynamic change of CaM activity was not dependent on light. When exogenous purified CaM was directly added to culture medium at different concentrations, CaM could promote GUS activity; the optimal concentration of CaM was about 10^{-7}-10^{-8} M. At the same concentration, BSA had no effect. Adding CaM antiserum and W7-agarose into the culture medium inhibited the endogenous CaM activity and also the GUS activity was reduced. This effect could be reversed by purified exogenous CaM in a dose dependent manner, and at 10^{-7} M CaM the effect was completely reversed. Histochemical-staining analysis also confirmed these results (manuscript in preparation).

Northern blot analysis was used to test the effect of apoplast CaM on gene expression of different subforms of *rbcS* in the tomato suspension culture cells. Total RNA was extracted from cells cultured for different days and hybridized to ^{32}P labelled *rbcS*3A or 3C oligonucleotide probe. The light-independent *rbcS*3A gene expression in dark conditions showed a dynamic change from 7-15 days of culture, the peak transcript level was achieved on day 13. CaM concentrations also showed a similar dynamic change though the response was seen earlier than *rbcS* gene expression. However, light-dependent *rbcS*3C transcript did not show any change in darkness. Adding 10^{-7} M of exogenous CaM into culture medium increased the low level of the expression of light-independent *rbcS*3A gene in total darkness. At the same concentrations, BSA and S-100 protein had no effect. Light-dependent *rbcS*3C transcript levels were not influenced by the exogenous CaM. CaM antibody also inhibited the high level expression of *rbcS*3A in darkness but the pre-immune serum had no effect. All these data suggested that CaM may be one of the major factors for the light-independent expression of *rbcS* gene (manuscript in preparation).

4. TRANSMEMBRANE MECHANISM OF APOPLAST CAM

If CaM could function even when localized at the outside of the cell, then there must exist a transmembrane mechanism for the intracellular signal transduction pathways. We have done a series of experiments to check this using pollen germination as a model system. As mentioned above, apoplast CaM has been demonstrated to have a role in the regulation of pollen germination and growth.

Using covalent cross-linking analysis with biotinylated CaM as a molecular probe, our preliminary experiments indicated presence of CaMBPs on the outer surface of lily pollen plasma membrane (Ma *et al* 1997). Houston *et al* (unpublished) also provided preliminary evidence that there are cell-surface receptor-like proteins (according to their binding characteristics) for CaM on animal (myeloid) cells.

Heterotrimeric G proteins are a key component in transmembrane signal transduction mechanism. Our experiments have proved that heterotrimeric G proteins are present on the plasma membrane of lily pollen. This was shown by two methods: (i) immunoblotting analysis indicated that two kinds of anti Gα antibodies recognized the same 41 kDa band. (ii) ADP-ribosylation analysis showed that only the 41 kDa band could be ADP-ribosylated specifically by PTX. This suggested that the 41 kDa polypeptide may be the heterotrimeric G proteins present at the pollen tube. Our work with membrane impermeable and permeable G protein regulators also showed that G protein may be involved in the regulation mechanism of pollen germination and tube growth. Microinjection of the membrane-impermeable agonist GTPγs stimulated pollen tube growth, whereas antagonists (GDPβs, anti-Gα antibody) inhibited the same process. Using membrane-permeable G protein agonist CTX (cholera toxin), G protein antagonist PTX (pertussis toxin), we also got similar results. Pharmacological experiments indicated that G protein might act downstream of the apoplast CaM in pollen germination and tube growth.

Using rightside–out plasma membrane vesicles as an experimental model system, we measured the GTPase activity of G protein by the spectrofluorometric method. We showed that adding purified CaM at the outside of vesicles could activate GTPase activity of G protein located inside, and this effects could be completely inhibited by PTX, GDPβ-S, GMPPNP (nonhydrolysable GTP analogs) and activated by CTX. Taken together, the above data give direct evidence for the role of apoplast CaM in activating the GTPase activity of heterotrimeric G proteins inside of the cell and then the activated heterotrimeric G protein may transduce the signal to promote pollen germination and tube growth (Ma *et al* 1999).

Beside above, the following pharmacological reagents were also microinjected to elucidate the downstream signal events:

- Ca^{2+} channel activator Bay K8644, -blocker verapamil and ionophore A23187,
- phospholipase C (PLC) inhibitor U73122, microinjection of antibody against PLCβ (1-4),
- caged IP_3 and microinjection of IP_3,
- IP_3 receptor antagonist heparin, microinjection of antibodies against IP_3 receptor (1-3)
- antagonists of protein kinase and measurements of protein phosphorylation activity.

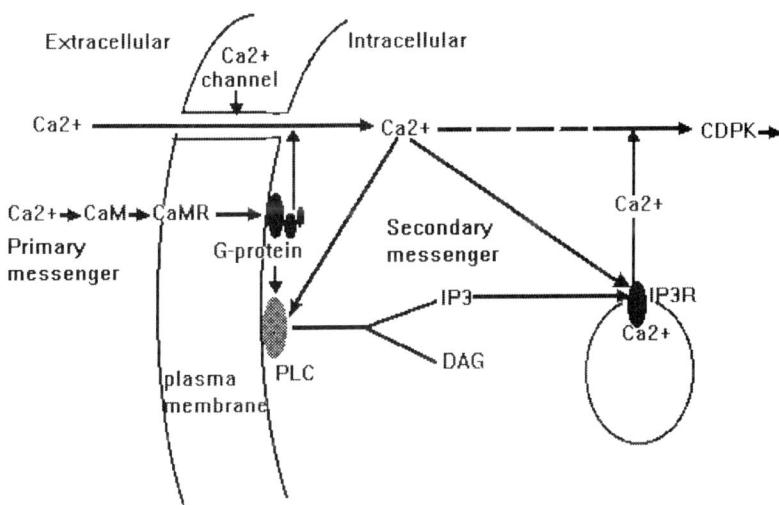

Figure 1. Proposed model of extracellular CaM signal transduction pathways.

Our preliminary experiments using the above reagents showed that both Ca^{2+} flux and phosphoinositide signaling pathways may be involved in the intracellular signal transduction of apoplast CaM to promote pollen germination and tube growth. From these experimental results, we suggest a model of apoplast CaM signal transduction pathway as shown in the Fig. 1.

Apoplast CaM binds to its receptor on the surface of a cell, and then activates it, which then activates the G protein, the G protein in turn activates Ca^{2+} channels, and PLC, which generates IP_3 that in turn releases Ca^{2+} into cytoplasm. The elevated cytoplasm Ca^{2+} signal may alter the protein phosphorylation status which either directly or *via* expression of genes would regulate the cell functions (Ma *et al* 1998 and unpublished data). Of course, we need to do more work to prove the above hypothesis conclusively.

5. PRESENCE OF APOPLAST CaM BINDING PROTEINS (CaMBPS)

We have also detected and localized the extracellular CaM binding proteins in several plants (Tang *et al* 1996) and animal species (Tang *et al* 1997) by biotinylated CaM gel overlay technique and immnogold-electronmicroscopic methods.

A major 21 kDa apoplast CaMBP was purified from wall protein of *A. dahurica* cells (Tang *et al* 1996). Using a 20 amino acids fragment from 21 kDa CaMBP N-terminal sequence as a probe, we isolated a cDNA encoding 21kDa CaMBP by PCR. The deduced sequence showed that 21 kDa CaMBP is a new number of citrate binding protein family (unpublished data). After making antibody, the function analysis indicated that 21 kDa CaMBP is not a CaM receptor-like protein, but it might be related to the regulation of the apoplast CaM activity (Mao *et al* 1999).

6. CONCLUSION

Since CaM is not only present at the apoplast area, but also has very important biological functions outside of the cell, and there exists transmembrane-intracellular signal transduction pathways, one needs to consider whether or not the apoplast CaM is a primary peptide messenger in the plant.

For a long time the plant apoplast was taken as an inert part of a cell, having only very narrow roles in the life of a cell (Strauss *et al* 1998). However, from our work and from the viewpoint of signal transduction, we think that the apoplast is a very important site for cell interaction with external environment and also in interaction between different cells. Any signal molecules, which are produced in one cell to influence the others, must pass through the apoplast.

Recently, some cell wall proteins were shown to have important biological or development cell functions, e.g. arabinogalactan (AGP) (Kreuger et al 1996), expansin (Fleming et al 1996) extensin etc. Some of these have indeed been shown to be the first-peptide messengers in plants, e.g. systemin and Nod factor 40 (Marx 1996). We think that apoplast CaM can be added to this list. Apoplast CaM also act as an extracellular signal in cAMP effects in *Dictyostelium* (Klein et al 1989) and extracellular Ca^{2+} effects in some animal cells (Brown et al 1993). They are intracellular signal molecules which also have developmental functions outside of the cell. From our apoplast CaM studies we have now learnt more about the role of apoplast in signal transduction and believe that research on apoplast will greatly contribute to the understanding of the total cell functions in plants.

REFERENCES

Biro, R.L., et al, 1984, Characterization of oat calmodulin and radioimmunoassay of its subcellular distribution. *Plant Physiol.* **75**: 382-386.
Brown, E.M., et al, 1993, Cloning and characterization of an extracellular Ca^{2+}-sensing receptor from bovine parathyroid. *Nature* **366**: 575-580.
Fleming, A.T., et al, 1996, Induction of leaf primordial by the cell wall protein expansin. *Science* **276**: 1415-1417.
Klein, J., et al, 1989, A chemoattractant receptor controls development in *Dictyostelium discoideum*. *Science* **241**: 1467-1472.
Kreuger, M., and Holst, G.J.V., 1996, Arabinogalactan proteins and plant differentiation. *Plant Mol. Biol.* **30**: 1077-1086.
Li, J.X., Liu, J.W., and Sun, D.Y., 1993, Immunoelectron microscopic localization of calmodulin in maize root. *Cell Res.* **3**: 11-19.
Ma, L.G., and Sun, D.Y., 1997, The effects of extracellular calmodulin on initiation of *Hippeastrum rutilum* pollen germination and tube growth. *Planta* **302**: 336-340.
Ma, L.G., and Sun, D.Y. et al, 2000, Does aluminum inhibit pollen germination via extracellular calmodulin. *Plant Cell Physiol.* **41**: 372-376.
Ma, L.G., Xu, X.D, Cui, S.J., et al, 1998, The involvement of phosphoinositide signaling pathway in the inhibitory effects of extracellular camodulin on pollen germination and tube growth. *Acta Photophysiologica Sinica* **24**: 198-200 (in Chinese).
Ma, L.G., 1997, Ph.D. Dissertation, *Studies on Extracellular Function and Molecular Evolution of Calmodulin*. China Agricultural University (in Chinese).
Ma, L.G., Xu, X.D., Cui, S.J., et al, 1999, The presence of heterotrimeric G protein and its role in signal transduction of extracellular calmodulin in pollen germination and tube growth. *Plant Cell* **11**: 1351-1363.
MacNeil, S., Dawson, R.A., and Crocker, G., et al, 1984, Effects of extracellular calmodulin and calmodulin antagonists on B16 melanoma cell growth. *J. Invest. Dermatol.* **83**: 15-19.
Mao, G.H., Tang, W.Q., and Sun, D.Y., 1999, Preliminary study on the physiological functions of extracellular 21 kDa calmodulin-binding protein from *Angelica dahurica*. *Acta Phytophysiologica Sinica* **25**: 165-170 (in Chinese).
Marx, J., 1996, Plants like animals may make use of peptide signals. *Science* **273**: 1338-1339.

Strauss, E., 1998, When wall can talk, plant biologists listen. *Science* **276**: 1415-1417.

Sun, D.Y., Li, H.B., and Chang, G., 1994, Extracellular calmodulin accelerates the proliferation of suspension cultured cells of *Angelica dahurica*. *Plant Sci.* **99**: 1-8.

Sun, D.Y., Bian, Y.Q., Zhao, B.H., *et al*, 1995, The effects of extracellular calmodulin on cell wall regeneration and cell division of protoplasts. *Plant Cell Physiol.* **36**: 133-138.

Sun, Y.U., Chen, J., and Sun, D.Y., 1998, Extracellular calmodulin stimulates the trans-plasmamembrane redox reaction of root protoplasm in *Zea mays*. *Acta Botanica Sinica* **40**: 437-441 (in Chinese).

Tang, J., Wu, S.P., Sun, D.Y., *et al*, 1996, Extracellular calmodulin-binding proteins in plants: Purification of a 21-kDa calmodulin-binding protein. *Planta* **98**: 510-516.

Tang, W.Q., Guo, Yi., Sun, D.Y., *et al*, 1997, Extracellular calmodulin-binding proteins in body fluids of animals. *J. Endocr.* **155**: 13-17.

Ye, Z.H., Sun, D.Y., and Guo, J.F., 1988, Preliminary study on wheat cell wall calmodulin. *Chin. Sci. Bull.* **33**: 624-626 (in Chinese).

CDPKs in Plant Signaling Networks
Progress in Research on a Groundnut CDPK

MAITRAYEE DAS GUPTA AND SUBHO CHAUDHURI
Department of Biochemistry, Calcutta University, 35, Ballygunge Circular Road, Calcutta 700019, India

1. INTRODUCTION

CDPKs or Calmodulin like Domain Protein Kinases are the most abundant members of the calcium regulated protein kinases found in signaling networks in plant cell (Roberts and Harmon 1992). They are biochemically distinct from protein kinase C (Nishizuka 1986) and Ca^{2+}/Calmodulin dependent protein kinases (Cohen 1985), since they require neither lipid nor calmodulin for activation. As the name implies the unique biochemistry of CDPKs appear to result from a novel structural arrangement within a single polypeptide; a catalytic domain of ser/thr kinase is joined to a calmodulin like regulatory domain containing four calcium binding helix loop helix EF hand motifs (Harper *et al* 1991). Presence of EF hand motifs qualify CDPKs to become a subfamily of EF hand containing calcium modulators in plant cells. They can sense the small transient changes in cellular calcium concentration and respond to the signal by change of conformation upon binding Ca^{2+} (Roberts and Harmon 1992). Atypical CDPKs containing significantly degenerate sequence in the calmodulin like domain are known as CDPK Related Protein Kinase or CRKs. They comprise another subfamily of calcium regulated protein kinases in plants (Furumoto *et al* 1996).

In plants, touch, low temperature, red light, hypoxia, gravity, and a number of growth factors have been demonstrated to have their signals mediated through change in cytoplasmic Ca^{2+} ion concentration (Roberts and Harmon 1992). These changes enormously vary in amplitude, kinetics and spatio-temporal distribution (Bush 1995). The cell distinguishes one stimulus from the other and generates variability in the response by using various modulators and/or target proteins. As classical homologues of CaM kinase or Protein Kinase C has not been reported in plants (Satterlee and Sussman 1998), CDPKs stand to be the most common modulator of calcium ion flux in plant cells. Each isoform of CDPK is believed to have a distinct role to play. They are found to be distributed in all parts of plant with each isotype having different expression patterns, and involved in any of every conceivable plant physiological event. In this report we have discussed the common properties of CDPKs, keeping the central focus on the association of CDPKs with various physiological events. Based on our work we suggest a CDPK from groundnut to be associated with desiccation related events during seed formation. The progress of our research on this CDPK from groundnut is described in this chapter.

2. BIOCHEMICAL PROPERTIES OF CDPKS

The shared biochemical properties of CDPKs have been comprehensively reviewed by Harmon and Roberts (1992), and almost a decade later though many more CDPKs have been included in the list, the scenario remains more or less the same. The salient features are mentioned here. In spite of having overall biochemical resemblance, the members of the CDPK family encoded by large multigene families have been found to significantly differ from each other.

2.1 Domain Organization and Size

From the sequences submitted in the Genbank it is clear that the domain arrangement of all members of the CDPK family is same (Fig. 1).

Figure1: Domain organization of CDPK

They have a catalytic domain resembling the catalytic domain of ser/thr protein kinase (gray box) and contain 255 amino acid residues in average. This domain is 33-42% identical to the catalytic domains of the family of Ca^{2+}/calmodulin dependent protein kinases (Hanks *et al* 1988) but is only 25% and 29% identical to PKCα (Parker *et al* 1986) and PKA respectively

(Shoji *et al* 1983). There is a regulatory domain (open box) resembling calmodulin, containing 4 helix-loop-helix EF hand motifs (black circles) comprising of 144 aa residues in average. This domain has 39% homology with spinach calmodulin sequence (Babu *et al* 1985). The catalytic and regulatory domains are joined together by a junction domain roughly consisting of 33 residues (black box). Together these 3 domains contribute 47 kDa (approx.) molecular weight to each CDPK. The molecular weight of CDPKs, which are all monomeric in nature, have been found to vary from 50 to 90 kDa, as per various biochemical reports and Genbank data. The extra molecular weight of various CDPK isotypes is contributed by the unique sequences of variable length that they have at their N and/or C terminal ends (lines).

2.2 Catalytic Properties

2.2.1 Choice of Substrates *in vitro*

Choice of substrates *in vitro*: Substrate specificities of most CDPKs as tested *in vitro* is broad and overlaps with that of protein kinase C and all Ca^{2+}/calmodulin dependent protein kinases (Roberts and Harmon 1992). With variable affinity various CDPK isotypes have been found to phosphorylate (i) a very specific substrate of PKC consisting of residues 4-14 of myelin basic protein, (ii) syntide II a good substrate of CaMKII and (iii) various histones. A number of other substrates commonly phosphorylated by protein kinase like glycogen synthase, ribosomal protein S6 also get phosphorylated by CDPKs. Substrates of cAMP dependent protein kinases are however very poorly phosphorylated by CDPKs. Some CDPKs are very specific in their choice of substrate. A CDPK characterised from *Mougotia* (Roberts 1989) and another from groundnut (DasGupta 1994) were found to specifically phosphorylate peptides derived from SkMLCs.

2.2.2 Autophosphorylation

Autophosphorylation: Ca^{2+} dependent autophosphorylation has been reported for several CDPKs and is believed to be a common property of these kinases (Roberts and Harmon 1992). The effect of autophosphorylation on CDPK activity varies from 'no effect' as in case of CDPKα from soybean (Putnam-Evans *et al* 1990) to an 'upregulatory' effect (Bogre *et al* 1988) or a 'downregulatory' effect (Saha and Singh 1995). In none of these cases however the enzyme was rendered Ca^{2+} independent, as would have been the case if compared with CaM kinases, where autophosphorylation makes the enzyme independent of Ca^{2+}/calmodulin (Lai *et al* 1986). In our laboratory, we noticed no effect of intramolecular autophosphorylation of thr residues of GnCDPK on its activity toward exogenous substrate. We reinvestigated the problem at suboptimal ATP concentrations to watch the reaction at slow motion. A pronounced lag time of 1 to 2 minutes followed by a linear increase of activity for 7,5 mins was seen in the initial rate of exogenous substrate phosphorylation under such conditions. Prior autophosphorylation completely abolished this lag phase and a sharp rise of exogenous substrate phosphorylation was seen from the first minute. Our results suggested autophosphorylation to be a prerequisite for the activation of the groundnut CDPK (Chaudhuri *et al* 1999). This mechanism is yet to be generalized.

2.2.3 Activation by Calcium

Calcium stimulates CDPKs by directly binding to them through EF hand calcium binding motifs in their calmodulin like domain. The concentration of calcium required for their half maximal activation ranges from 0.1-2 µM (Roberts and Harmon 1992). CDPK isotypes considerably differ in their calcium binding properties depending on the presence or absence of substrate (Lee *et al* 1998). CaM antagonists like phenothiazines, a napthalene sulphonamide (W7), calmidazolium *etc* interact with all CDPKs through their calmodulin like domain and inhibits their activity (Chaudhuri *et al* 1999). Compounds that inhibit PKC e.g. sphingosine and staurosporine also has been found to inhibit CDPKs (Roberts and Harmon 1992).

2.2.4 Mechanism of activation of CDPKs

Interestingly, the junction domain of CDPKs which is one of the most highly conserved regions in this family of kinases, has homology with the autoinhibitory pseudosubstrate domains of the Ca^{2+}/calmodulin dependent protein kinases of the animal system e.g. CaMKII (Bennet and Kennedy 1987), SmMLCK (Shoemaker *et al* 1990) and SkMLCK (Takio *et al* 1986). The homology is more pronounced with skMLCK and smMLCK and is centered on the calmodulin binding subdomain sequence. The autoinhibitory properties of junction domains of CDPKα from soybean and AK1 from *Arabidopsis* have been well documented (Harper *et al* 1994, Harmon *et al* 1994). In consistence with the homology of the junction domains with calmodulin binding domains, it has also been demonstrated with these CDPKs, that the junction domains undergo a Ca^{2+} dependent intramolecular interaction with the calmodulin like domain of the kinase leading to its activation (Huang *et al* 1996, Yoo and Harmon 1996).

3. PHYSIOLOGICAL RELEVANCE OF CDPK DEPENDENT PHOSPHORYLATION EVENTS

The relationship of CDPK dependent phosphorylations with certain physiological events has been suggested in several cases though the claims are not definitive. Different approaches have been made to ascertain the role of these kinases. Attempts have been made towards (i) identification of endogenous substrates, (ii) monitoring expression of the kinase in different tissues and stages of plant life cycle, (iii) checking inducibility of the kinase under specific defined conditions, (iv) prediction of function from sequence homology.

3.1 Identification of Endogenous Substrates

The first endogenous substrate for a CDPK was demonstrated to be nodulin-26, a nodule specific membrane protein in soybean (Weaver and Roberts 1992). Nod-26 is homologous to numerous transport and channel proteins in membranes. There are other evidences of CDPKs phosphorylating membrane channel proteins. A potential substrate of an oat plasma membrane CDPK was found to be a H^+ATPase (Schaller and Sussman 1988). In beet roots, CDPK dependent phosphorylation of H^+ATPase is found to inhibit ATP hydrolysis and proton transport activity of the channel (Lino *et al* 1998). Another example of CDPK phosphorylating a channel protein came from the observation made in guard cells of *Vicia faba*, where it has been shown that vacuolar chloride channels are activated by CDPK dependent phosphorylation (Pei *et al* 1996). A potassium channel KATI in guard cells have also been found to be phosphorylated by CDPK (Li *et al* 1998). Collectively these evidences suggest a general mechanism of CDPK dependent membrane transport regulation in plant cells.

CDPKs also use some key regulatory enzymes of plant metabolism as substrate. They have been demonstrated to phosphorylate sucrose synthase (Zhang and Chollet 1997) and phosphoenol pyruvate carboxylase (Ogawa *et al* 1998) in soybean nodule, indicating their regulatory role in dark reactions of photosynthesis. Nitrate reductase is phosphorylated by a CDPK and the phosphorylated protein is found to bind a 14-3-3 protein and thereby gets inhibited (Douglas *et al* 1998). This phenomenon has been demonstrated in spinach and shows the involvement of CDPK dependent protein phosphorylation in regulation of nitrogen assimilation in plants. Role of CDPK dependent phosphorylation in lipid metabolism is demonstrated in wheat plants where small molecular wt proteins homologous to lipid transfer proteins (LTPs) has been found to be phosphorylated by a CDPK (Neumann *et al* 1994). Small and large chains of napin like basic protein fractions from radish and castor bean seeds, who were found to act as calmodulin antagonist, are also very good substrates of CDPK (Neumann *et al* 1996). CDPK here is believed to be tuning the signals generated by Ca^{2+}. Fundamental processes like translation are also found to be regulated by CDPK dependent phosphorylation. In lily anther, CDPK has been found to phosphorylate the eukaryotic elongation factor 1α in multiple sites indicating its role in regulating translation (Wang and Poovaiah 1999).

3.2 Tissue Specific Expression of CDPK and its Subcellular Localization

Tissue specific expression of CDPK was reported first in maize (Estruch *et al* 1994). A CDPK isotype in this plant was found to be specifically expressed in pollen grains in late stages of its development. It has been demonstrated that this pollen specific maize CDPK is required for germination of pollen and the pollen tube growth. Another example of tissue specific expression of CDPK comes from rice where a CDPK expression has been shown to be spatially and temporally regulated during seed development (Kawasaki *et al* 1993). In our laboratory, we have shown in case of a groundnut CDPK that its expression is specifically restricted to the cotyledons. Similar to the rice CDPK its expression is spatio-temporaly regulated during seed development. In addition we have shown that the expression of the enzyme decays with the onset of germination. Seeds which fail to germinate show no such sign of decay in expression of the CDPK (Chaudhuri *et al* communicated). It is possible that these CDPKs have some role in initiating and protecting the state of dormancy in the seeds. Together these cases clearly point to the involvement of these kinases in tissue specific physiological processes. In contrast other CDPKs have been found to be expressed all throughout the plant from leaves to roots (Putnam-Evans *et al* 1990). Immunocytochemical localization of CDPK proteins with a soybean CDPK specific monoclonal antibody, indicated association of CDPKs with F-actin (Putnam-Evans *et al* 1989), chromatin (Roberts and Harmon 1992) and plasma membrane (Schaller *et al* 1992).

3.3 Induction of CDPK by Various Signals

Evidences are now in abundance showing expression of specific CDPK types to be induced in presence of different phytohormones, indicating these inducible CDPKs to have unique role in specific calcium dependent signal transduction pathways. A rice seed plasma membrane CDPK has been found to be highly induced following treatment of the seed with gibberellin for 10 mins (Abo-el-Saad *et al* 1995). ABA has been found to induce a particular CDPK in *Arabidopsis* (PKAB1) (Sheen 1996). Expression of a CDPK in tobacco leaves has been found to be stimulated in presence of phytohormones like ABA, GA, cytokinin, and methyl jasmonate where it was otherwise found to be absent (Yoon *et al* 1999). The same CDPK was also induced in leaves by wounding, fungal elicitors, chitosan and NaCl. A membrane bound CDPK type from tobacco was found to be induced in leaves after treating them with sucrose (Iwata *et al* 1998). Induction of a CDPK was also noticed in presence of 14-3-3 protein (Camoni *et al* 1998) indicating the importance of CDPK dependent phosphorylation in crosstalk between signaling pathways.

3.3.1 Induction of CDPKs in presence of stress

Association of specific CDPK types with specialized physiological events has been best examplified in *Arabidopsis* with relation to their expression in response to stress (Urao *et al* 1994). In an attempt to characterize the protein kinases that might be participating downstream to calcium signals mediating stress perception, (Urao *et al* 1994), we identified two different CDPKs, ATCDPK1 and ATCDPK2 that were specifically expressed in drought induced conditions. The sequence of ATCDPK1 and ATCDPK2 was 59% homologous to each other. The expression of both these genes start within 10 minutes of the start of dehydration. The expression reaches a maxima by 1 hour in case of ATCDPK1 and 20 minutes in case of ATCDPK2. Neither of these kinases were found to be expressed in high or low temperature or in presence of the phytohormone ABA which is known to mediate stress signal in plants.

This role of CDPKs in stress signaling was further analyzed by Sheen in her work on *Arabidopsis* CDPKs (Sheen 1996). Her work has illuminated the fact that though both ATCDPK1 and ATCDPK2 are stress induced, they have completely different signals to transduce. She has shown that constitutive expression of ATCDPK1 (and a related kinase ATCDPK1a), but not ATCDPK2 (and five other isotypes of CDPKs) can bypass stress signals and activate a stress and ABA inducible promoter HVA1. This activation of HVA1 promoter by ATCDPK1 could be partially blocked by a constitutively active protein phosphatase 2C. The same treatment however could completely abolish ABA dependent activation of HVA1 promoter. The ATCDPK1 mediated pathway thus seem to partially overlap with the ABA dependent pathway in a stress activated signaling network.

3.4 Analysis of CDPK Sequences

The first cDNA clone (SK5) for a CDPK was reported from soybean (Harper *et al* 1991). Today there are almost fifty CDPK type sequences reported in the GenBank. All the CDPKs as mentioned earlier contain three domains; a catalytic domain, a junction domain and a calmodulin like domain. The extension in the N and C terminal ends attribute unique domains to the kinases for their specialized function. Keeping these unique domains aside, the relative homology of CDPKs in the three conserved domains varies from 50 to 60%. The homology increases in case of certain members indicating parallelism in their behavior. The situation is best exemplified in *Arabidopsis* where several CDPKs has been characterized (Sheen 1996). Two of them, namely ATCDPK1 and ATCDPK1a share 96% homology with each other. As is expected, all the functional behavior of ATCDPK1 (discussed in section 3.3.1) is found to be mimicked by ATCDPK1a. A CDPK from groundnut (Acc. no. Y18055) which is specifically expressed during seed development, shows 88% homology with ATCDPK1. Because the upstream activator of ATCDPK1 expression is known to be drought, it is very attractive to postulate the physiological relevance of groundnut CDPK dependent phosphorylation. It is possible that the groundnut kinase is associated with the physiological desiccation events during seed development.

4. CONCLUSION

The discovery and onward progress of research on CDPKs have established a very important signaling module in plants. Calcium is already found to be an important second messenger in plant signal transduction pathways and CDPKs seem to be the most common translator of cytoplasmic calcium flux to protein phosphorylation. Collectively all evidences say that CDPKs are attached to every conceivable physiological event in plants. What is not clear yet is why plant signalling network has been forced to abandon the diffusible calmodulin mediated kinases and adapt a module where a calcium modulator and a kinase has been fused together.

REFERENCES

Abo-el-saad, M., and Wu, R., 1995, A rice membrane calcium-dependent protein kinase is induced by gibberellin. *Plant Physiol.* **108**: 787-793.

Babu, Y.S., Sach, J.C., Bugg, C.E. *et al*, 1985, Three dimesional structure of calmodulin. *Nature* **315**: 37-40.

Bogre, L., Olah, Z., and Dudits, D., 1988, Calcium dependent protein kinase from alfalfa (*Medicago varia*): Partial purification and autophosphorylation. *Plant Sci.* **58**: 135-144.

Bush, S.D., 1995, Calcium regulation in plant cells and its role in signalling. *Annu. Rev. Plant Physiol. Plant Mol. Biol.* **46**: 95-122.

Chaudhuri, S., Seal, A., and DasGupta, M., 1999, Autophosphorylation dependent activation of a calcium-dependent protein kinase from groundnut. *Plant Physiol.* **120**: 13-18.

Chaudhuri, S., Seal, A., and DasGupta, M., 2000, Expression of a draught induced calcium dependent protein kinase in fruit formation stages of a groundnut *Arachis hypogea* plant. Communicated.

Cohen, P., 1985, The role of a protein phosphorylation in the hormonal control of the enzyme activity. *Eur. J. Biol.* **151**: 439-448.

Camoni, L., Harper, J.F., and Palmgren, M.G., 1998, 14-3-3 proteins activate a plant calcium-dependent protein kinase (CDPK). *FEBS Lett.* **430** (3): 381-384.

DasGupta, M., 1994, Characterisation of a calcium dependent protein kinase from *Arachis hypogea* (Groundnut) seeds. *Plant Physiol.* **104**: 961-969.

Douglas, P., Moorhead, G., Hong, Y., Morrice, N., and MacKintosh, C., 1998, Purification of a nitrate reductase kinase from *Spinacea oleracea* leaves, and its identification as a calmodulin-domain protein kinase. *Planta* **206** (3): 435-442.

Ellard-Ivey, M., Hopkins, R.V., White, T.J., and Lomax, T.L., 1999, Cloning, expression and N-terminal myristoylation of CpCPK1, a calcium-dependent protein kinase from zucchini (*Cucurbita pepo L.*). *Plant Mol. Biol.* **39** (2): 199-208.

Esteruch, J.J., Kadwell, S., Merlin, E., and Crossland, L., 1994, Cloning and characterization of a maize pollen-specific calcium-dependent calmodulin-indipendent protein kinase. *Proc. Natl. Acad. Sci. USA* **91**: 8837-8841.

Furumoto, T., Ogawa, N., Hata, S., and Izui, K., 1996, Plant calcium dependent protein kinase-related kinases (CRKs) do not require calcium for their activities. *FEBS Lett.* **396**: 147-151.

Farmer, P.K., and Choi, J.H., 1999, Calcium and phospholipid activation of a recombinant calcium dependent protein kinase (DcCPK1) from carrot. *Biochem. Biophys. Acta* **1434** (1): 6-17.

Hanks, S.K., Quinn, A.M., and Hunter, T., 1988, The protein kinase family: conserved features and deduced phylogeny of the catalytic domains. *Science* **241**: 42-52.

Harmon, A.C., Yoo, B.C., and McCaffery, C., 1994, Pseudosubstrate inhibition of CDPK, a protein kinase with a calmodulin-like domain. *Biochemistry* **33**: 7278-7287.

Harper, J.F., Jing-Feng Huang, J., and Lioyd, S., 1994, Genetic identification of an autoinhibitor in CDPK, a protein kinase with a calmodulin-like domain. *Biochemistry* **33**: 7267-7277.

Harper, J.F., Sussman, M.R., Schaller, G.E., Putnam-Evans, C., and Harmon, A.C., 1991, A calcium dependent protein kinase with a regulatory domain similar to calmodulin. *Science* **252**: 951-954.

Harper, J.F., Binder, B.M., and Sussman, M.R., 1993, Calcium and lipid regulation of an *Arabidopsis* protein kinase expressed in *Escherichia coli*. *Biochemistr* **32**: 3282-3290.

Hung, J.F., Teyton, L., and Harper, J.F., 1996, Activation of a Ca^{2+} dependent protein kinase involves intramolecular binding of a calmodulin-like regulatory domain. *Biochemistry* **35**: 13222-30.

Iwata, Y., Kuriyama, M., Nakakita, M., Kojima, H., Ohto, M., and Nakamura, K., 1998, Characterization of a calcium-dependent protein kinase of tobacco leaves that is associated with the plasma membrane and is inducible by sucrose. *Plant Cell Physiol.* **39** (11): 1176-1183.

Kawasaki, T., Hayashida, N., Baba, T., Shinozaki, K., and Shimada, H., 1993, The gene encoding a calcium dependent protein kinase located near the sbe 1 gene encoding starch branching enzyme specifically expressed in developing rice seeds. *Gene* **129**: 183-189.

Kawasaki, T., Okumura, S., Kishimoto, N., Shimada, H., Higo, K., and Ichikawa, N., 1999, RNA maturation of the rice SPK gene may involve trans-splicing. *Plant J.* **18** (6): 625-632.

Lai, Y., Narin, A.C., and Greengard, P., 1986, Autophosphorylation reversibly regulates the Ca^{2+}/Calmodulin dependence of a Ca^{2+}/Calmodulin-dependent protein kinase II. *Proc. Natl. Acad. Sci. USA* **83**: 4253-4257.

Lee, J.Y., Yoo, B.C., and Harmon, A.C., 1998, Kinetic and calcium-binding properties of three calcium-dependent protein kinase isoenzymes from soybean. *Biochemistry* **37**: 6801-6809.

Li, J., Jee, Y.R., and Assmann, S.M., 1998, Guard cells poses a calcium-dependent protein kinase that phosphorylates the KAT1 potassium channel. *Plant Physiol.* **116** (2): 785-795.

Lino, B., Baizabal-Aguirre,V.M., and Gonzalez de la Vara, L.E., 1998, The plasma-memmbrane H(+)-ATPase from beet root is inhibited by a calcium dependent phosphorylation. *Planta* **204** (3): 352-359.

Neumann, G.M., Condron, R., Thomas, I., and Polya, G.M., 1994, Purification and sequencing of a family of wheat lipid transfer protein homologues phosphorylated by plant calcium dependnet protein kinase. *Biochem. Biophys. Acta* **1209** (2): 183-190.

Neumann, G.M., Condron, R., Thomas, I, and Polya, G.M., 1996, Purification and sequencing of multiple forms of *Brassica napus* seed napin small chains that are calmodulin antagonists and substrates for plant calcium-dependent protein kinase. *Biochem. Biophys. Acta* **1295** (1): 23-33.

Neumann, G.M., Condron, R., and Polya, G.M., 1996, Purification and sequencing of napins like proteins and large chains from *Momordica charantia* and *Ricinus communis* seeds and determination of sites phosphorylated by plant Ca^{2+}-dependent protein kinase. *Biochem. Biophys. Acta* **1298** (2): 223-240.

Nishizuka, Y., 1986, Studies and perspectives of Protein Kinase C. *Science* **232**: 305-312.

Nishiyama, R., Mizuno, H., Okada, S., Yamaguchi, T., Takenaka, M., and Ohyama, K., 1999, Two mRNA species encoding calcium-dependent protein kinases are differentially expressed in sexual organs of Marchantia polymorpha through alternative splicing. *Plant Cell Physiol.* **40** (2): 205-212.

Ogawa, N., Yabuta, N., Ueno, Y., and Izui, K., 1998, Characterization of a Maize Ca^{2+}dependent protein kinase phosphorylating Phosphoenol pyruvate carboxylase. *Plant Cell Physiol.* **39** (10): 1010-1019.

Parker, P.L., Coussens, L., Totty, N., Rhee, L., and Young, S., 1986, The complete primary structure of PKC- the major phorbol ester receptor. *Science* **233**: 853-859.

Patil, S., Takezawa, D., and Poovaiah, B.W., 1995, A chimeric Ca^{2+}/calmodulin-dependent protein kinase (CcaMK) gene characterized by a catalitic domain, a calmodulin-binding domain, and a neural visinin-like Ca^{2+}-binding domain was recently cloned from plants. *Proc. Natl. Acad. Sci. USA* **92**: 4797-4801.

Pei, Z.M., Ward, J.M., Harper, J.F., and Schroeder, J.I., 1996, A novel chloride channel in Vicia fava guard cell vacuoles activated by the ser-thr kinase CDPK. *EMBO J.* **15** (23): 6564-6574.

Putnam-Evans, C., Harmon, A.C., and Cormier, M., 1990, Purification and characterization of a novel calcium dependent protein kinase from soybean. *Biochemistry* **29**: 2488-2495.

Putnam-Evans, C., Harmon, A.C., Palevitz, B.A., Fechheimer, M., and Cormier, M.J., 1989, Calcium-dependent protein kinase is localised with F-actin in plant cells. *Cell Motil. Cytoskel.* **12**: 12-22.

Roberts, D.M., 1989, Detection of a Calcium activated protein kinase in *Mougeotia* by using synthetic peptide substrate. *Plant Physiol.* **91**: 1613-1619.

Roberts, D.M., and Harmon, A.C., 1992, Calcium modulated proteins: targets of intracellular calcium signals in higher plants. *Annu. Rev. Plant. Physiol. Plant Mol. Biol.* **43**: 375-414.

Saha, P., and Singh, M., 1995, Characterization of a winged bean (*Psophocarpus tetragonolobus*) protein kinase with calmodulin like domain: regulation by autophosphorylation. *Biochem. J.* **305**: 205-210.

Satterlee, J.S., and Sussman, M.R., 1998, Unuasual membrane-associated protein kinases in higher plants. *J. Membr. Biol.* **164** (3): 205-13.

Schaller, G.E., and Sussman, M.R., 1988, Phosphorylation of the plasma-membrane H^+-ATPase of oat roots by a calcium-stimulated protein kinase. *Planta* **173**: 509-518.

Schaller, G.E., Harmon, A.C., and Sussman, M.R., 1992, Characterization of a Calcium and Lipid dependent protein kinase associated with the plasma membrane of Oat. *Biochemistry* **31**: 1721-1727.

Sheen, J., 1996, Calcium dependent protein kinases and stress signal transduction in plants. *Science* **274**: 1900-1902.

Shoji, S., Ericssion, L.H., Walsh, K.A., Fischer, E.H., and Titani, K., 1983, Amino acid sequence of the catalytic subunit of bovine type II adenosine cyclic 3',5'-phosphate dependent protein kinase. *Biochemistry* **22**: 3702-3709.

Urao, T., Katagiri, T., Mizoguchi, T., Yamaguchi-Shinozaki, K., Hayashida, N., and Shinozaki, K., 1994, Two genes that encode Ca^{2+}-dependent protein kinases are induced by drought and high-salt stresses in *Arabidopsis thaliana*. *Mol. Gen. Genet.* **224**: 331-340.

Wang, W., and Poovaiah, B.W., 1999, Interaction of plant chimeric calcium/calmodulin-dependent protein kinase with a homolog of eukaryotic elongation factor-1 alpha. *J. Biol. Chem.* **274** (17): 12001-12008.

Weaver, C.D., and Roberts, D.M., 1992, Determination of the site of phosphorylation of Nodulin 26 by the calcium dependent protein kinase from soybean nodules. *Biochemistry* **31**: 8954-8959.

Yoo, B.C., and Harmon, A.C., 1996, Intramolecular binding contributes to the activation of CDPK, a protein kinase with a calmodulin-like domain. *Biochemistr* **35**:12029-37.

Yoon, G.M., Cho, H.S., Ha, H.J., Liu, J.R., and Lee, H.S., 1999, Characterization of NtCDPK1, a Calcium-dependent protein kinase gene in *Nicotiana tabacum* and the activity of its encoded protein. *Plant Mol. Biol.* **39** (3): 991-1001.

Zang, X.Q., and Chollet, R., 1997, Seryl-phosphorylation of soybean nodule sucrose synthase (nodulin-100) by a Ca^{2+}-dependent protein kinase. *FEBS Lett.* **410** (2-3): 126-130.

The Function of the Maize CRINKLY4 Receptor-like Kinase in a Growth Factor Like Signaling System

[1,2]PHILIP W. BECRAFT, [1]MEENA R. CHANDOK, [2]PING JIN, [1]TAO GUO
[1]YVONNE ASUNCION-CRABB AND [1]YUAN ZHANG
[1]*Zoology & Genetics Department, Iowa State University, Ames, IA;* [2]*Agronomy Department, Iowa State University, Ames, IA, USA*

1. INTRODUCTION

One of the key questions in developmental biology is how cell fates are decided. In animal systems, complex signaling systems are often involved in establishing the cellular pattern of tissues. Much less is known about how cellular pattern and cell fates are established in plant systems. Studies have shown that, in general, cell fate is determined by position rather than by lineage (Dawe and Freeling 1991, Irish 1991). This implies that plant cells also signal one another during development to organize the differentiation of various cell types in the appropriate pattern for a given tissue. Recent evidence for the importance of receptor protein kinases in plant development support this (Becraft 1998). The maize aleurone provides an excellent system for studying plant cell fate acquisition; it is a simple system, with only a single cell fate decision between aleurone and starchy endosperm. Also, the genetically controlled anthocyanin pigmentation in the aleurone provides a visible marker that allows convenient screening for mutants disrupted in aleurone formation.

Mutants in the *crinkly4* (*cr4*) gene disrupt aleurone differentiation, causing a mosaic aleurone phenotype and the defects are much more prevalent on the abgerminal face of the kernel (Becraft *et al* 1996). The mutants also have plant phenotypes with stunted plants due to shortened internodes and crinkled leaves. The cellular defects indicate that *Cr4* has a complex developmental function, controlling cell proliferation, expansion, differentiation and pattern formation. The epidermis shows particularly strong effects on cell morphology and proliferation. Cells are abnormally large and irregularly shaped. Ultrastructural defects in epidermal cells include irregular deposition of cell wall material, lack of vacuoles and the presence of extraneous membrane vesicles. The irregular cell shape leads us to speculate that cytoskeletal functions are effected but tests of this hypothesis are not yet complete. The gene was isolated by transposon tagging and it encodes a receptor kinase containing 901 amino acids.

The complex array of developmental processes controlled by CR4 and the identity of CR4 as a predicted receptor kinase suggest analogy to growth factor receptors of animal systems. The processes controlled by many growth factor receptors, including epidermal growth factor receptor and transforming growth factor-β receptor include a list much like CR4; they regulate cell fate, proliferation, differentiation, patterning and are receptor protein kinases (Massague 1998, Schweitzer and Shilo 1997).

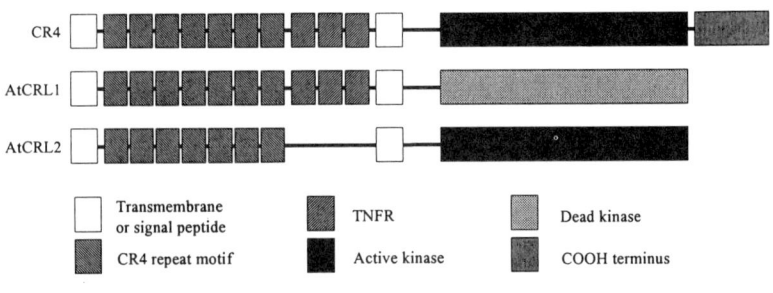

Figure 1. Diagramatic representation of the CR4 protein and the predicted proteins encoded by two *Arabidopsis* genes.

2. THE CR4 PROTEIN

The presumptive cytoplasmic domain contains a serine/threonine kinase with an approximately 40 amino acid juxtamembrane region and 119 amino acid carboxy terminus. Bacterially expressed CR4 cytoplasmic domain is capable of autophosphorylation and this activity is abolished by site directed mutagenesis of the invariant Asp^{652} to Ala, verifying the functionality of the kinase domain. The extracellular domain contains a region similar to the cysteine-rich repeats of tumor necrosis factor receptor. This similarity leads us to speculate that the signal ligand for CR4 is a peptide, perhaps related to TNF. In addition, there are seven repeats of a motif containing around 37 amino acids. These repeats are critical for CR4 function because the *cr4-624* mutant allele results from an aspartic acid to asparagine substitution in the 4th repeat.

Several genes encoding proteins with similar extracellular motifs occur in *Arabidopsis*. At least two have the predicted structures of receptor kinases. One contains the seven novel repeats and three TNFR-like repeats in the predicted extracellular domain. The cytoplasmic domain contains a kinase that is most likely inactive because it has a deletion of part of the 7th and 8th subdomains, including an invariant proline-glutamic acid pair. The second gene has a kinase domain that appears to be intact by sequence analysis. The extracellular domain contains the novel repeat motif but lacks a region of similarity to TNFR. It is not yet known whether either of these are functional homologs of CR4.

3. CR4 SIGNALS A CELL-AUTONOMOUS DIFFERENTIATON RESPONSE

To better understand the developmental role of CR4 signaling, a genetic mosaic analysis was conducted. The recessive *cr4-R* allele was marked with the linked, semi-dominant *Oy-700* allele. In a hemizygous state, *Oy-700* confers an albino phenotype. Seeds heterozygous for *Oy-700, cr4-R/++* in coupling linkage were treated with γ-rays to induce chromosome breakage. Loss of the short arm of chromosome 10 carrying the wild type alleles uncovers the mutant *cr4* allele and these genetically mutant cells are recognizable by their albino phenotype. The net effect is to generate albino sectors of *cr4* mutant tissue in otherwise normal plants. The results of these experiments showed that the *cr4* mutant phenotype was cell autonomous; wild type cells were not able to rescue neighboring mutant cells. This indicates that the cellular processes regulated by CR4 signal transduction are strictly intracellular. Furthermore, it was found that CR4 is required in both epidermal and internal cells throughout the shoot.

4. MUTANTS IDENTIFYING CANDIDATES FOR OTHER COMPONENTS CR4 SIGNALING

Mutations in genes encoding other components of the CR4 signaling system are expected to produce phenotypes similar to *cr4*. The *defective kernel1* (*dek1*) gene is a candidate for such a gene. Recessive mutations in *dek1* eliminate the aleurone layer from the endosperm and also cause arrested embryo development (Cone *et al* 1989). A weak allele, *dek1-Dooner*, causes a phenotype very reminiscent of *cr4* mutants. It causes a mosaic aleurone with defects more prevalent on the abgerminal face. Surface cells of the aleurone have the attributes of starchy endosperm cells indicating that cell fate has not been properly specified. Mutant plants have short internodes and crinkled leaves. The epidermal cells appear enlarged and irregularly shaped in transverse section.

Sectors of *dek1* mutant cells in the endosperm have fuzzy borders suggesting that the gene acts non-cell-autonomously (Neuffer 1995). This caused us to speculate that *dek1* might encode the signal ligand for the CR4 receptor. A genetic mosaic analysis was repeated, with an independent albino marker, *vp5*, linked in *cis* to the *dek1-792* mutant allele. Heterozygous seeds were irradiated, generating albino sectors that were genetically mutant for *dek1*, in otherwise normal plants. It was clear that wild type cells were not able to rescue neighbouring mutant cells as would be expected for a non-autonomous loss of function mutant. There were some suggestions that mutant cells could confer a mutant phenotype to adjacent wild type cells. This is unexpected for a recessive mutation and is not currently understood. In general however, the gene appears to function cell-autonomously, reducing the likelihood that *dek1* encodes the CR4 ligand.

Transposon tagging was undertaken to isolate the *dek1* gene. The *dek1-Dooner* mutant was used to pollinate wild type lines containing active *Mu* transposons. Four thousand ears of F1 kernels were screened and five putatively transposon tagged mutant alleles were recovered. A *Mu3* element was found to cosegregate with the *dek1-34* allele and efforts are currently underway to clone the genomic fragment.

Several other mutants that represent additional candidate components of the CR4 signal transduction system have been isolated from *Mu* lines. The *crinkly dwarf* mutant has a plant phenotype that resembles *cr4* while the kernel phenotype appears normal. Conversely, several *mosaic aleurone* mutants, of as yet undefined complementation, have kernel phenotypes similar to *cr4*, but normal plant phenotypes. These mutants represent potential plant specific and endosperm specific signal transduction components, respectively.

5. BIOCHEMICAL STUDIES OF CR4 SIGNALING

5.1 MAP Kinase Signalling

Because MAP kinase cascades commonly function downstream of receptor protein kinases, we investigated the possibility that CR4 regulates a MAP kinase. Protein extracts from wild type and *cr4-651* mutants were compared. The *cr4-651* allele produces a strong phenotype and contains a premature stop in subdomain 10 of the kinase catalytic domain. The extracts were run on SDS-PAGE and immunoblotted with human ERK1 antibodies. Wild type extracts contained an immunoreactive protein of approximately 44 kDa. The abundance of this protein was greatly reduced in the mutant. Extracts were then run on a myelin basic protein (MBP) substrate gel. When the gels were incubated with γ^{32}P-ATP, MAP kinase activity was detected at 44 kDa. Despite having lower levels of immunoreactive protein, the mutant contained a higher level of MBP phosphorylating activity. To verify that the MBP phosphorylating activity was due to the same MAP kinase, ERK1 antibodies were used for immunoprecipitation. The immunoprecipitated proteins were run on MBP substrate gels and again the mutant extract contained a higher level of activity than wild type. These results suggest that CR4 signaling negatively regulates a MAP kinase cascade.

5.2 Proteins that Bind CR4

Several affinity chromatography experiments were conducted to purify proteins that bind to the cytoplasmic domain of CR4. Two columns were prepared, one from the complete CR4 cytoplasmic domain expressed and purified from bacteria and another of a synthetic peptide from the juxtamembrane region. In one experiment, designed to isolate potential substrates of CR4, proteins were labelled *in vivo* with ^{32}P orthophosphate and the extracts passed over a column containing the CR4 cytoplasmic domain. Two phospho-proteins have been purified and N-terminal peptide sequence obtained. In another experiment, unlabeled extracts were passed over the juxtamembrane peptide column and N-terminal sequence has been obtained for 4 purified proteins. To date, the identity of all these purified proteins is unknown. Efforts are currently underway to identify and clone the genes encoding these factors. However, the affinity fractions have proven valuable for testing the interactions of other proteins described below.

5.2.1 Interaction with thioredoxin-h

Two thioredoxin-h-like proteins and an arm repeat-containing protein were found to interact in a yeast 2-hybrid screen with the cytoplasmic domain of the S receptor kinase, involved in *Brassica* self-incompatibility

(Bower et al 1996, Gu et al 1998). These prey plasmids were obtained from Daphne Goring and found to interact in yeast 2-hybrid assays with a bait plasmid containing the cytoplasmic domain of CR4. Antibodies to wheat thioredoxin-h (THX) were obtained from Robert Buchanan and immunoreactivity was detected in the affinity chromatography fractions from the complete cytoplasmic domain, indicating that maize thioredoxin also could bind CR4. Autoradiography and immunoblotting indicated that an *in vivo* labelled phosphoprotein and an anti-THX immunoreactive protein, present in the affinity fractions, both migrated at approximately 13 kDa in SDS-PAGE, suggesting that the phosphorylated THX binds CR4.

To test whether THX could act as a substrate for the CR4 kinase, THX was immunoprecipitated from maize seedling extracts and incubated with the bacterially expressed CR4 kinase domain in the presence of γ^{32}P-ATP. THX became phosphorylated in the presence of wild type CR4 kinase but not with a mutant form where the essential Asp^{652} had been substituted by Ala. This demonstrated that THX could act as a substrate for CR4 *in vitro*. Co-immunoprecipitation experiments were conducted to test *in vivo* interactions between THX and CR4. Proteins were immunoprecipitated with antibodies raised against a juxtamembrane peptide from CR4. An anti-THX immunoreactive protein of 13 kDa was pulled down along with CR4. When this mixture was incubated with γ^{32}P-ATP, both CR4 and THX became phosphorylated. The net result of these experiments is that THX binds to the cytoplasmic domain of CR4 *in vivo* and can act as a substrate for the CR4 kinase.

THX is encoded by a multigene family and it is not yet know which specific members interact with CR4 *in vivo*. Yeast 2-hybrid was used to test the interactions of THXs encoded by specific genes. Several maize ESTs encoding THX proteins were obtained. Two were cloned into a 2-hybrid prey plasmid and both proteins interacted with CR4. Other THX proteins remain to be tested.

Cytosolic thioredoxin activity was assayed in wild type and mutant plant tissues. Wild type tissues contained higher levels of activity than mutants. Immunoblots of cytosolic proteins indicated that mutants also contained lower levels of THX protein, however *thx* transcripts were unchanged indicating that the differential regulation is post-transcriptional. These results suggest that CR4 signaling positively regulates thioredoxin-h activity. It is still unknown what role THX phosphorylation plays. Possibilities include that phosphorylation could modulate thioredoxin enzymatic activity, could modulate interaction with CR4 or other proteins, or could induce THX translocation to the nucleus or other cellular location. These possibilities will be tested.

5.2.2 Possible interaction with ROP GTPase

Small GTPases of the Ras superfamily function downstream of many receptor protein kinases in animal systems. A Rho-like GTPase of plants (ROP) was found associated with the CLAVATA1 receptor kinase complex in *Arabidopsis* (Trotochaud 1999). Interactions between CR4 and ROP were tested in collaboration with John Fowler. Affinity fractions from the juxtamembrane peptide column contained a protein that was recognized by antibodies to a maize ROP protein. A ROP immunoreactive protein was pulled down by immunoprecipitation with anti-CR4 antibodies indicating that this interaction occurred *in vivo*. There is a discrepancy between the molecular weight of the CR4 interacting protein and that reported for ROP. ROPs are generally in the neighborhood of 20 kDa while the protein observed in these experiments was approximately 40 kDa. It remains to be determined whether this apparent size difference is due to a technical difference in extract preparation or whether a different protein with a related epitope interacts with CR4.

6. CONCLUSION

CR4 signaling regulates an array of developmental processes including cell fate, patterning, proliferation, shape and differentiation. The function of the CR4 receptor-like kinase therefore appears to be analogous to a growth factor receptor from animal systems. The developmental processes regulated by CR4 are cell autonomous indicating that the *cr4* mutant phenotype does not result from the lack of a secondary signal controlled by CR4. Several other mutants, including *dek1*, have similar phenotypes suggesting they may also function in CR4 signaling. CR4 contains a novel extracellular domain with a region of similarity to TNFR and a region containing 7 novel repeats. These motifs are also found in proteins encoded by the *Arabidopsis* genome. The cytoplasmic domain contains a serine/threonine kinase. Thioredoxin h binds to the CR4 cytoplasmic domain and appears to function in CR4 signal transduction. Thioredoxin h can be phosphorylated by CR4 and *cr4* mutants have reduced levels of cytosolic thioredoxin activity suggesting that thioredoxin h is positively regulated by CR4 signaling. Several other proteins have been affinity purified and one is immnologically related to ROPs although the apparent molecular weight is different. MAP kinase activity is increased in *cr4* mutants suggesting that CR4 signaling negatively regulates a MAP kinase cascade. Continued genetic and biochemical analyses will further elucidate this signalling system.

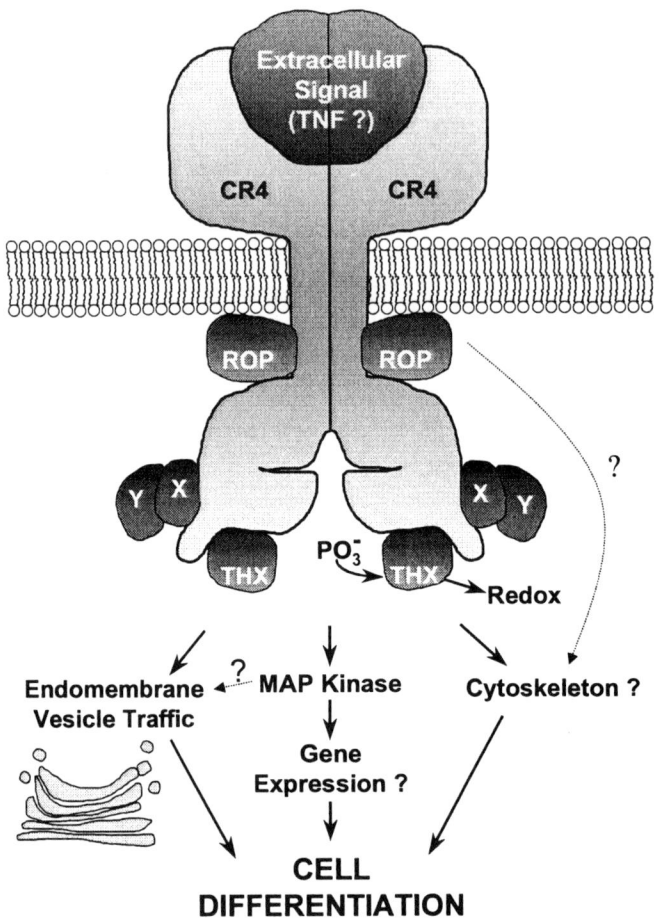

Figure 2. Model for CR4 signaling in the control of cell differentiation

7. CONCLUSION

CR4 signaling regulates an array of developmental processes including cell fate, patterning, proliferation, shape and differentiation. The function of the CR4 receptor-like kinase therefore appears to be analogous to a growth factor receptor from animal systems. The developmental processes regulated by CR4 are cell autonomous indicating that the *cr4* mutant phenotype does not result from the lack of a secondary signal controlled by CR4. Several other mutants, including *dek1*, have similar phenotypes suggesting they may also function in CR4 signaling. CR4 contains a novel extracellular domain with a region of similarity to TNFR and a region containing 7 novel repeats. These motifs are also found in proteins encoded by the *Arabidopsis* genome. The cytoplasmic domain contains a serine/threonine kinase. Thioredoxin h binds to the CR4 cytoplasmic domain and appears to function in CR4 signal transduction. Thioredoxin h can be phosphorylated by CR4 and *cr4* mutants have reduced levels of cytosolic thioredoxin activity suggesting that thioredoxin h is positively regulated by CR4 signaling. Several other proteins have been affinity purified and one is immnologically related to ROPs although the apparent molecular weight is different. MAP kinase activity is increased in *cr4* mutants suggesting that CR4 signaling negatively regulates a MAP kinase cascade. Continued genetic and biochemical analyses will further elucidate this signalling system.

REFERENCES

Becraft, P. W., 1998, Receptor kinases in plant development. *Trends Plant Sci.* **3**: 384-388.
Becraft, P. W., Stinard, P. S., and McCarty, D. R., 1996, CRINKLY4: a TNFR-like receptor kinase involved in maize epidermal differentiation. *Science* **273**: 1406-1409.
Bower, M. S., Matias, D. D., Fernandes-Carvalho, E., Mazzurco, M., Gu, T., Rothstein, S. J., and Goring, D. R., 1996, Two members of the thioredoxin-h family interact with the kinase domain of a Brassica S locus receptor kinase. *Plant Cell* **8**: 1641-1650.
Cone, K. C., Frisch, E. B., and Phillips, T. E., 1989, *dek1* interferes with aleurone differentiation. *Maize Genet. Coop. Newslett.* **63**: 67-68.
Dawe, R. K., and Freeling, M., 1991, Cell lineage and its consequences in higher plants. *Plant J.* **1**: 3-8.
Gu, T., Mazzurco, M., Sulaman, W., Matias, D. D., and Goring, D. R., 1998, Binding of an arm repeat protein to the kinase domain of the S-locus receptor kinase. *Proc. Natl. Acad. Sci. USA* **95**: 382-387.
Irish, V. F., 1991, Cell lineage in plant development. *Curr. Opin. Genet. Dev.* **1**: 169-173.
Massague, J., 1998, TGF-β Signal Transduction. *Annu. Rev. Biochem.* **67**: 753-791.
Neuffer, M. G., 1995, Chromosome breaking sites for genetic analysis in maize. *Maydica* **40**: 99-116.
Schweitzer, R., and Shilo, B. Z., 1997, A thousand and one roles for the Drosophila EGF receptor. *Trends Genet.* **13**: 191-196.
Trotochaud, A. E., Hao, T., Wu, G., Yang, Z., and Clark, S. E., 1999, The CLAVATA1 receptor-like kinase requires CLAVATA3 for its assembly into a signaling complex that includes KAPP and a Rho-related protein. *Plant Cell* **11**: 393-406.

Novel Calcium/Calmodulin-modulated Proteins
Chimeric Protein Kinase and Small Auxin Up-regulated RNA

B.W. POOVAIAH, WUYI WANG, AND TIANBAO YANG
Laboratory of Plant Molecular Biology and Physiology, Department of Horticulture, Washington State University, Pullman, WA 99164-6414, USA; Present address of Wuyi Wang: Department of Botany, University of Wisconsin, Madison, WI 53706-1381, USA

1. INTRODUCTION

Ca^{2+}/CaM-regulated protein phosphorylation is believed to play a pivotal role in amplifying and diversifying the action of Ca^{2+}-mediated signals (Poovaiah and Reddy 1993). Although Ca^{2+}/CaM-dependent protein phosphorylation is implicated in regulating a number of cellular processes in plants, not much is known about Ca^{2+}/CaM-dependent protein kinases and their role in Ca^{2+} signaling. A chimeric Ca^{2+}/CaM-dependent protein kinase (CCaMK) gene with a neural visinin-like Ca^{2+}-binding domain was cloned and characterized (Patil *et al* 1995). CCaMK is characterized by the presence of a catalytic domain, a CaM-binding domain, and a neural visinin-like Ca^{2+}-binding domain in a single polypeptide, making it distinctly different from other protein kinases (Takezawa *et al* 1996a, Ramachandiran *et al* 1997, Poovaiah *et al* 1999).

Recent evidence indicates that there is a close relationship between the mechanism of auxin action and Ca^{2+}/CaM-mediated signaling, but the interaction between them is not clear. We have isolated and characterized a novel CaM-binding protein which is encoded by a corn homolog of *SAURs*, designated as *ZmSAUR1*. *SAURs* belong to one group of the early auxin-response genes (Abel and Theologis 1996). Initially isolated from soybean (McClure and Guilfoyle 1987), *SAUR* genes have also been characterized from several dicots, such as mung bean (Yamamoto *et al* 1992), *Arabidopsis* (Gil *et al* 1994) and apple (Watillion *et al* 1998). In all cases examined, *SAURs* genes encode short transcripts with highly conserved open reading frames that accumulate rapidly and specifically after auxin treatment. Soybean *SAUR* gene transcription can be detected as soon as 2.5 min after auxin application (McClure and Guilfoyle 1987). We have demonstrated that the corn *ZmSAUR1* is a rapid auxin-responsive gene and the results discussed here suggest the involvement of Ca^{2+}/CaM messenger system in auxin action.

2. STRUCTURAL FEATURES OF CCAMK

CCaMK contains a catalytic kinase domain, followed by a CaM-binding domain and a neural visinin-like Ca^{2+}-binding domain in a single polypeptide (Fig. 1). The catalytic domain contains all 11 conserved subdomains characteristic of serine/threonine protein kinases and shares high homology to mammalian Ca^{2+}/CaM-dependent protein kinases (CaM kinases). Homology modeling of the kinase domain and CaM-binding domain using the X-ray structure of CaM kinase I suggests high conservation in these two domains between CCaMK and CaM kinase I (Poovaiah *et al* 1997). The two regulatory domains of CCaMK (the CaM-binding domain and the visinin-like Ca^{2+}-binding domain) share similarities with the CaM-binding domain of animal Ca^{2+}/CaM-dependent protein kinases and neural visinin-like proteins, respectively. The neural visinin-like proteins belong to a family of Ca^{2+}-sensitive regulator. The presence of a visinin-like Ca^{2+}-binding domain is unique to CCaMK. The structural features of CCaMK suggest that it has evolved from a fusion of two genes that are functionally different and phylogenetically diverse in origin.

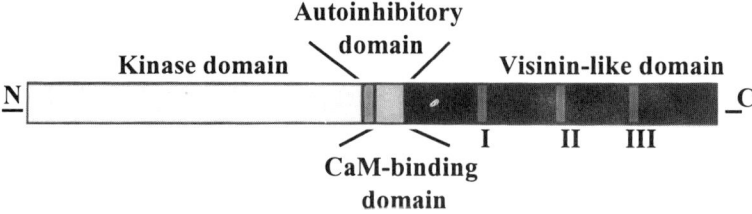

Figure 1. Schematic representation of CCaMK showing various structural features.

3. BIOCHEMICAL CHARACTERISTICS OF CCAMK

CCaMK binds to CaM in a Ca^{2+}-dependent manner (kDa ~55 nM). The CaM-binding domain has been mapped to a stretch of 19 amino acids between the catalytic domain and the visinin-like domain (Takezawa *et al* 1996a). CCaMK phosphorylates various protein and peptide substrates in a Ca^{2+}/CaM-dependent manner and the autophosphorylation at the threonine residue is Ca^{2+}-dependent. Schematic diagram of wild-type and truncated mutants of CCaMK, which were used for ^{35}S-CaM binding assays are shown in Fig. 2A.

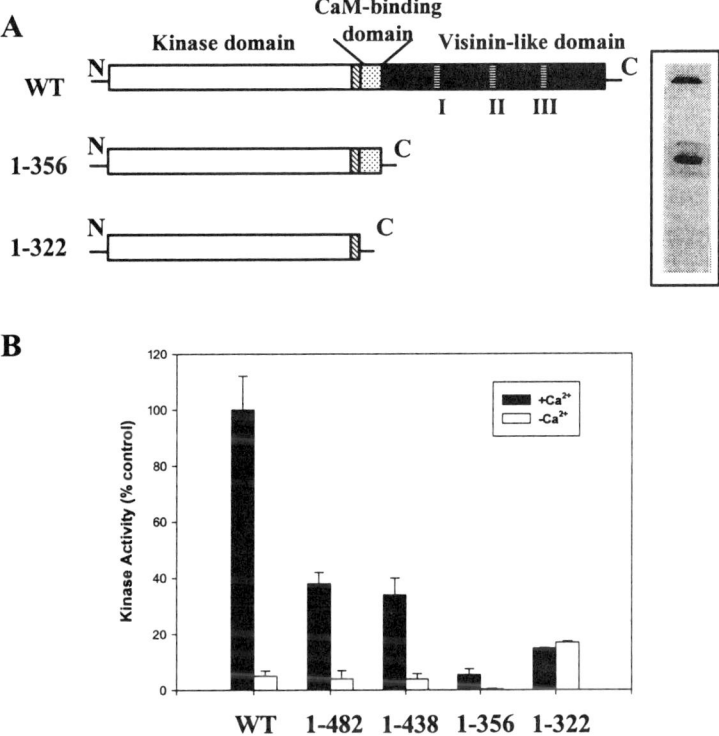

Figure 2. (A) Identification of CaM-binding site of CCaMK. Schematic diagram of wild-type and truncated mutants of CCaMK, which were used for ^{35}S-CaM binding assays are shown on the left. The mutants 1-356 and 1-322 represent CCaMK lacking the visinin-like domain and both visinin-like and CaM-binding domains. *E. coli*-expressed wild-type and mutant CCaMKs were electrophoresed on SDS-polyacrylamide gel and transferred onto nitrocellulose filter. The excised bands containing the expressed proteins were subjected to ^{35}S-CaM binding assay. The autoradiogram is shown on the right of each diagram (boxed area). The radioactivity (cpm) of bound ^{35}S-CaM was 11,600 for wild-type, 12,500 for the mutant 1-356, and 99 for the mutant 1-322, respectively (from *J. Biol. Chem.*, **271**: 8126-8132, 1996). (B) Ca^{2+}/CaM-dependent and independent activity of the wildtype and mutants of lily CCaMK.

The mutants 1-482, 1-438, 1-356, and 1-322 lack EF hand III, both EF hands II and III, all three EF hands; and all three EF-hands and the CaM-binding domain, respectively. The kinase activity of CCaMK and its mutants were assayed in the presence of 0.5 mM $CaCl_2$ plus 1 μM CaM (solid bars), or 2.5 mM EGTA (open bars). The mean values and standard deviations were calculated from three independent experiments (from *J. Biochem.* **121**: 984-990, 1997).

In order to study the role of the neural visinin-like domain in the regulation of CCaMK activity, various deletion mutants were created. Mutants lacking one, two, or all three Ca^{2+} binding EF hands within the visinin-like domain were created. The results show that all of the mutants, except the mutant 1-322, were Ca^{2+}/CaM-dependent for their kinase activities (Fig. 2B). Disrupting portions of the visinin-like domain by deleting EF hands III, II, and I gradually reduced the total activity of CCaMK, indicating that the Ca^{2+}-binding EF hands are crucial for the full Ca^{2+}/CaM-dependent activity of CCaMK (Ramachandiran *et al* 1997). Ca^{2+}-dependent mobility shift of wildtype and site-directed mutants revealed that Ca^{2+} induces conformational changes in the visinin-like domain (Takezawa *et al* 1996a). These results confirm the existence of a Ca^{2+}/CaM-dependent protein kinase in plants that is modulated by both Ca^{2+} and Ca^{2+}/CaM.

4. CCaMK-INTERACTING PROTEIN: EF-1α

Identification of the substrate and interacting proteins of CCaMK is pivotal to understanding its role in Ca^{2+}/CaM-mediated signal transduction. The yeast two-hybrid system was used to isolate genes encoding CCaMK substrates/interacting proteins (Wang and Poovaiah 1999). One of the cDNA clones obtained from the screening (*LlEF-1α1*) has high similarity with the eukaryotic elongation factor-1α (EF-1α). CCaMK phosphorylated the C-terminal region of LlEF-1α1 in a Ca^{2+}/CaM-dependent manner. In the presence of EGTA or Ca^{2+}, CCaMK phosphorylated the LlEF-1α1 only at the basal level. The phosphorylation level was stimulated up to around 25-fold by adding Ca^{2+} and CaM. *In vitro* binding assays revealed that CCaMK binds to LlEF-1α1 in a Ca^{2+}-independent manner. Dissociation of CCaMK from EF-1α by Ca^{2+} and subsequent phosphorylation by CCaMK in a Ca^{2+}/CaM-dependent manner suggest that these interactions may play a role in regulating the biological functions of EF-1α.

The phosphorylation site (Thr-257) for CCaMK in LlEF-1α1 has been identified using site-directed mutagenesis. When Thr-257 in the C-terminal region was mutated into Ala, it was no longer phosphorylated by CCaMK (Fig. 3B). The phosphorylation site of CCaMK is consistent with the consensus phosphorylation site sequence (R-X-X-S/T) of the mammalian CaMK II. Interestingly Thr-257 is located in the putative tRNA-binding region of LlEF-1α1. To investigate whether LlEF-1α1 can also be phosphorylated by Ca^{2+}-dependent but CaM-independent protein kinases (CDPK), one of CDPK isoforms (CRPK2) isolated from corn root tips was expressed and purified from *E. coli* (Takezawa *et al* 1996). Unlike CCaMK, CRPK2 phosphorylated multiple sites of LlEF-1α1 in a Ca^{2+}-dependent, but CaM-independent manner, and its phosphorylation sites are different from that for CCaMK (Fig. 3C). This suggests that these two kinases have some structural similarities, yet they differ in their regulation of kinase activity. The phosphorylation of EF-1α by these two kinases may have different functional significance.

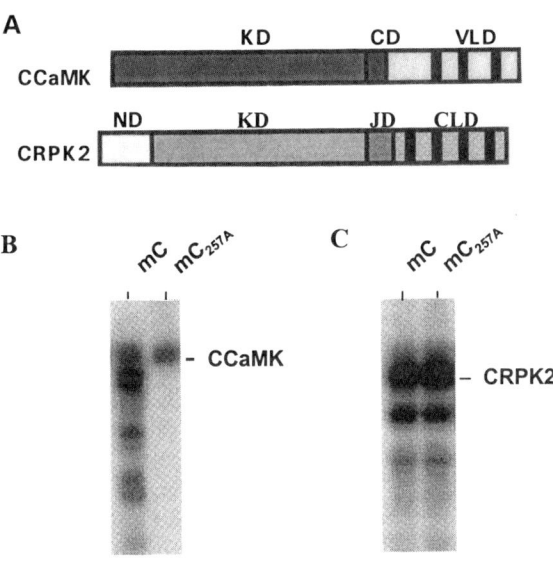

Figure 3. Phosphorylation of the C-terminal region (mC) of LlEF-1α1 (amino acids 204-447) and its site-directed mutant by CCaMK and CRPK2. (A) Schematic diagram compares the functional domain organization of CCaMK and CRPK2. Domains of CCaMK: kinase domain (KDA); CaM-binding domain (CD); and neural visinin-like domain (VLD). The three EF-hand Ca^{2+}-binding sites in VLD are marked. Domains of CRPK2: N-terminal domain (ND); kinase domain (KDA); junction domain (JD); and CaM-like domain (CLD). The four EF-hand Ca^{2+}-binding sites in CLD are marked. (B) Ca^{2+}/CaM-dependent phosphorylation of the mC and the mC_{257A} by CCaMK. The mC_{257A} is the mutant of mC where T_{257} has been mutated into A_{257}. (C) Ca^{2+}-dependent phosphorylation of the mC and the mC_{257A} by CRPK2 (from *J. Biol. Chem.* **274**: 12001-12008 1999).

5. CROSS-TALK BETWEEN Ca^{2+}/CaM-MEDIATED SIGNALING AND AUXIN-MEDIATED SIGNALING

Corn SAUR homolog, designated as *ZmSAUR1*, was cloned using a CaM-binding screening approach. The cDNA codes for a polypeptide of 147 amino acids with the calculated molecular mass and the isoelectric point 16.6 kDa and 7.22, respectively. The deduced amino acid sequence of *ZmSAUR1* is aligned in Fig. 4 with those of soybean *SAUR* 10A5, 15 A (McClure et al 1989), a mung bean *SAUR ARG7* (Yamamoto et al 1992), and *Arabidopsis SAUR-AC1* (Gil et al 1994). The size of ZmSAUR1 is larger than other SAURs. However, searching the *Arabidopsis* genomic sequences, a ZmSAUR1 homolog, with even larger molecular weight, named as SAUR-A2 was found (Fig. 4). The difference lies in the N-terminal 54 amino acids and about 30 amino acids in the C-terminus of ZmSAUR1, where soybean and other plant SAURs have less similarity. In contrast, the sequences are highly similar within the central portion (from 55-117 in ZmSAUR1) in all SAURs. Between these residues, ZmSAUR1 is 70.6% similar (58.8% identical) to the soybean 10A5 and 72.5% similar (54.9% identical) to *Arabidopsis* SAUR-AC1. Thus it seems likely that the central conserved portion of these proteins is most important for their function.

To further study the properties of ZmSAUR1, ZmSAUR1 protein was expressed in *E. coli*, using the pET-14b expression vector. The recombinant protein was present mainly in the soluble fraction. The following two experiments proved that CaM binds to the ZmSAUR1 protein. First, the 18.8 kDa fusion protein [16.6 kDa ZmSAUR1 plus 2.2 kDa N-terminal $(His)_6$ tag] was purified by CaM affinity chromatography to near homogeneity as judged by SDS-PAGE. Second, ^{35}S-labeled CaM binds to ZmSAUR1 protein only in the presence of Ca^{2+}. After adding 2 mM EGTA, no CaM-banding was observed, suggesting that CaM-binding to ZmSAUR1 is Ca^{2+}-dependent. The proteins from *E. coli* transformed with the pET-14b vector did not show any CaM-binding.

To map the CaM-binding region of ZmSAUR1, two mutants were prepared (Fig. 5). The mutant mΔC lacks the C-terminal 81 amino acids residues which includes the conserved central portion, while the mutant mΔN lacks the N-terminal 66 residues. The wild type ZmSAUR1 and the two mutants were expressed in *E. coli* and purified as described (Yang and Poovaiah 2000). These proteins were used for ^{35}S-CaM-binding assays in the presence and absence of Ca^{2+}. The binding of CaM to wild type and mutant mΔC were similar, whereas, CaM did not bind to the mutant mΔN (Fig. 5), indicating that a CaM-binding region is restricted to the 66 amino acids of the N-terminal, where SAURs showed the least similarity. CaM binding to wild type and mΔC of ZmSAUR1 was prevented by the addition of 2 mM

EGTA (data not shown), indicating an absolute requirement of Ca^{2+} for CaM-binding.

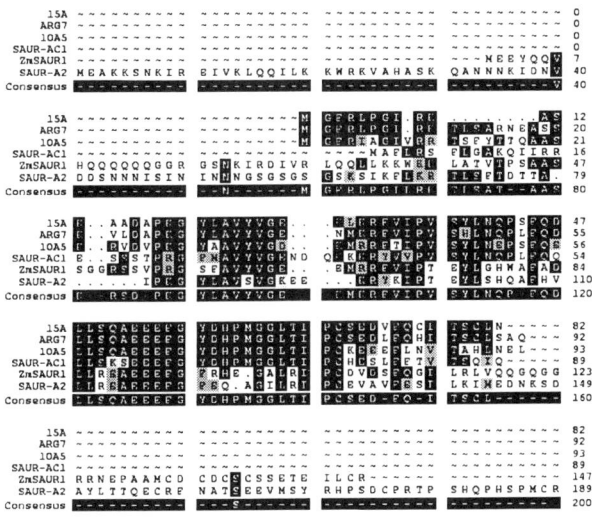

Figure 4. Comparison of the deduced amino acid sequence of ZmSAUR1 to soybean, *Arabidopsis* and mung bean SAURs. The accession numbers of soybean SAUR 10A5, 15A, *Arabidopsis* SAUR-AC1, SAUR-A2 and mung bean SAUR ARG7 are P33079, P33081, S70188, AL021633 and D14414, respectively.

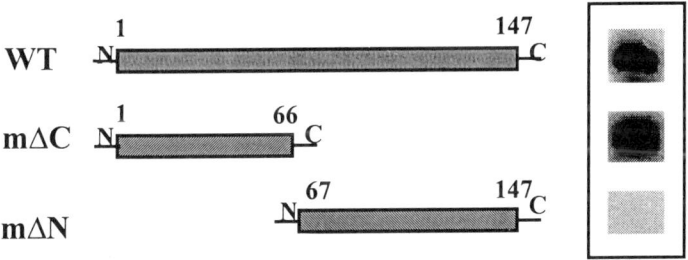

Figure 5. CaM binds to the N-terminal portion of ZmSAUR1. Wild type (WT) and two deletion mutants of ZmSAUR1 were used for ^{35}S-CaM-binding assays. Mutant mΔC and mΔN represent ZmSAUR1 lacking the C-terminal 81 amino acids and N-terminal 66 amino acids. *E coli*-expressed wild type and mutants were electrophoresed on SDS-PAGE and transferred onto PVDF membrane. The excised bands containing the expressed proteins were subjected to ^{35}S-CaM binding assay. The autoradiogram is shown on the right of each diagram (*boxed area*).

CaM is a protein capable of recognizing the basic amphiphilic α-helical domain of the target proteins. Helical wheel projection of the peptide sequences predicted that the CaM-binding region was further restricted to amino acids 20-45 of ZmSAUR1. The amino acid residues 32-45 formed a typical basic amphiphilic α-helix, similar to CCaMK (Takezawa et al 1996a). A peptide with 26 residues corresponding to the amino acids 20-45 was incubated with bovine CaM, and complex formation was assessed by nondenaturing PAGE in the presence or absence of Ca^{2+}. The results showed that the peptide is capable of forming a stable complex with CaM in the presence of Ca^{2+} but not in the absence of Ca^{2+}.

Based on the primary structure of SAURs shown in Fig. 4, it seems that ZmSAUR1 is very divergent isoform with a longer N-terminal domain. However, using helical wheel projection method, we found that the N-terminal portions in all the SAURs listed in Fig. 4 as well as other SAURs in data base can form a basic amphiphilic α-helix, suggesting that SAURs in general are CaM-binding proteins. CaM-binding affinity of ZmSAUR1 was studied using different concentrations of ^{35}S-labeled CaM. Binding of labelled CaM to ZmSAUR1 saturated at concentrations above 100 nM, indicating the presence of a saturable high affinity binding site in ZmSAUR1. From the Scatchard plot analysis of the saturation curve, the dissociation constant (KDa) of CaM for ZmSAUR1 was estimated to be about 15 nM. The binding of CaM to ZmSAUR1 was completely blocked in the presence of 2 mM EGTA. The Scatchard analysis also indicated that ZmSAUR1 has a single CaM-binding site.

Corn coleoptile segments floated in 10 μM α-naphthaleneacetic acid (NAA) solution showed significant increase in length from that of control without NAA application. After about 2 hours incubation, the elongation of coleoptiles was detectable. After 16 hours, the coleoptile elongation increased by more than 50%. However, the coleoptiles in the medium without NAA application elongated only about 5%. To study the auxin induction kinetics of *ZmSAUR1* expression, the corn coleoptile segments were collected at different times after incubation in the medium with 10 μM NAA for RNA preparation. Northern analyses indicated the level of *ZmSAUR1* is undetectable if NAA was not applied (Fig. 6). Treatment with NAA led to a significant induction of the *ZmSAUR1* with a size of ~ 0.8 kb, which coincides with the cDNA size of *ZmSAUR1*. The induction began within 10 min, a sharp increase occurred between 20-60 min with half maximal after 30 min, and saturation in 60 min. This demonstrates that *ZmSAUR1* is indeed an early auxin responsive gene.

6. CONCLUSION

Ca^{2+} has unique regulatory roles in plant growth and development. CaM, a primary intracellular Ca^{2+} receptor, plays a crucial role in transducing the Ca^{2+} signal. During the past decade, a number of Ca^{2+}/CaM-modulated proteins have been identified. These Ca^{2+}/CaM -modulated proteins play an important role in Ca^{2+} signal amplification and diversification. Functional analysis and continued identification of CaM target proteins will further enhance our understanding of the molecular mechanism of Ca^{2+}/CaM-mediated signal transduction.

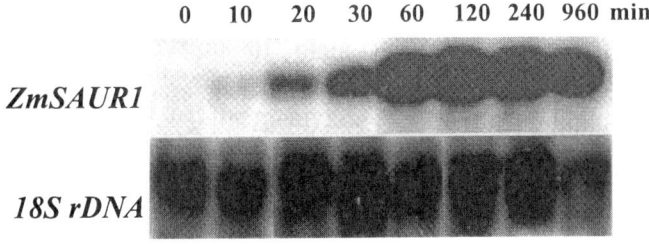

Figure 6. Effect of auxin (10 µM NAA) on the induction of ZmSAUR1 in corn coleoptiles. Three sets of 10 coleoptile segments were floated in test solution in the dark for time periods indicated on the top (min). Autoradiograms of Hybond N$^+$ filter hybridized successively with ^{32}P-labeled *ZmSAUR1* cDNA fragment 37-480 and *18S rDNA* fragment 158-1669 (accession number X16077).

ACKNOWLEDGEMENT

Part of this work was supported by the National Science Foundation Grant MCB 96-30337, and the National Aeronautics and Space Administration Grant NAG-10-0061 to BWP.

REFERENCES

Abel, S,. and Theologis, A., 1996, Early genes and auxin action. *Plant Physiol.* **111**: 9-17.
Gil, P., Liu, Y., Orbovic, V., Verkamp, E., Poff, K.L., and Green, P.J., 1994, Characterization of the auxin inducible SAUR-AC1 gene for use as a molecular genetic tool in *Arabidopsis. Plant Physiol.* **104**: 777-784.
McClure, B.A., and Guilfoyle, T.J., 1987, Characterization of a class of small auxin-inducible soybean polyadenylated RNAs. *Plant Mol. Biol.* **9**: 611-623.

McClure, B.A., Hagen, G., Brown, C.S., Gee, M.A., and Guilfoyle, T.J., 1989, Transcription, organization and sequence of an auxin-regulated gene cluster in soybean. *Plant Cell* **1**: 229-239.

Patil, S., Takezawa, D., and Poovaiah, B.W., 1995, Chimeric plant calcium/calmodulin-dependent protein kinase gene with a neural visinin-like calcium-binding domain. *Proc. Natl. Acad. Sci. USA* **92**: 4897-4901.

Poovaiah, B.W., and Reddy, A.S.N., 1993, Calcium and signal transduction in plants. *CRC Crit. Rev. Plant Sci.* **12**: 185-211.

Poovaiah, B.W., Wang, W., Takezawa, D., Sathyanarayanan, P.V., and An, G., 1997, Calcium-mediated signaling in plants: calmodulin and Ca^{2+}/calmodulin-dependent protein kinase. *J. Plant Biol.* **40**: 190-197.

Poovaiah, B.W., Xia, M., Liu, Z., Wang, W., Yang, T., and Sathyanarayanan, P.V., 1999, Developmental regulation of the gene for chimeric calcium/calmodulin-dependent protein kinase in anthers. *Planta* **209**: 161-171.

Ramachandiran, S., Takezawa, D., Wang, W,. and Poovaiah, B.W., 1997, Functional domains of plant chimeric calcium/calmodulin-dependent protein kinase: regulation by autoinhibitory and visinin-like domains. *J. Biochem.* **121**: 984-990.

Reddy, A. S. N., Takezawa, D., Fromm, H., and Poovaiah, B.W., 1993, Isolation and characterization of two cDNAs that encode for calmodulin-binding proteins from corn root tips. *Plant Sci.* **94**: 109-117.

Takezawa, D., Ramachandiran, S., Paranjape, V., and Poovaiah, B.W., 1996a, Dual regulation of a chimeric plant serine/threonine kinase by calcium and calcium/calmodulin. *J. Biol. Chem.* **271**: 8126-8132.

Takezawa, D., Patil, S., Bhatia, A., and Poovaiah, B.W., 1996b, Two Ca^{2+}-dependent protein kinase genes from corn roots. *J. Plant Physiol.* **149**: 329-335.

Wang, W., and Poovaiah, B.W., 1999, Interaction of plant chimeric calcium/calmodulin-dependent protein kinase with a homolog of eukaryotic elongation factor-1. *J. Biol. Chem.* **274**: 12001-12008.

Wang, W., Takezawa, D., Narasimhulu, S.B., Reddy, A.S.N., and Poovaiah, B.W., 1996a, A novel kinesin-like gene with a calmodulin-binding region within the motor domain. *Plant Mol. Biol.* **31**: 87-100.

Wang, W., Takezawa, D., and Poovaiah, B.W., 1996b, A potato cDNA encoding a homolog of mammalian multidrug resistant P-glycoprotein. *Plant Mol. Biol.* **31**: 683-687.

Watillon, B., Kettmann, R., Arredouani, A., Hecquet, J.F., Boxus, P., and Burny, A., 1998, Apple messenger RNAs related to bacterial lignostilbene dioxygenase and plant SAUR genes are preferentially expressed in flowers. *Plant Mol. Biol.* **36**: 909-915.

Yamamoto, K.T., Mori, H., and Imaseki, H., 1992, cDNA cloning of indole-3-acetic acid-regulated genes: Aux22 and SAUR from mung bean (*Vigna radiata*) hypocotyl tissue. *Plant Cell Physiol.* **33**: 93-97.

Yang, T., and Poovaiah, B.W., 2000, Molecular and biochemical evidence for the involvement of calcium/calmodulin in auxin action. *J. Biol. Chem.* **275**: 3137-3143.

A Novel Ca^{2+}/CaM-regulated Microtubule Motor Protein from Plants: Role in Trichome Morphogenesis and Cell Division

A.S.N. REDDY
Department of Biology and Program in Cell and Molecular Biology, Colorado State University, Fort Collins, Colorado 80523, USA

1. INTRODUCTION

In recent years Ca^{2+} has emerged as a key messenger in transducing many hormonal and environmental signals in plants (Hepler and Wayne 1985, Poovaiah and Reddy 1987, Poovaiah and Reddy 1993, Bowler and Chua 1994, Trewavas and Malho 1997, Sander *et al* 1999). However, the biochemical and molecular events involved in the mode of Ca^{2+} action are poorly understood. Ca^{2+} is believed to control many biochemical and molecular processes by interacting with several proteins either directly or through calmodulin (CaM), a highly conserved, multifunctional regulatory protein in eukaryotes (Roberts *et al* 1986, Snedden and Fromm 1998). Calmodulin action in regulating biochemical and molecular events and ultimately physiological processes involves its interaction with other proteins called CaM-binding proteins. The effect of this interaction usually results in regulation of enzymatic activity/function of the binding protein. In animal systems a large number of CaM-binding proteins have been characterized (Rhoads and Friedberg 1997). In plants, little is known about the number, localization, identity, function and regulation of CaM-binding proteins (Snedden and Fromm 1998, Zielinski 1998). In our attempts to identify interacting protein partners of CaM, we screened expression libraries of *Arabidopsis* and other plants with labeled CaM. This screening has resulted in isolation of several CaM-binding proteins. Among these we identified a novel kinesin-like calmodulin-binding protein (KCBP) that is regulated by Ca^{2+}/CaM.

Kinesin, a microtubule (MT) motor protein, was first identified and biochemically characterized from squid giant axons in 1985 as an ATPase involved in transport of vesicles and the gene encoding the conventional kinesin was isolated in 1988 (Brady 1985, Vale *et al* 1985, Yang *et al* 1988). This kinesin (also referred as conventional kinesin) is a tetramer consisting of two light chains and two heavy chains. The heavy chain has three structural domains: a motor domain that is located in the amino-terminal region, and contains conserved ATP-and MT-binding sites, a central stalk region that forms an alpha-helical coiled-coil region involved in dimerization, and a globular tail that binds to two light chains. The motor activity is associated with the motor domain of the heavy chain (Hirokawa 1998). During the last ten years there has been an explosion in the identification of kinesin-like proteins (KLPs) in eukaryotes. A large number (over 140 full-length sequences) of KLPs have been identified in plants and animals (http://www.blocks.fhcrc.org/~kinesin). The members of the kinesin superfamily, which consists of conventional kinesin and KLPs, perform numerous force-generating tasks such as transport of vesicles and organelles, mitotic and meiotic spindle formation and elongation, chromosome segregation, MT dynamics, germplasm aggregation, and intraflagellar transport (Barton and Goldstein 1996, Endow 1999). The common feature among the members of the kinesin superfamily is a highly conserved motor domain of about 340 amino acid residues. Outside the motor domain there is little or no sequence similarity between the members of different subfamilies (Moore and Endow 1996). Phylogenetic analysis of kinesins and KLPs using the conserved motor domain, groups the majority of known kinesins into one of ten subfamilies. However, some KLPs do not fall into these subfamilies and could represent prototypes of additional subfamilies (Reddy 2000). Members of one of the subfamilies (a C-terminal subfamily) possess a C-terminal motor domain with minus-end motility while the rest (N-terminal or internal motor domain kinesin subfamilies) are plus-end motors (Moore and Endow 1996). Six subfamilies have been shown to be involved in some aspects of cell division and three subfamilies are involved in organelle/vesicle transport. The presence of a superfamily of kinesins suggests that the motor domain performs many diverse MT-based transport functions by being fused to unique domains that are specific to the cargo that they transport. The nonconserved globular tail domains confer diversity to interact with specific cargoes and may function in different cellular processes.

2. KCBP IS A NOVEL MEMBER OF KINESIN FAMILY

The cDNA encoding a kinesin-like protein that we isolated in a protein-protein interaction-based screening with labeled CaM showed binding to ^{35}S-labeled or biotinylated CaM only in the presence of Ca^{2+} but not in the presence of EGTA, a Ca^{2+} chelator (Fordham-Skelton *et al* 1994, Reddy *et al* 1996b). The predicted protein has an estimated molecular mass of 143 kDa. An approximately 340 amino acid region in the C-terminus of this protein showed significant sequence similarity with the motor domain of kinesins and KLPs (Reddy *et al* 1996b). Because of the similarity of this CaM-binding protein with kinesins, this protein was designated as kinesin-like calmodulin-binding protein (KCBP). The region amino terminal to the motor domain is composed of two domains. First, amino acid residues from 610 to 890 form a coiled-coil structure, suggesting that the native KCBP may form a dimer. Bacterially expressed motor domain with part of the coiled-coil region forms a dimer, confirming that this region is involved in dimerization (Deavours *et al* 1998). Second, the N-terminal tail region that extends from the beginning of the protein to the coiled-coil region has no sequence similarity with known kinesin heavy chains or KLPs. However, the tail region of KCBP shows significant similarities to two domains that are present in some animal myosins (see below).

To date, five KLPs (KatA, KatB, KatC, KatD and KCBP) have been characterized at the molecular level from *Arabidopsis* (Reddy 2000). In addition, as of September 1999, I have identified seventeen other KLPs in a search of the *Arabidopsis* genome database suggesting that there are at least twenty-two KLPs in *Arabidopsis* (Reddy 2000). This number is likely to increase as the genome sequence is completed. Based on the number of KLPs in *Arabidopsis*, the number of cellular processes that are regulated by kinesins in plants is likely to match or exceed the number of processes controlled by these proteins in animals. Of the total 22 KLPs, a majority (thirteen out of twenty-two) of them appear to have their motor located at the N-terminus whereas nine KLPs have C-terminal motors. It is interesting to find that in *Arabidopsis* members of only two subfamilies (C-terminal and BimC) are represented. Three *Arabidopsis* KLPs belong to the BimC and nine belong to the C-terminal subfamily. A majority of the rest (ten KLPs) falls into a new subfamily. None of the *Arabidopsis* kinesins fall into a conventional kinesin subfamily.

4. CALMODULIN-BINDING DOMAIN IN KCBP IS LOCATED IN A STRETCH OF 23 AMINO ACIDS ADJACENT TO THE MOTOR DOMAIN

KCBP is unique among kinesins and KLPs in having a CaM-binding domain (CBD). To map the CBD we expressed different regions of the protein and tested for their ability to bind CaM. These studies have shown that the CBD is located in a 52 amino acid stretch at the C-terminus (Reddy *et al* 1996b). To further narrow down the region involved in CaM-binding, a 23 amino acid long peptide was synthesized and used in binding studies. The synthetic peptide bound to CaM indicating that this region is involved in CaM-binding. To test if the binding of the peptide to CaM is Ca^{2+}-dependent, we performed binding studies in solution using the synthetic peptide and CaM in the presence of Ca^{2+} or EGTA. The binding of synthetic peptide to CaM was judged by gel mobility shift assay in polyacrylamide gels containing 4M urea. These studies have shown that the KCBP peptide binds to CaM in a Ca^{2+}-dependent manner at concentrations as low as 1 µM (Reddy *et al* 1996b). In *Arabidopsis*, there are six *Cam* genes (*AtCam* 1-6) and four *Cam*-related genes which encode four CaM isoforms and four CaM-like proteins, respectively (Zielinski 1998). KCBP binds to bovine CaM and three isoforms of *Arabidopsis* CaM (CaM-2, -4 and -6) that have been tested (Reddy *et al* 1999). CaM-2 isoform showed two-fold higher affinity for KCBP than CaM-4 and -6. It is not known if any of the CaM-related proteins in *Arabidopsis* interact with KCBP.

4. KCBP HAS MYOSIN- AND KINESIN-LIKE FEATURES

Interestingly, KCBP has some domains that are present in MT- and actin-based motors, suggesting that this class of motors is unique in having kinesin and myosin features. The N-terminal tail of KCBP shows significant similarities to the MyTH4 and talin-like regions present in myosins VIIa and X (Reddy *et al* 1996b, Reddy and Reddy 1999). The MyTH4 and the talin-like domains of KCBP showed the highest similarity with the myosin VIIa tail (29% identity and 45% similarity in the MyTH4 domain and 23% identity and 37% similarity in the talin-like region) (Reddy and Reddy 1999). The MyTH4 domain and talin-like region have not been found in any other known members of the kinesin superfamily, suggesting that KCBP is a molecular hybrid consisting of a motor domain from MT-based motors and a tail region of actin-based motors. The existence of a talin-like region together with a MyTH4 domain in KCBP, myosin VIIa and X is interesting and likely to have some functional significance. It is likely that KCBP and myosins could interact with some unknown common protein(s) through their tail homology regions to either cross-bridge MTs and actin filaments or facilitate cargo exchange between these two types of molecular motors. Some recent reports have provided evidence for the direct interaction of actin- and MT-based motors (Brown 1999). On the other hand, the talin-like region in these motors may be involved in anchoring the MTs and actin filaments to plasma membrane since talin has been shown to bind membranes (Reddy 2000).

5. KCBP IS A MINUS-END DIRECTED MICROTUBULE MOTOR

We have tested KCBP motor for its ability to bind MTs to a glass surface and induce gliding of MTs. The truncated protein used in motility assays contained about 70 residues of the central α-helical coiled-coil domain and the entire motor domain including the CaM-binding residues that lie C-terminal to the conserved motor domain. The KCBP fusion protein bound MTs to a coverslip surface and induced gliding of MTs across the glass surface with a velocity of 8-10 µm/min (Song *et al* 1997). The polarity of the KCBP motor was determined by assaying gliding of axoneme-MT complexes on KCBP bound to coverslips. These assays have shown that KCBP is a minus end-directed motor. The velocity of KCBP movement on MTs is similar to that of Ncd. In the presence of Ca^{2+} and CaM, MTs did not bind to KCBP, suggesting that CaM regulates KCBP binding to MTs in the cell by causing the motor to release from MTs.

6. KCBP CONTAINS TWO MICROTUBULE BINDING SITES

Analysis of the interaction of various truncated versions of KCBP with MTs indicates that there are two MT binding domains, one located at the C terminus and the second one located at the N terminus (Narasimhulu and Reddy 1998). Unlike the MT binding domain in the C terminus, the N-terminal region of KCBP bound MTs both in the presence and absence of ATP, indicating that the MT binding domain in the N terminus is insensitive to ATP. The presence of ATP-independent MT binding regions in the non-motor domains of KAR3, Ncd, and the kinesin heavy chain have been reported (Reddy 2000). However, there is no sequence similarity between the N-terminal tail region of KCBP and known KLPs. The motor domain of KCBP binds to both alpha- and beta-tubulin subunits also and this binding is regulated by Ca^{2+}/CaM (Narasimhulu *et al* 1997).

7. KCBP IS UBIQUITOUS IN PLANTS

A homologue of KCBP has been isolated from potato, tobacco and maize (Abdel-Ghany and Reddy 2000, Reddy et al 1996a, Wang et al 1996). Furthermore, antibodies specific to KCBP detected an expected size protein in *Haemanthus*, *Tradescantia* and *Lilium* (Smirnova et al 1998, Voss et al 1999), suggesting ubiquitous presence of KCBP in phylogenetically divergent plant species of dicots and monocots. However, a homologue of KCBP has not been found in the completely sequenced genomes of *S. cerevisiae* and *C. elegans* .It is of interest to determine if KCBP is present in non-flowering plants and the origin of KCBP. A CaM-binding C-terminal kinesin (kinesin C) was cloned recently from sea urchin (Rogers et al 1999). The CaM-binding domain of kinase C shared 35% sequence identity with the CaM-binding domain in KCBP. The existence of CaM-binding KLPs in both plants and sea urchin suggests that the origin of this group of KLPs predates the divergence of plants and animals from a common ancestor about 1.5 billion years ago (Wang et al 1999). Alternatively, CaM-binding KLPs may have evolved independently in plant and animals after they diverged from a common ancestor. Kinesin C differs considerably from KCBP and does not show any sequence similarity in the stalk and tail regions (Reddy 2000). The amino-terminal tail and stalk regions of KCBPs from different plant systems are highly conserved and contain myosin tail homology (MyTH4) and talin-like regions that are not present in kinesin C (Reddy 2000). These data suggest that CaM-binding KLPs may constitute a distinct group within the C-terminal subfamily and KCBP is a prototype for that group. Phylogenetic analysis of kinesins using the motor domain sequence grouped KCBPs and kinesin C with other known carboxy-terminal MT motor proteins. However, *Arabidopsis* KCBP together with its homologs from potato, tobacco and maize constitute a distinct group within the carboxy-terminal subfamily of motors (Reddy 2000) which is consistent with structural and functional features of KCBP. The presence and conservation of the N- and C-terminal regions of KCBPs in monocots and dicots indicates that KCBP originated prior to the divergence of monocots and dicots from a common ancestor which is believed to have occurred about 130 to 200 million years ago (Wolfe et al 1989, Crane et al 1995).

8. KCBP IS CODED BY A SINGLE GENE AND IS EXPRESSED HIGHLY IN TISSUES CONTAINING DIVIDING CELLS

A single transcript of about 4 kb was found to hybridize with the cDNA (Reddy et al 1996b). Flowers, shoot tips and suspension culture had the highest level of expression. *Arabidopsis* and other plants from which KCBP is cloned appear to contain a single gene encoding KCBP. *Arabidopsis* KCBP gene contains twenty-one exons and twenty introns (Oppenheimer et *al* 1997, Reddy *et al* 1998). The first nine exons contain the coding region for the tail region. The coiled-coil region is coded by exons ten to fifteen and the conserved motor domain is coded by exons sixteen to twenty. The CaM binding domain unique to KCBP is coded by the last exon, suggesting that KCBP may have evolved by fusion of an exon coding for a CaM binding domain with a kinesin-like protein gene (Reddy *et al* 1998). A genomic clone encoding KCBP was recently isolated from a monocot (Abdel-Ghany and Reddy 2000). The number and location of introns is highly conserved between dicot and monocot KCBPs. However, comparison of KCBP gene structure with *Drosophila* Ncd, a kinesin-like protein with a C-terminal minus-end motor, and *Chlamydomonas* KHP1, a KLP with an amino-terminal motor, did not reveal conservation of intron number or location.

9. KCBP IS INVOLVED IN CELL DIVISON

In plants, the transition from interphase to mitosis and progression through cell division involves several noteworthy differences from that of animal cells. In late G_2, just prior to prophase, plant cortical MTs rearrange to form a band of MTs called the preprophase band (PPB), that encircles the cell just below the cell membrane (Staehelin and Hepler 1996). The site of the PPB accurately predicts the future location of the cell plate but the exact mechanism whereby this occurs is not known. Another distinctive feature of plant mitosis is the formation of a bipolar spindle in the absence of centrosomes (Franklin and Cande 1999). Furthermore, cytokinesis, the process that produces two daughter cells following the completion of nuclear division, occurs in plant and animal cells by two distinct mechanisms. Cytokinesis in higher plant cells occurs *via* the formation of a polysaccharide cell plate, which expands centrifugally to join a predetermined zone of the plasma membrane of the dividing cell (Staehelin and Hepler 1996). Formation and growth of the cell plate between daughter nuclei is controlled by the phragmoplast, which is composed of two disks of parallel MTs with their plus ends toward the equatorial zone (Staehelin and Hepler 1996). The phragmoplast is implicated in transporting vesicles containing cell plate materials to the cell plate (Staehelin and Hepler 1996). The association of vesicles with phragmoplast MTs indicate that MT-based motors move the vesicles to the cell plate (Staehelin and Hepler 1996). Microinjection studies with fluorescently labeled tubulin indicate that the initial phragmoplast is formed from preexisting MTs and the later stages of centrifugal growth is derived by new MTs (Zhang *et al* 1993). These differences between plants and animals point to the possibility that plants may contain novel motor(s) to perform these function. As described below, KCBP appears to be involved in forming plant specific mitotic arrays and/or functions associated with these structure.

We localized KCBP in *Arabidopsis* and tobacco by indirect immunofluorescence microscopy using affinity-purified anti-KCBP antibody. The KCBP was localized to the preprophase band, the mitotic spindle and the phragmoplast (Bowser and Reddy 1997). Similar localization patterns were observed both in *Arabidopsis* and tobacco. Localization of KCBP in *Haemanthus* also showed localization to mitotic MT arrays with some differences. For example, in *Haemanthus* anti-KCBP staining is seen almost exclusively at the spindle poles in late anaphase. Additionally, anti-KCBP staining is observed in association with the cell plate in *Haemanthus* but is absent from the midzone region of the phragmoplast in BY-2 and *Arabidopsis* suspension cells (Smirnova *et al* 1998). The association of KCBP with MT arrays in dividing cells suggests that this minus-end-directed MT motor protein is likely to be involved in the formation of these MT arrays and/or functions associated with these structures. Furthermore, the formation of the spindle in the absence of centrosomes in plants may require novel motors such as KCBP. Using an antibody that activates the motor function of KCBP, it was shown that this minus-end kinesin is differentially active during the various phases of cell division in stamen hair cells of *Tradescantia* (Voss *et al* 1999). Injection of KCBP-antibody results in the premature breakdown of the nuclear envelope and early onset of pro-metaphase, subsequently the cells are arrested in late pro-metaphase or metaphase. Injection later during anaphase causes aberrant phragmoplast and cell plate formation and the completion of cytokinesis is delayed or inhibited. Anaphase and interphase are not affected by KCBP antibody injection. Although KCBP transcripts are detected in every tissue that is tested, high level KCBP expression is found in tissues with meristematic cells and in suspension cultures (Reddy *et al* 1996a, 1996b). High levels of KCBP mRNA in meristematic regions and suspension culture and localization of KCBP to the preprophase band, the mitotic apparatus and the phragmoplast suggest that KCBP may be abundant in M phase of the cell cycle. We tested whether or not this was the case using tobacco BY-2 cells, which are amenable to synchronization. These studies have shown that KCBP was abundant in M-phase cells and declined as the cells entered interphase (Bowser and Reddy 1997).

Two C-terminal motors (KCBP and Kat A) and one N-terminal motor (TKRP-125) from plants localize to mitotic MT arrays (PPB, spindle and phragmoplast), suggesting their involvement in cell division (Liu *et al* 1996, Asada *et al* 1997, Bowser and Reddy 1997, Smirnova *et al* 1998). The assembly of the acentriolar spindle in plants may involve convergence of MT minus-ends leading to the formation of spindle poles (Smirnova *et al* 1998, Franklin and Cande 1999). The fact that KCBP is a minus-end-directed motor (Song *et al* 1997) and localizes to the spindle suggests that it may be involved in acentriolar spindle formation in plants, perhaps in a role similar to the one played by Ncd in the formation of anastral meiotic spindles in *Drosophila* oocytes (Matthies *et al* 1996). Based on the presence of an ATP-insensitive MT binding domain in the amino-terminus and another MT binding site in the C-terminal motor domain, it is likely that KCBP cross-links MTs and its minus-end motor activity allows it to focus MTs. Immunolocalization studies suggest that between early and late anaphase, the localization of KCBP shifts toward the spindle pole, supporting our hypothesis for a role for KCBP in cell division and in the formation of a converging bipolar spindle (Smirnova *et al* 1998). Localization of minus-end (KatA and KCBP) motors and a plus-end (TKRP125) motor in the spindle suggests their involvement in counterbalancing forces generated by plus- and minus-end motors to stabilize the spindle. Minus-end motors such as KCBP could cross-link and slide anti-parallel MTs in the spindle midzone and provide a plateward force on the spindle poles whereas the plus-end motors such as TKRP could provide pole-separating forces that antagonize the plateward force. Such antagonizing forces are likely to be important in maintaining the spindle shape. In some cases, plus- and minus-end motors have been shown to function together in an antagonistic way (Reddy 2000).

Numerous forces are at play in the development, maintenance, and function of the phragmoplast. Golgi-derived vesicles are transported to the forming cell plate where the plus-ends of phragmoplast MTs interdigitate. It is likely that MT motors are involved in powering the movement of the vesicles along the MTs. The polarity of MTs and microfilaments in the phragmoplast (Staehelin and Hepler 1996) indicates that plus-end motors are likely to be involved in the transport of vesicles to the cell plate. Of the three KLPs (TKRP125, KCBP and Kat A) that localize to the phragmoplast, only TKRP125 is a plus-end motor, indicating its involvement in vesicle transport to the cell plate. It is unlikely that KCBP and KatA play a role in vesicle transport to the cell plate since these are known to be minus-end motors and the MTs are oriented with their plus ends facing the cell plate. One intriguing possible function of minus end motors could be recycling of Golgi vesicle membranes from the expanding cell plate, a process which would require a minus-end directed motor protein. Functional analysis of TKRP125 by Asada and coworkers indicates its role in phragmoplast MT organization (Asada *et al* 1997). Using antibody against a short peptide from the motor domain of TKRP125, GTP- and ATP-dependent translocation of phragmoplast MTs was inhibited in permeabilized tobacco BY2 cells (Asada *et al* 1997). However, immunostaining of the phragmoplast with the kinesin antibodies is not punctate raising the possibility that other kinesins may be involved in transport and the ones that have been studied are primarily involved in organization of the phragmoplast. Phragmoplast associated kinesins may also play a role in the MT reorganization associated with its growth. The finding that the amino-terminus of KCBP possesses a functional ATP-insensitive MT-binding site and the ability of KCBP to bundle MTs lend support to this idea (Narasimhulu and Reddy 1998, Kao et al 2000). The motor domain of KCBP with and without CaM-binding has been shown to bundle MTs (Kao et al 2000). However, MT bundles induced by motor domain with CaM-binding domain are dissociated in the presence of Ca^{2+}/CaM (Kao et al 2000). These results, together with the KCBP localization data, suggest the involvement of KCBP in establishing mitotic MT arrays during different stages of cell division and that Ca^{2+}/CaM regulates the formation of these MT arrays.

10. KCBP IS ESSENTIAL FOR NORMAL DEVELOPMENT OF TRICHOMES

In wild-type *Arabidopsis*, trichomes are single epidermal cells with a stalk and three or four branches (Heulskamp et al 1994, Oppenheimer 1998). In zwichel (*zwi*) mutants, the trichomes have a short stalk and only two branches. Recently, Oppenheimer et al 1997 have cloned the ZWICHEL gene which encodes a kinesin-like CaM-binding protein (KCBP), suggesting that this novel MT motor protein is required for elongation of the stalk and normal branching probably *via* reorganization of the cortical MT cytoskeleton. KCBP is expressed in other tissues besides developing leaves, and in cultured cells in a cell cycle regulated manner. Furthermore, as discussed above immunolocalization data and microinjection studies indicate a mitotic role also for KCBP. However, all of the *zwi* mutants appear to grow normally with only the defect in trichome morphogenesis (Oppenheimer et al 1997, Krishnakumar and Oppenheimer 1999). The lack of a phenotype in other tissues suggests that either KCBP is nonessential or another motor with overlapping functions may substitute for ZWI function in other tissues. In *Arabidopsis* there are many KLPs (more than 22) including several C-terminal motors. In trichomes the other kinesins with functional overlap may not be expressed or they may not substitute for ZWI function in trichomes. A number of reports indicate that the functions of many kinesins in non-plant systems are redundant (Goldstein and Philip 1999). Four alleles of *zwi* have been characterized. However, none of the *zwi* mutants are null mutants and the N-terminal region of the protein is not affected in any of the known mutants. Hence, *zwi* mutants have either truncated or low levels of KCBP protein. This raises the possibility that the N-terminal region of ZWI can perform some functions in other tissues and cells other than trichomes whereas trichome development requires the entire protein. KCBP, like some other motors, may have several functional domains (Porter et al 1993, 1995). Since the motor domain of KCBP is affected in all mutants, it is likely that motor domain of KCBP function is essential for trichome morphogenesis whereas the N-terminal region alone may still perform others functions in other parts of the plant. Although localization of KCBP to cortical MTs has not been reported, genetic studies with trichomes strongly indicate a role for KCBP in cell expansion.

How does KCBP influence trichome branching which requires localized cell wall expansion? In plant cells, cortical MT array are known to play a critical role in cell expansion by controlling the orientation of newly synthesized cellulose microfibrils (Lloyd 1994, Nicol and Hofte 1998). Hence the reorientation of MTs which could be accomplished by either depolymerization/polymerization of MTs or movement of existing MTs could play a role in trichome branching. Both these processes may involve motors and the fact the KCBP, a minus-end MT motor, is required for branching indicates that it may be involved in reorientation of MTs thereby directing the deposition of MTs. Two MT binding sites on KCBP and its ability to bundle MTs supports the idea that it participates in MT reorganization (Narasimhulu and Reddy 1998, Kao et al 2000). Since the motor domain of KCBP, which is regulated by Ca^{2+}/CaM, is essential for trichome branching it is likely that Ca^{2+} through CaM regulates trichome branching. It would be interesting to look at the organization of MTs and localization of KCBP in trichomes of wild type and *zwi* mutants during different stages of trichome development.

11. Ca^{2+}/CALMODULIN REGULATES THE INTERACTION OF KCBP WITH MICROTUBULES

Because KCBPs differ from all other kinesins and KLPs in containing a CaM binding domain at the C terminus, we investigated the role of this domain in regulating KCBP activity. To determine the effect of Ca^{2+}/CaM, we used cosedimentation assays to analyze the interaction of the 1.5 C (MD plus CBD) protein with MTs in the presence of Ca^{2+}, CaM, or Ca^{2+}/CaM (Narasimhulu and Reddy 1998). The results of these experiments indicate that the motor protein cosediments with MTs in the presence of Ca^{2+} alone or CaM alone. However, KCBP remained in the supernatant when MTs were supplemented with Ca^{2+} and CaM, suggesting that in the presence of Ca^{2+}/CaM the protein is released from MTs. This is consistent with the data on the effect of Ca^{2+}/CaM on MT gliding (Song et al 1997). Further experiments were conducted to determine if Ca^{2+}/CaM could dissociate pre-formed AMP-PNP-induced KCBP/MT complexes (Deavours et al 1998). Resuspension of pelleted rigor complexes with ATP caused release of KCBP from MTs while resuspension of rigor complexes with AMP-PNP maintained the rigor complex. Resuspension with AMP-PNP and Ca^{2+}, or

AMP-PNP and CaM, also maintained the rigor complex. However, addition of both Ca^{2+} and CaM caused KCBP to dissociate from MTs even in the presence of AMP-PNP (Deavours et al 1998). The extent of KCBP dissociation was dependent on the concentration of CaM with greater dissociation associated with a higher concentration of CaM. The MT-dependent ATPase activity is also inhibited by Ca^{2+}/CaM (Deavours et al 1998). To determine whether the CaM binding domain is necessary for Ca^{2+}/CaM regulation, MT pelleting assays were carried out with the 1.0 C protein, which lacks the CaM-binding domain, in the presence of Ca^{2+}, CaM and Ca^{2+}/CaM. A predominant portion of the 1.0 C protein remained in the pellet in all three cases, suggesting that the Ca^{2+}/CaM effect observed with the 1.5 C protein is no longer evident with the motor domain protein lacking the CaM-binding domain. These studies have shown that binding of activated CaM to KCBP inhibits its interaction with MTs via the CaM-binding domain. It appears that the binding of CaM to KCBP affects MT binding regions resulting in inhibition of KCBP binding to MTs and dissociation of KCBP/MT complex. Structural studies with KCBP in free and CaM bound forms are needed to verify these speculations. Based on *in vitro* studies with KCBP it is reasonable to speculate that spatial and temporal changes in free cytosolic Ca^{2+} levels in response to signals are likely to regulate KCBP activity in the cell.

Among motor proteins, heavy chains of unconventional myosins also bind CaM. The CaM binding sites in the myosins consist of highly basic repeats with a core consensus sequence of IQXXXRGXXXR, known as the IQ motif (Wolenski 1995), present near the motor-stalk or motor-tail junction. The effect of Ca^{2+} and CaM on myosin function has been best described for brush border myosin I. Brush border myosin has three "IQ" motifs and binds to three molecules of CaM (Wolenski 1995). Unlike most CaM-binding proteins, myosins bind CaM in the absence of Ca^{2+}. In an *in vitro* motility assay, Ca^{2+} has been shown to inhibit the motility and actin-dependent ATPase activity of myosin I (Wolenski 1995). Ca^{2+} causes dissociation of one of the bound CaMs leading to inhibition of movement. The motility can be restored by adding back purified CaM, indicating that brush border myosin I motility in *in vitro* assays is dependent on bound CaM. This recovery of motility is likely due to the association of exogenous CaM with the heavy chain and points to an atypical mode of CaM regulation; dissociation from the heavy chain in the Ca^{2+}-bound state and activation of enzymatic activity in the Ca^{2+}-free state (Wolenski 1995). The inhibition of the enzymatic activity and motility of myosin by Ca^{2+} is similar to the effect of Ca^{2+} on a CaM binding MT-dependent C-terminal motor from plants (KCBP), although the mechanism of inhibition is very different between these proteins.

In the *Arabidopsis* genome there are at least seventeen myosin-like proteins (Reddy 2000). All *Arabidopsis* myosin-like proteins that have been identified so far possess two to six putative CaM-binding "IQ" motifs. However, interaction with, or regulation by CaM has not yet been demonstrated for these proteins. The presence of these motifs suggests an important regulatory role for Ca^{2+} and CaM in regulating the function of myosins. Two myosin heavy chains (170 and 175 kDa) in plants associate with CaM, indicating that CaM serves as a light chain for these myosins. Yokota *et al* (1999a 1999b) using highly purified plant myosins have provided the first evidence that Ca^{2+} inhibits the myosin motor activity as well as actin-dependent ATPase activity in plants. Ca^{2+} at concentrations higher than 1 µM inhibited reversibly the F-actin dependent ATPase activity as well as the motor activity. The inhibition of myosin activity in the presence of Ca^{2+} appears to be due to dissociation of CaM from the heavy chain since Ca^{2+} inhibition of motor activity can be restored by exogenous addition of CaM. Since the genes encoding these myosins are not cloned the structural features of these myosins are not known.

12. INTERACTION OF KCBP WITH OTHER PROTEINS

Krishnakumar and Oppenheimer (1999) isolated three extragenic suppressors (*suppressor of zwichel*-3; *suz1, suz2* and *suz3*) that rescued trichome branch number defect. These genetic studies (Oppenheimer *et al* 1997, Krishnakumar and Oppenheimer 1999) strongly suggest that KCBP interacts with several proteins and may function as a multiprotein complex. We found that the tail region of KCBP interacts with a protein kinase (KIPK, KCBP interacting protein kinase) in the yeast two-hybrid system (Day *et al* 2000). The association of KIPK with KCBP suggests regulation of KCBP or KCBP-associated proteins by phosphorylation and/or KCBP is involved in targeting KIPK to a proper cellular location.

13. CONCLUSION

KCBP is a ubiquitous MT motor protein in plants. It is unique among KLPs in having a CBD and myosin homology regions. The affect of activated calmodulin on KCBP suggests that calcium, through calmodulin, acts as a molecular switch and down-regulates the activity of KCBP. The evidence so far suggests a role for KCBP in cell division and trichome morphogenesis.

ACKNOWLEDGEMENTS

I thank members of my laboratory and all my collaborators who have contributed to the KCBP project; and Irene Day for her comments on the manuscript. Work on KCBP was funded by a grant from National Science Foundation to ASNR.

REFERENCES

Abdel-Ghany, S., and Reddy, A.S.N., 2000, A novel calcium/calmodulin-regulated microtubule motor protein is highly conserved between monocots and dicots. In preparation.

Asada, T., Kuriyama, R., and Shibaoka, H. 1997, TKRP125, a kinesin-related protein involved in the centrosome-independent organization of the cytokinetic apparatus in tobacco BY-2 cells. *J. Cell Sci.* **110**: 179-189.

Barton, N.R., and Goldstein, L.S.B., 1996, Going mobile: Microtubule motors and chromosome segregation. *Proc. Natl. Acad. Sci. USA* **93**: 1735-1742.

Bowler, C., and Chua, N.-H., 1994, Emerging themes of plant signal transduction. *Plant Cell* **6**: 1529-1541.

Bowser, J., and Reddy, A.S.N., 1997, Localization of a kinesin-like calmodulin-binding protein in dividing cells of *Arabidopsis* and tobacco. *Plant J.* **12**: 1429-1438.

Brady, S.T., 1985, A novel brain ATPase with properties expected for the fast axonal transport motor. *Nature* **317**: 73-75.

Brown, S.S., 1999, Cooperation between microtubule- and actin-based motor proteins. *Annu. Rev. Cell Dev. Biol.* **15**: 63-80.

Crane, P.R., Fris, E.M., and Pedersen, K.R., 1995, The origin and early diversification of angiosperms. *Nature* **374**: 27-33.

Day, I.S., Miller, C., Golovkin, M., and Reddy, A.S.N., 2000, Interaction of a kinesin-like calmodulin-binding protein with a protein kinase (submitted).

Deavours, B.E., Reddy, A.S., and Walker, R.A., 1998, Ca^{2+}/calmodulin regulation of the *Arabidopsis* kinesin-like calmodulin-binding protein. *Cell Motil. Cytoskeleton* **40**: 408-416.

Endow, S.A., 1999, Microtubule motors in spindle and chromosome motility. *Eur. J. Biochem.* **262**: 12-18.

Fordham-Skelton, A.P., Safadi, F., Golovkin, M., and Reddy, A.S.N., 1994, A non-radioactive method for isolating complementary DNAs encoding calmodulin-binding proteins. *Plant Mol. Biol. Reptr.* **12**: 355-363.

Franklin, A.E., and Cande, W.Z., 1999, Nuclear organization and chromosome segregation. *Plant Cell* **11**: 523-534.

Goldstein, L.S.B., and Philip, A.V., 1999, The road less traveled: Emerging principles of kinesin motor utilization. *Annu. Rev. Cell Dev. Biol.* **15**: 141-183.

Hepler, P.K., and Wayne, R.O., 1985, Calcium and plant development. *Annu. Rev. Plant Physiol.* **36**: 397-439.

Hirokawa, N., 1998, Kinesin and dynein superfamily proteins and the mechanism of organelle transport. *Science* **279**: 519-526.

Heulskamp, M., Misra, S., and Jürgens, G., 1994, Genetic dissection of trichome cell development in *Arabidopsis*. *Cell* **76**: 555-566.

Kao, Y.-L., Deavours, B.E., Phelps, K.K., Walker, R., and Reddy, A.S.N., 2000, Bundling of microtubules by motor and tail domains of a kinesin-like calmodulin-binding protein from

Arabidopsis: Regulation by Ca^{2+}/calmodulin. *Biochem. Biophys. Res. Commun.* **267**: 201-207.

Krishnakumar, S., and Oppenheimer, D.G., 1999, Extragenic suppressors of the *Arabidopsis* zwi-3 mutation identify new genes that function in trichome branch formation and pollen tube growth. *Development* **126**: 3079-3088.

Liu, B., Cyr, R.J., and Palevitz, B.A., 1996, A kinesin-like protein, KatAp, in the cells of *Arabidopsis* and other plants. *Plant Cell* **8**: 119-132.

Lloyd, C., 1994, Why should stationary plant cells have such dynamic microtubules? *Mol. Biol. Cell* **5**: 1277-80.

Matthies, H.J.G., McDonald, H.B., Goldstein, L.S.B., and Theurkauf, W.E., 1996, Anastral meiotic spindle morphogenesis: role of the nonclaret disjunctional kinesin-like protein. *J. Cell Biol.* **134**: 455-464.

Moore, J.D., and Endow, S.A., 1996, Kinesin proteins: a phylum of motors for microtubule-based motility. *BioEssays* **18**: 207-219.

Narasimhulu, S.B., Kao, Y.-L., and Reddy, A.S.N., 1997, Interaction of *Arabidopsis* kinesin-like calmodulin-binding protein with tubulin subunits: Modulation by Ca^{2+}-calmodulin. *Plant J.* **12**: 1139-1149.

Narasimhulu, S.B., and Reddy, A.S.N., 1998, Characterization of microtubule binding domains in the *Arabidopsis* kinesin-like calmodulin-binding protein. *Plant Cell* **10**: 957-965.

Nicol, F., and Hofte, H., 1998, Plant cell expansion: scaling the wall. *Curr. Opin. Plant Biol.* **1**: 12-17.

Oppenheimer, D.G., 1998, Genetics of plant cell shape. *Curr. Opin. Plant Biol.* **1**: 520-4.

Oppenheimer, D.G., Pollock, M.A., Vacik, J., Szymanski, D.B., Ericson, B., Feldmann, K., and Marks, D., 1997, Essential role of a kinesin-like protein in *Arabidopsis* trichome morphogenesis. *Proc. Natl. Acad. Sci. USA* **94**: 6261-6266.

Poovaiah, B.W., and Reddy, A.S.N., 1987, Calcium messenger system in plants. *CRC Cri. Rev. Plant Sci.* **6**: 47-103.

Poovaiah, B.W., and Reddy, A.S.N., 1993, Calcium and signal transduction in plants. *CRC Cri. Rev. Plant Sci.* **12**: 185-211.

Porter, J.A., Minke, B., and Montell, C., 1995, Calmodulin binding to Drosophila NinaC required for termination of phototransduction. *EMBO J.* **14**: 4450-4459.

Porter, J.A., Yu, M., Doberstein, S.K., Pollard, T.D., and Montell, C., 1993, Dependence of calmodulin localization in the retina on the NINAC unconventional myosin. *Science* **262**: 1038-1042.

Reddy, A.S.N., 2000, Molecular motors and their functions in plants. *Intl. Rev. Cytol. Cell Biol.* (in press).

Reddy, A.S.N., Narasimhulu, S.B., and Day, I.S., 1998, Structural organization of a gene encoding a novel calmodulin-binding kinesin-like protein from *Arabidopsis*. *Gene* **204**: 195-200.

Reddy, A.S.N., Narasimhulu, S.B., Safadi, F., and Golovkin, M., 1996a, A plant kinesin heavy chain-like protein is a calmodulin-binding protein. *Plant J.* **10**: 9-21.

Reddy, A.S.N., Safadi, F., Narasimhulu, S.B., Golovkin, M., and Hu, X., 1996b, A novel plant calmodulin-binding protein with a kinesin heavy chain motor domain. *J. Biol. Chem.* **271**: 7052-7060.

Reddy, V., and Reddy, A.S.N., 1999, A plant calmodulin-binding motor is part kinesin and part myosin. *Bioinformatics* (in press).

Reddy, V., Safadi, F., Zielinski, R.E., and Reddy, A.S.N., 1999, Interaction of a kinesin-like protein with calmodulin isoforms from *Arabidopsis*. *J. Biol. Chem.* **274**: 31727-31733.

Rhoads, A.R., and Friedberg, F., 1997, Sequence motifs for calmodulin recognition. *FASEB J.* **11**: 331-340.

Roberts, D.M., Lukas, T.J., and Watterson, D.M., 1986, Structure, function, and mechanism of action of calmodulin. *CRC Cri. Rev. Plant Sci.* **4**:, 311-339.

Rogers, G.C., Hart, C.L., Wedman, K.P., and Scholey, J.M., 1999, Identification of kinesin-C, a calmodulin-binding carboxy-terminal kinesin in animal (*Strongylocentrotus purpuratus*) cells. *J. Mol. Biol.* **294**: 1-8.

Sander, D., Brownlee, C., and Harper, J., 1999, Communicating with calcium. *Plant Cell* **11**: 691-706.

Smirnova, E., Reddy, A.S.N., Bowser, J., and Bajer, A.S., 1998, A minus end-directed kinesin-like motor protein, KCBP, localizes to anaphase spindle poles in Haemanthus endosperm. *Cell Motil. Cytoskeleton* **41**: 271-280.

Snedden, W.A., and Fromm, H., 1998, Calmodulin, calmodulin-related proteins and plant responses to the environment. *Trends Plant Sci.* **3**: 299-304.

Song, H., Golovkin, M., Reddy, A.S.N., and Endow, S.A., 1997, In vitro motility of AtKCBP, a calmodulin-binding kinesin-like protein of *Arabidopsis*. *Proc. Natl. Acad. Sci. USA* **94**: 322-327.

Staehelin, L.A., and Hepler, P.K., 1996, Cytokinesis in higher plants. *Cell* **84**, 821-824.

Trewavas, A.J., and Malho, R., 1997. Signal perception and transduction: The origin of the phenotype. *Plant Cell* **9**: 1181-1195.

Vale, R.D., Reese, T.S., and Sheetz, M.P., 1985, Identification of a novel force-generating protein, kinesin, involved in microtubule-based motility. *Cell* **42**: 39-50.

Voss, J., Safadi, F., Reddy, A.S.N., and Hepler, P.K., 1999, Kinesin-like calmodulin binding protein is necessary in plant cell division. *Mol. Biol. Cell* **10**: 373a.

Wang, D.Y., Kumar, S., and Hedges, S.B., 1999, Divergence time estimates for the earlyhistory of animal phyla and the origin of plants, animals and fungi. *Proc. R.. Soc. Lond. B Biol. Sci.* **266**: 163-171.

Wang, W., Takezawa, D., Narasimhulu, S.B., Reddy, A.S.N., and Poovaiah, B.W., 1996, A novel kinesin-like protein with a calmodulin-binding domain. *Plant Mol. Biol.* **31**, 87-100.

Wolenski, J.S., 1995, Regulation of calmodulin-binding myosins. *Trends Cell Biol.* **5**: 310-316.

Wolfe, K.H., Guoy, M., Ynag, Y.W., Sharp, P., and Li, W.H., 1989, Date of monocot-dicot divergence estimated from chloroplast DNA sequence data. *Proc. Natl. Acad. Sci. USA* **86**: 6201-6205.

Yang, J.T., Saxton, W.M., and Goldstein, L.S.B., 1988, Isolation and characterization of the gene encoding the heavy chain of *Drosophila* kinesin. *Proc. Natl. Acad. Sci. USA* **85**: 1864-1868.

Yokota, E., Muto, S., and Shimmen, T., 1999a, Inhibitory regulation of higher-plant myosin by Ca^{2+} ions. *Plant Physiol.* **119**: 231-40.

Yokota, E., Yukawa, C., Muto, S., Sonobe, S., and Shimmen, T., 1999b, Biochemical and immunocytochemical characterization of two types of myosins in cultured tobacco bright yellow-2 cells. *Plant Physiol.* **121**: 525-534.

Zhang, D.H., Wadsworth, P., and Hepler, P.K., 1993, *Cell Motil. Cytoskel.* **24**, 151-155.

Zielinski, R.E., 1998, Calmodulin and calmodulin-binding proteins in plants. *Annu. Rev. Plant Physiol. Plant Mol. Biol.* **49**: 697-725.

Progress Towards the Identification of Cytokinin Receptors

RICHARD HOOLEY

IACR-Long Ashton Research Station, University of Bristol, Department of Agricultural Sciences, Long Ashton, BRISTOL BS41 9AF, UK

1. INTRODUCTION

Since the discovery of cytokinins (Miller *et al* 1955) it has become clear that, in addition to promoting plant cells to divide in culture, this class of plant hormones influences a wide range of events during growth and development (Taiz and Zeiger 1998). For example, cytokinins inhibit leaf senescence, root elongation and lateral root initiation. They are involved in promoting leaf and cotyledon expansion, auxiliary meristem growth, nutrient mobilisation, root hair growth and stomatal opening. Cytokinins do not function in isolation of other plant hormones. Their stimulation of cell division in tissue culture requires the presence of auxin, and it is evident that in many responses cytokinins act both in synergy and antagonism with auxin, ethylene and gibberellin. Similarly, cytokinin and light signalling pathways interact in the regulation of chloroplast development and anthocyanin biosynthesis. There also appear to be strong interactions between cytokinin, nitrate and sugar signals. The importance of this class of hormones in plant growth and development has prompted many studies of their mechanism of action (see recent reviews of Hare and van Stadon 1997, Schmülling *et al* 1997, Brault and Maldiney 1999) and these have led to the identification of potential components of cytokinin signalling pathways and targets of them. In this review I will consider the progress that has been made towards the identification of cytokinin binding proteins and genes encoding receptors that are implicated in cytokinin signalling and discuss whether or not any of these may be cytokinin receptors.

2. STRUCTURE ACTIVITY RELATIONS AND SITES OF ACTION

In spite of the fact that numerous cytokinin compounds have been found in plants we have only a rudimentary understanding of the biosynthetic pathways leading to them (Prinsen *et al* 1997). Furthermore, because cytokinins applied to plants are metabolised it is difficult to be certain about which cytokinins have inherent biological activity. Most naturally occurring cytokinins that are thought to be active are N^6-substituted adenine molecules with a branched 5-carbon side chain (Taiz and Zeiger 1998). The most abundant of these is *trans*-zeatin, though the less abundant isopentenyl adenine is also highly active. Riboside and ribotide derivatives tend to be less active than the free bases and the abundant *N*-and *O*-linked glucoside are mostly inactive. Cytokinins with a hydroxylated benzyl ring at N^6 such as *meta*-topolin are naturally occurring and highly active (Strnad 1997). Kinetin, a highly active synthetic aminopurine derivative, has a furfuryl ring at N^6. Taken together these structure activity relationships suggest that recognition of active cytokinins by receptors involves features associated with the adenine ring and with isoprenoid or aromatic N^6 substitutions to this.

Our understanding of the molecular mechanisms underlying plant hormone perception and signal transduction has advanced rapidly as hormone response mutants in *Arabidopsis* have been isolated and characterised. Certainly in the case of ethylene this approach has led to the identiflcation of a bona fide ethylene receptor (Bleecker and Schaller 1996). Compared with other classes of plant hormones however, only very few cytokinin response mutants bave been isolated (Deikman 1997) and, as yet, none of these have helped to elaborate a molecular understanding of how cytokinins are perceived by plant cells.

Are cytokinin receptors soluble or integral membrane proteins? The cellular site of action of cytokinins is far from clear. Nevertheless, some evidence points to them acting at the plasma membrane and possibly regulating calcium flux (reviewed by Brault and Maldiney 1999). Silverman *et al* (1998) demonstrated that kinetin can induce a specific hyperpolarization of the plasma membrane of alfalfa root hairs when presented to the outside of the cell, but not when it was microinjected into the cytoplasm. This suggests that the kinetin is perceived at the external face of the plasma membrane. Other cytokinins induce a similar hyperpolarization response and promote alfalfa root hair growth. Furthermore, the ion transport inhibitor diisothiocyanostilbene-2,2'-disulfonic acid (DIDS) prevents both the hyperpolarization and growth responses, suggesting that they are linked. Thus, in alfalfa root hairs, cytokinin responses may be mediated by perception of the hormone by a receptor at the plasma membrane. In moss, cytokinins have been shown to promote calcium entry into cells, possibly *via* a G protein regulated voltage-dependent calcium channel (Schumaker and Gizinski 1996). Such a mechanism is most likely to involve perception of cytokinin at the plasma membrane. Nevertheless, these observations do not preclude the existence of cytokinin receptors at other sites within plant cells.

3. CYTOKININ BINDING PROTEINS

Protein extracts from a variety of plant tissues and species have been employed in searches for cytokinin binding proteins. A range of techniques have been applied and these include conventional binding assays, photoaffinity labelling, anti-idiotypic antibodies and affinity chromatography (Brinegar 1994). Only a small number of binding activities have been characterised in any detail and in no case has a clear role been demonstrated in a specific cytokinin response. The wheat cytokinin binding protein CBF-1 has been most thoroughly characterised (Brinegar 1994). CBF-1 is a vicilin-type seed storage protein present in abundance in embryo protein bodies as a 165 kDa homotrimeric protein. CBF-1 is not a cytokinin receptor even though it does bind the 2-azido-benzyladenine photo affinity probe, and a number of other active cytokinin derivatives (Brinegar 1994). One possibility is that CBF-1 sequesters cytokinins in embryos in such a way that they can be rapidly mobilised perhaps early after germination.

Mitsui *et al* (1996) identified a protein complex of 130 kDa in extracts from tobacco leaves that has cytokinin binding activity. One of the subunits, CBP57, of this protein complex has been cDNA cloned and found to share sequence similarity with S-adenosyl-L-homocysteine (SAH) hydrolase. This enzyme is known, in other organisms, to bind and be regulated by purines such as adenosine and cAMP. It does seem possible therefore that in plants cytokinins may perform an analogous function in regulating the activity of the enzyme (Mitsui *et al* 1996, 1997). Likely methylation substrates of SAH hydrolase include proteins and DNA and these might reasonably be components or targets of cytokinin signalling.

Certain phenylurea derivates are potent cytokinin antagonists and as such may be useful tools with which to identify and characterise cytokinin receptors. One such compound has been developed as a photoaffinity probe and used to identify and characterised a glutathione S-transferase as a cytokinin binding protein (Gonneau *et al* 1998). However, because the cytokinin binding properties of this enzyme do not correlated well with their biological activity the glutathione S-transferase is thought unlikely to have a central role in cytokinin signalling, though it may be involved in metabolism (Gonneau *et al* 1998).

Using anti-idiotypic antibodies, Kulaeva *et al* (1995) partially purified a soluble cytokinin binding protein from barley leaf cytosol and showed that, in the presence of *trans*-zeatin, it could stimulate elongation of nascent RNA transcripts in both isolated nuclei and a chromatin-associated RNA-polymerase I system. The identity of this protein is not currently known. Nevertheless, it has also been reported that in nuclei isolated from *Arabidopsis* leaves that have been treated with kinetin, steady-state levels of accurately initiated rRNA transcripts are higher than in nuclei from untreated leaves (Gaudino and Pikaard 1997).

In a number of very early reports cytokinin binding activity was detected in membrane fractions from higher plants (Sussman and Kende 1978, Kobayashi *et al* 1981) and moss protonema (Gardner *et al* 1978). While none of these have been characterised in any detail, or studied further in subsequent reports, they may support a membrane location for cytokinin receptors. A recent novel approach to detecting cytokinin binding activity in *Arabidopsis* membranes (Brault *et al* 1999) could renew further research in this area.

4. A G PROTEIN-COUPLED RECEPTOR HOMOLOGUE INVOLVED IN CYTOKININ SIGNALLING

There is a growing body of evidence that plants have some of the components of the G protein signalling pathway characteristic of many eukaryotic organisms and that this, or a derivative of it, operates in plants and transduces certain plant hormones (Hooley 1999). GCR1 is the first plant homologue of the G protein-coupled receptor (GPCR) superfamily and could potentially be a cytokinin receptor (Plakidou-Dymock *et al* 1998). GCR1 is a 326 amino acid polypeptide that has up to 23% amino acid identity (53% similarity) to known GPCRs, and a number of amino acid motifs that are conserved in some classes of GPCRs. Hydropathy analysis indicates that GCR1 has 7 potential transmembrane spanning domains and membrane topology prediction algorithms support a structure characteristic of GPCRs. *GCR1* is a single copy gene in *Arabidopsis* and is expressed at very low levels. Using RT-PCR, transcripts can be detected in 4 day seedlings and in roots, and vegetative tissues of 2-7 week plants. Evidence that GCR1 is involved in cytokinin signalling has come from studies of transgenic *Arabidopsis* containing antisense *GCR1* behind the constitutive CaMV*35S* promoter. These have reduced sensitivity to the cytokinin benzyl adenine in roots and shoots, yet respond normally to all other plant hormones.

The ligand for GCR1 needs to be identified and efforts to do this are focussing on expression of GCR1 in yeast. While GCR1 may be a cytokinin receptor it could alternatively be a downstream component in cytokinin signalling, or a receptor for a different ligand, the signalling pathway for which interacts with cytokinin signalling.

5. A TWO-COMPONENT RECEPTOR INVOLVED IN CYTOKININ SIGNALLING

Another integral membrane receptor with a potential role in cytokinin signalling is CKI1, identified in an activation tagging screen in *Arabidopsis* because it conferred cytokinin-independent shoot organogenesis in callus (Kakimoto 1996). Overexpression of receptors can lead to ligand independent signalling and this could explain the constitutive cytokinin response observed in CKI1 overexpressing callus. Alternatively, higher levels of the CKI1 receptor may enable cells to detect and give a greater response to their endogenous cytokinins. Other plausible interpretations of the effect of overexpressing CKI1 exist and distinguishing between all these possibilities requires further research.

CKI1 has significant sequence similarity to histidine kinase two-component response-regulators and it contains conserved motifs that are found in both the histidin kinase and response-regulator domains of this family of signalling proteins. Two-component response-regulators are activated either by a ligand, or by interacting with a transport or receptor protein (Chang and Stewart 1998). This leads to autophosphorylation of a highly conserved histidine residue located in the sensory kinase domain of the protein, and subsequent transfer of the phosphate group to a conserved aspartate in the response-regulator or receiver domain. Further phosphorelay can occur to additional response-regulator proteins as part of a signalling cascade.

If CKI1 is activated in this way then a family of *Arabidopsis* response-regulator proteins, and a maize response-regulator protein, are potential recipients of phosphotransfer reactions involving CKI1 (Brandstatter and Kieber 1998; Sakakibara *et al* 1998, Taniguchi *et al* 1998). The fact that genes encoding these plant response regulator proteins have been shown to be rapidly upregulated in leaves by cytokinins and by nitrate treatment of roots of nitrogen-starved plants, a treatment thought to cause an increase in foliar cytokinins, argues in favour of such a mechanism.

6. SUMMARY

Cytokinin receptors have been sought in plants for more than 20 years using biochemical, and more recently, molecular approaches. These have succeeded in identifying cytokinin binding proteins that include a number of enzymes, and two genes encoding unrelated integral membrane receptors. While some of these are good candidate cytokinin receptors, none can be assigned this role with any confidence at the present time. Future studies need to address the identity of ligands that interact with the cloned receptors and to acquire robust data demonstrating the involvement of candidate receptors in specific cytokinin responses.

ACKNOWLEDGEMENTS

IACR receives grant-aided support from the Biotechnology and Biological Sciences Research Council (BBSRC) of the United Kingdom. Work in R Hooley's lab is supported by the BBSRC Integration in Cellular Responses Programme and European Union Framework IV Programme.

REFERENCES

Bleecker, A.B., and Schaller, E.G., 1996, The mechanism of ethylene perception. *Plant Physiol.* **111**: 653-660.

Brandstatter, I., and Kieber, J.J., 1998, Two genes with similarity to bacterial response regulators are rapidly and specifically induced by cytokinin in *Arabidopsis*. *Plant Cell* **10**: 1009-1019.

Brault, M., and Maldiney, R., 1999, Mechanisms of cytokinin action. *Plant Physiol. Biochem.* **37**: 403-412.

Brault, M., Caiveau, O., Pedron, J., Maldiney, R., Sotta, B., and Miginiac, E., 1999, Detection of membrane-bound cytokinin binding proteins in *Arabidopsis thaliana* cells. *Eur. J. Biochem.* **260**: 512-519.

Brinegar, A.C., 1994, Cytokinin binding proteins and receptors. In *Cytokinins: Chemistry, Activity, and Function* (D.W.S. Mok and M.C. Mok, eds), CRC Press, Boca Raton, pp. 217-232.

Chang, C., and Stewart, R.C., 1998, The two component system. Regulation of diverse signalling pathways in prokaryotes and eukaryotes. *Plant Physiol.* **117**: 723-731.

Deikman, J., 1997, Elucidating cytokinin response mechanisms using mutants. *Plant Growth Regul.* **23**: 33-40.

Gardner; G., Sussman, M., and Kende, H., 1978, In vitro cytokinin binding to a particulate cell fraction from protonema of *Funaria hygrometrica*. *Planta* **143**: 67-73.

Gaudino, R.J., and Pikaard, C.S., 1997, Cytokinin induction of RNA polymerase I transcription in *Arabidopsis thaliana*. *J. Biol. Chem.* **272**: 6799-6804.

Gonneau, M., Mornet, R., and Laloue, M., 1998, A *Nicotiana plumbaginifolia* protein labeled with an azido cytokinin agonist is a glutathione S-transferase. *Physiol. Plant.* **103**: 114-124.

Hare, P.D., and van Staden, J., 1997, The molecular basis of cytokinin action. *Plant Growth Regul.* **23**: 41-78.

Hooley, R., 1999, A role for G proteins in plant hormone signalling? *Plant Physiol. Biochem.* **37**: 393-402.

Kakimoto, T., 1996, CKI1, a histidine kinase homolog implicated in cytokinin signal transduction. *Science* **274**: 982-985.
Kulaeva, O.N., Karavaiko, N.N., Selivankina, S.Y., Zemlyachenko, Y.V., and Shipilova, S.-V., 1995, Receptor of *trans*-zeatin involved in transcription activation by cytokinins. *FEBS Lett.* **366**: 26-28.
Kobayashi, K., Zbell, B., and Reinert, J., 1981, A high affinity binding site for cytokinin to a particulate fraction in carrot suspension cells. *Protoplasma* **106**: 145-155.
Miller, C.O., Skoog, F., von Saltza, M.H., and Strong, M., 1955, Kinetin, a cell devision factor from deoxyribonucleic acid. *J. Am. Chem. Soc.* **77**: 1329-1334.
Mitsui, S., Wakasugi, T., and Sugiura, M., 1996, A cytokinin-binding protein complex from tobacco leaves. The 57 kDa subunit has high homology to S-adenosyl-L.homocysteine hydrolase. *Plant Growth Regul.* **18**: 39-43.
Mitsui, S., Wakasugi, T., Hanano, S., and Sugiura, M., 1997, Localization of a cytokinin-binding protein CBP57/S-adenosyl-L-homocysteine hydrolase in a tobacco root. *J. Plant Physiol.* **150**: 752-754.
Plakidou-Dymock, S., Dymock, D., and Hooley, R., 1998, A higher plant seven transmembrane receptor that influences sensitivity to cytokinins. *Curr. Biol.* **8**: 315-324.
Prinsen, E., Kaminek, M., and van Onckelen, H.A., 1997, Cytokinin biosynthesis: a black box? *Plant Growth Regul.* **23**: 3-15.
Sakakibara, H., Suzuki, M., Takei, K., Deji, A., Taniguchi, M., and Sugiyama, T., 1998, A response-regulator homologue possibly involved in nitrogen signal transduction mediated by cytokinin in maize. *Plant J.* **14**: 337-344.
Schmülling, T., Schafer, S., and Romanov, G., 1997, Cytokinins as regulators of gene expression. *Physiol. Plant.* **100**: 505-519.
Schumaker, K.S., and Gizinski, M.J., 1996, G proteins regulate dihydropyridine binding to moss plasma membranes. *J. Biol. Chem.* **271**: 21292-21296.
Silverman, F.P., Assiamah, A.A., and Bush, D.S., 1998, Membrane transport and cytokinin action in root hairs of *Medicago sativa*. *Planta* **205**: 23-31.
Strnad, M., 1997, The aromatic cytokinins. *Physiol. Plant.* **101**: 674-688.
Sussman, M., and Kende, H., 1978, In vitro cytokinin binding to a particulate fraction of tobacco cells. *Planta* **140**: 251-259.
Taiz, L., and Zeiger, E., 1998, *Plant Physiol., Second Edition.* Sinauer Associates, Inc, Sunderland, MA, USA.
Taniguchi, M., Kiba, T., Sakakibara, H., Ueguchi, C., Mizuno, T., and Sugiyama, T., 1998, Expression of *Arabidopsis* response regulator homologs is induced by cytokinins and nitrate. *FEBS Lett.* **429**: 259-262.

Salicylic Acid- And Nitric Oxide-Mediated Signal Transduction In Disease Resistance

[1]DANIEL F. KLESSIG, [1]JÖRG DURNER, [1]ROY NAVARRE, [1]DHIRENDRA KUMAR, [1]JYOTI SHAH, [1]JUN MA ZHOU, [1]SHUQUN ZHANG, [1]DAVID WENDEHENNE, [1]PRADEEP KACHROO, [1]HERMAN SILVA, [1]KEIKO YOSHIOKA, [1]YOUSSEF TRIFA, [2]DOMINIQUE PONTIER, [2]ERIC LAM, [1]ZHIXIANG CHEN, [1]MARC ANDERSON AND [1]HE DU

[1]*Waksman Institute, Rutgers, The State University of New Jersey, 190 Frelinghuysen Road, Piscataway, NJ 08854, USA,* [2]*AgBiotech Center, Rutgers, The State University of New Jersey, 59 Dudley Road, New Brunswick, NJ 08901, USA.*

1. INTRODUCTION

Salicylic acid (SA) has been shown to play a critical role in signaling the activation of plant defense responses following pathogen infection. These responses include the induction of cell death and disease resistance in the infected leaf, as well as the activation of systemic disease resistance in the uninfected portions of the plant. SA has been shown to induce these phenomena through a variety of mechanisms that may include altering the activity or synthesis of certain enzymes, increasing the expression of various defense genes and/or potentiating the activation of certain defense responses.

2. RESULTS/DISCUSSION

2.1 SA-Interacting Proteins in Tobacco

Several proteins in tobacco, including catalase, ascorbate peroxidase and SA binding protein 2 (SABP2), have been shown to interact with SA (Chen *et al* 1993, Durner and Klessig 1995, Du and Klessig 1997). Catalase and ascorbate peroxidase are the major hydrogen peroxide (H_2O_2) scavenging enzymes in plant cells. SA inhibits their activity by serving as a one electron-donating substrate; through this process, a phenolic free radical is generated (Durner and Klessig 1996). This free radical may then activate defense gene expression by causing increased levels of lipid peroxidation. In tobacco suspension cells, SA treatment has previously been shown to induce lipid peroxidation. Moreover, lipid peroxides, which are the products of lipid peroxidation, have been shown to activate pathogenesis-related (*PR*)-*1* gene expression (Anderson *et al* 1998). In addition to lipid peroxides, SA's interaction with catalase and ascorbate peroxidase may lead to increased levels of H_2O_2, another compound that has been shown to activate *PR-1* expression (Chen *et al* 1993).

In contrast to catalase and ascorbate peroxidase, the function of SABP2 is unknown. However, this soluble, low molecular weight protein binds SA with a kDa of approximately 90 nM, which is 150 fold lower than that exhibited by catalase. SABP2 exhibits an even greater affinity for the functional SA analog benzothiadiazole (BTH); this correlates with the fact that BTH is more effective than SA at activating *PR* gene expression and enhanced disease resistance. BTH has also been shown to inhibit the activity of catalase and ascorbate peroxidase (Wendehenne *et al* 1998).

2.2 MAP Kinases are Associated with Disease Resistance in Tobacco

Protein phosphorylation and/or dephosphorylation has been implicated in the SA-mediated activation of defense responses in tobacco (Conrath *et al* 1997). A 48 kDa SA-induced protein kinase (SIPK) was identified in tobacco suspension cells (Zhang and Klessig 1997). Subsequent purification of this kinase and isolation of its encoding gene revealed that it is a member of the tobacco mitogen-activated protein (MAP) kinase family. Another member of this MAP kinase family is encoded by the wounding-induced protein kinase (*WIPK*) gene. *WIPK* is transcriptionally activated by wounding and, based on this observation, it was proposed to encode a wounding-activated kinase (Seo *et al* 1995). However, recent studies have shown that this wounding-induced kinase is actually encoded by *SIPK* (Zhang and Klessig 1998a). In conjunction with various collaborators, we have assessed the effects of wounding, pathogen infection and pathogen-derived elicitors on the activity of SIPK and WIPK (Zhang *et al* 1998, Zhang and Klessig 1998a, b, Romeis *et al* 1999). Interestingly, both kinases are activated in resistant tobacco cultivars following TMV infection, but not in susceptible cultivars (Zhang and Klessig 1998b). Their mechanism of activation in TMV-infected plants, however, is quite distinct. SIPK is activated exclusively via phosphorylation by the upstream MAP kinase kinase. In contrast, large increases in *WIPK* mRNA and protein precede a dramatic increase in the activity of this MAP kinase.

2.3 Factors that Directly or Indirectly Interact with NPR1 of *Arabidopsis*

Through analysis of mutant *Arabidopsis*, we have identified several potential components of the SA-mediated pathway leading to disease resistance. One class of mutants constitutively expresses *PR* genes (*cep*). These plants also spontaneously develop lesions, contain elevated levels of SA and exhibit enhanced disease resistance (Silva *et al* 1999). A second class of mutants is salicylate insensitive (*sai1*; Shah *et al* 1997). Plants from this class fail to express *PR* genes following SA or BTH treatment and have reduced resistance to pathogen attack. The *sai1* mutation was shown to be allelic with *npr1* (Cao *et al* 1997) and *nim1* (Ryals *et al* 1997); thus, it was renamed *npr1-5*. Isolation of the *NPR1/NIM1* gene revealed that it encodes a 65 kDa protein which contains ankyrin repeats and shares limited homology with IκBα (Cao *et al* 1997, Ryals *et al* 1997). We have also identified suppressors of the salicylate insensitive phenotype of *npr1-5* (*ssi*; Shah *et al* 1999). Analysis of the *ssi1* mutant suggests that the wild type protein functions as a switch that modulates cross talk between the SA/NPR1 pathway regulating *PR-1* expression and the jasmonate/ethylene-regulated defense pathway(s) leading to defensin gene (*PDF1.2*) expression (Shah *et al* 1999).

In order to identify proteins that interact with NPR1, the yeast two-hybrid screen was utilized. Several NPR1-interacting proteins were identified including two members of the TGA/OBF family of bZIP transcription factors (Zhou *et al* 2000). Six members of this family were shown to differentially bind NPR1. This family of transcription factors was previously implicated in activating the promoters of several SA-inducible genes. Two members of this family, TGA2 and TGA3 exhibited the strongest binding to NPR1. Strikingly, point mutations in *NPR1* that block SA signaling (*npr1-1*, *npr1-2*, *npr1-5* and *nim1-2*) abolished TGA2 and TGA3 binding. Moreover, both TGA2 and TGA3 bound a TGACG element in the *PR-1* promoter. This element was shown to be required for SA inducibility *in planta* (Lebel *et al* 1998). Thus, our results argue that the SA/NPR1 pathway induces *PR-1* expression via members of the TGA/OBF transcription factor family (Fig. 1).

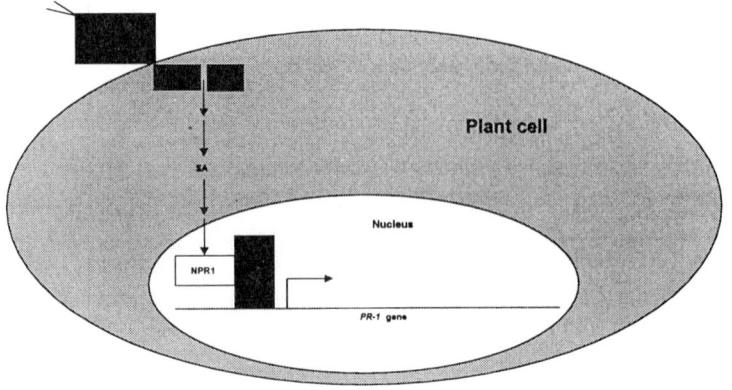

Figure 1. Simplified model of SA-mediated activation of the *Arabidopsis* PR-1 gene by a pathogen.

2.4 Nitric Oxide (NO) and the Activation of Plant Defense Responses

NO has been shown to function as an important redox-active signal in the induction of innate immunity in animals. To determine whether it serves a similar function in plants, NO synthase (NOS) activity was monitored in TMV-infected resistant and susceptible tobacco plants. NOS activity was observed to increase in resistant but not susceptible plants (Fig. 2; Durner *et al* 1998). Moreover, treating tobacco suspension cells with recombinant mammalian NOS or various NO donors increased the expression of *PR-1* and phenylalanine ammonia lyase (*PAL*), while the NOS inhibitor L-NMMA blocked *PR-1* expression after TMV infection. The expression of *PR-1* and *PAL* was also shown to be induced by cyclic ADP ribose (cADPR) and cyclic GMP (cGMP); these molecules are known to serve as second messengers for NO in mammals. Further evidence that cGMP is involved in the NO-induced expression of plant defense genes came from the observation that cGMP levels exhibited a transient but dramatic increase in NO-treated tobacco and that the guanylate cyclase (GC) inhibitors LY 83583 and ODQ suppressed NO-induced *PAL* expression. Using NahG tobacco plants, which are unable to accumulate SA, the NOS-induced expression of *PR-1* was shown to be SA dependent, while that for *PAL* was SA independent.

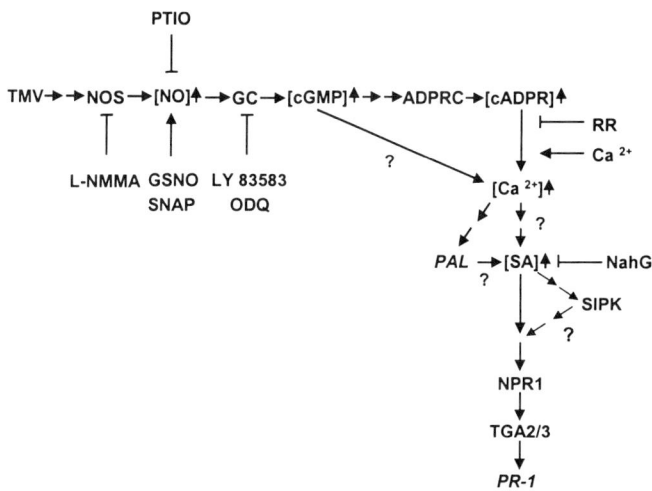

Figure 2. Proposed SA- and NO-mediated pathway for activation of defense genes PAL and PR-1.

To further compare the NO signaling pathways in plants and mammals, we assessed whether NO treatment affected the activity of aconitases and MAP kinases in tobacco; these enzymes are known to be regulated by NO in mammals. Injecting tobacco leaves with recombinant NOS or treating suspension cells with various NO donors was observed to activate SIPK transiently (Kumar *et al* 2000). By contrast, no activation of WIPK was detected. Since SIPK activation was blocked in NOS-treated NahG tobacco, it appears to be dependent on SA. Treating tobacco extracts with NO donors was also shown to inhibit aconitase activity (Navarre *et al* 2000). This result is consistent with studies in mammals, which have demonstrated that aconitase is an important NO and redox sensor. In the presence of NO, the activity of mammalian cytosolic aconitase is inhibited and the enzyme is converted into an mRNA binding protein known as the iron regulatory protein (IRP). The IRP regulates iron homeostasis by binding and thereby altering the stability and translatibility of transcripts for proteins that regulate iron levels. A cytosolic aconitase gene (NtACO1) was cloned from tobacco and shown to exhibit 61% identity and 76% similarity with the amino acid sequence of the human IRP1. Moreover, NtACO1 contained conserved residues thought to be involved in mRNA binding by IRP1; this result suggests that the tobacco enzyme may also function as an IRP.

Taken together, our results argue that several critical NO signaling components in animals are also involved in the NO-mediated activation of defense responses in plants.

3. CONCLUSION

SA and NO are two key signaling molecules involved in activating a variety of plant defense responses to pathogens. Both appear to act through a number of effector proteins to activate these defense responses.

ACKNOWLEDGEMENTS

This work was supported by Grants MCB 9723952 and MCB 9514239 from the National Science Foundation and Grant 9802200 from the U.S., Department of Agriculture.

REFERENCES

Anderson, M., Chen, Z., and Klessig, D.F., 1998, Possible involvement of lipid peroxidation in salicylic acid-mediated induction of *PR-1* gene expression. *Phytochemistry* **47**: 555-566.

Cao, H., Glazebrook, J., Clarke, J.D., Volko, S., and Dong, X., 1997, The *Arabidopsis NPR1* gene that controls systemic acquired resistance encodes a novel protein containing ankyrin repeats. *Cell* **88**: 57-63.

Chen, Z., Silva, H., and Klessig, D.F., 1993, Active oxygen species in the induction of plant systemic acquired resistance by salicylic acid. *Science* **262**: 1883-1886.

Conrath, U., Silva, H., and Klessig, D.F., 1997, Protein dephosphorylation mediates salicylic acid-induced expression of *PR-1* genes in tobacco. *Plant J.* **11**: 747-757.

Durner, J., Wendehenne, D., and Klessig, D.F., 1998, Defense gene induction in tobacco by nitric oxide, cyclic GMP and cyclic ADP ribose. *Proc. Natl. Acad. Sci. USA* **95**: 10328-10333.

Durner, J., and Klessig, D.F., 1995, Inhibition of ascorbate peroxidase by salicylic acid and 2,6-dichloroisonicotinic acid, two inducers of plant defense responses. *Proc. Natl. Acad. Sci. USA* **92**: 11312-11316.

Durner, J., and Klessig, D.F., 1996, Salicylic acid is a modulator of tobacco and mammalian catalases. *J. Biol. Chem.* **271**: 28492-28501.

Du, H., and Klessig, D.F., 1997, Identification of a soluble, high affinity salicylic acid-binding protein in tobacco. *Plant Physiol.* **113**: 1319-1327.

Kumar, D., Zhang, S., and Klessig, D.F., 2000, Differential induction of tobacco MAP kinases by the defense signals nitric oxide, salicylic acid, ethylene and jasmonic acid. *Mol. Plant-Microbe Interact.* In press.

Lebel, E., Heifetz, P., Thorne, L., Uknes, S., Ryals, J., and Ward, E.,1998, Functional analysis of regulatory sequences controlling *PR-1* gene expression in *Arabidopsis*. *Plant J.* **16**: 223-233.

Navarre, R., Wendehenne, D., Durner, J., Noad, R., and Klessig, D.F., 2000, Nitric oxide modulates the activity of tobacco aconitase. *Plant Physiol.* In press.

Romeis, T., Piedras, P., Zhang, S., Klessig, D.F., Hirt, H., and Jones, J.D.G., 1999, Rapid, Avr9- and Cf-9-dependent, activation of MAP kinases in tobacco cell cultures and leaves: Convergence in resistance gene, elicitor, wound and salicylate responses. *Plant Cell* **11**: 273-287.

Ryals, J.A., Weymann, K., Lawton, K., Friedrich, L., Ellis, D., Steiner, H.Y., Johnson, J., Delaney, T.P., Jesse, T., Vos, P., and Uknes, S., 1997, The *Arabidopsis NIM1* protein shows homology to the mammalian transcription factor inhibitor IκB. *Plant Cell* **9**: 425-439.

Seo, S., Okamoto, M., Seto, H., Ishizuka, K., Sano, H., and Ohashi, Y., 1995, Tobacco MAP kinase: A possible mediator in wound signal transduction pathways. *Science* **270**: 1988-1992.

Shah, J., Kachroo, P., and Klessig, D.F., 1999, The *Arabidopsis ssi1* mutation restores pathogenesis-related gene expression in *npr1* plants and renders defensin gene expression salicylic acid dependent. *Plant Cell* **11**:191-206.

Silva, H., Yoshioka, K., Dooner, H.K., and Klessig, D.F., 1999, Characterization of a new *Arabidopsis* mutant exhibiting enhanced disease resistance. *Mol. Plant-Microbe Interact.* **12**: 1053-1063.

Wendehenne, D., Durner, J., Chen, Z., and Klessig, D.F., 1998, Benzothiadiazole, an inducer of plant defenses, inhibits catalase and ascorbate peroxidase. *Phytochemistry* **47**: 651-657.

Zhang, S., Du, H., and Klessig, D.F., 1998, Activation of the tobacco SIP kinase by both a cell wall-derived carbohydrate elicitor and purified proteinaceous elicitins from *Phytophthora* spp. *Plant Cell* **10**: 435-449.

Zhang, S., and Klessig, D.F., 1997, Salicylic acid activates a 48-kDa MAP kinase in tobacco. *Plant Cell* **9**: 804-824.

Zhang, S., and Klessig, D.F., 1998a, The tobacco wounding-activated mitogen-activated protein kinase is encoded by SIPK. *Proc. Natl. Acad. Sci. USA* **95**: 7225-7230.

Zhang, S., and Klessig, D.F., 1998b, Resistance gene *N*-mediated *de novo* synthesis and activation of a tobacco mitogen-activated protein kinase by tobacco mosaic virus infection. *Proc. Natl. Acad. Sci. USA* **95**: 7433-7438.

Zhou, J.M., Trifa, Y., Silva, H., Pontier, D., Lam, E., Shah, J., and Klessig, D.F., 2000, NPR1 differentially interacts with members of the TGA/OBF family of transcription factors which bind an element of the *PR-1* gene required for induction by salicylic acid. *Mol. Plant-Microbe Interact.* In press.

Involvement of ROS and Caspase-like Proteases During Cell Death Induction by Plant Pathogens
Signaling in Plant Cell Death

ERIC LAM*, OLGA DEL POZO, AND DOMINIQUE PONTIER
Biotech Center, Rutgers, The State University of New Jersey, Foran Hall, 59 Dudley Rd., New Brunswick, NJ 08903, USA.

1. INTRODUCTION

Programmed cell death (PCD) is now recognized widely as an important process conserved among multicellular eukaryotes. In contrast to abiotically-induced cell deaths resulting from physical injuries, PCD is a highly organized event that requires the active participation of critical components from the dying cells. Together with cell proliferation by mitosis, the rate of PCD determines the number of cells at a given time as well as the sculpting of the size and shape of complex organs. In the past 10 years, work in animal systems has identified one conserved switch through which the most well-studied form of PCD called apoptosis is regulated. The central component of this death switch is a specialized family of cysteine proteases called caspases (Cryns and Yuan 1998). These proteases share similar structural features around their active sites and they cleave after an aspartate residue in their target sequence (i.e. the P1 position). Two classes of caspases are distinguished by their position in a protease cascade that leads to PCD activation. The "initiator" caspases are controlled by interaction with a number of signal sensors/pathways through protein/protein interaction. These interactions catalyze the necessary cleavage of the initiator procaspases resulting in activation of their latent protease activities, which in turn leads to the cleavage and activation of the downstream "effector" or "executioner" caspases. To date, three distinct signaling pathways in animal cells are known to channel through different caspases (reviewed in Mehmet

2000): caspase-8 mediates PCD activation by membrane localized receptors containing the so-called "death domain", caspase-9 is activated by Apaf-1 upon cytochrome c and dATP binding to the latter, and caspase-12 is essential for PCD activation through the ER pathway. Caspase-3 is a common executioner protease downstream from these initiator caspases. The critical substrates downstream from caspase-3 that orchestrate the highly ordered cell death program remain to be defined. One recently discovered target of caspase-3 is the inhibitor protein called Icad which regulates the activity of a nuclease that is responsible for nuclear DNA fragmentation (Sakahira *et al* 1999).

In plants the mechanism of PCD remains largely unknown. It is a phenomenon that occurs throughout the life cycle of various species, from the differentiation of tracheary elements to the controlled degeneration of the tapetal cell layer in the anthers. Due to the absence of engulfment of the dying cells by their neighbors because of the physical barrier imposed by the cell wall, it is anticipated that PCD in plants may be different in some key aspects from those of apoptosis (Mittler and Lam 1996). Thus, the detailed characterization of plant PCD could shed light on the evolution of this important process and may help us to identify key components in the death pathway that would be conserved between plant and animal cells. One of the most well-characterized pathways of PCD in plants is the hypersensitive response (HR) cell death that is commonly induced during incompatible interactions between host plants and pathogens. Together with the activation of defense gene expression and production of compounds such as phytoalexins, the HR is thought to be important for the efficient restriction of pathogen spread within the host. This chapter will review our recent progress in the study of cell death signaling during the HR. In particular, the possible roles of reactive oxygen species (ROS) and caspases will be discussed.

2. RESULTS / DISCUSSION

2.1 Requirement for Reactive Oxygen Species (ROS) for Cell Death Induction During Plant-Pathogen Interaction

The role of the oxidative burst in plant defense has been speculated for some time since its initial discovery (reviewed in Lamb and Dixon 1997). Although there are good evidence that ROS generation is required for activation of HR cell death, it is probably insufficient on its own to do so. This is exemplified by the isolation of mutants in bacterial pathogens which can activate ROS generation in the host without triggering cell death (Glazener *et al* 1996). Nevertheless, the requirement of ROS as one of the critical signals for HR cell death induction can be demonstrated by the application of inhibitors for NADPH oxidase such as DPI (diphenylene iodonium) or the suppression of ROS accumulation by catalase or other ROS scavengers (Lamb and Dixon 1997). More recently, the role of NO (nitric oxide) as a synergistic activator of HR cell death with ROS has emerged (see chapter by Klessig *et al* this Proceedings). In addition, the demonstration of the role for a small GTP-binding protein Rac, which may regulate ROS generation from the NADPH oxidase, to control cell death and disease resistance in transgenic rice is also consistent with the critical role that ROS is likely to play in HR cell death (Kawasaki *et al* 1999).

Work from our laboratory several years ago with tobacco has also demonstrated the critical importance of ROS for cell death induction during plant-pathogen interactions (Mittler *et al* 1996). In this work, we showed that depletion of oxygen by flushing the inoculated leaf tissues with nitrogen suppressed HR cell death induction by either viral or bacterial pathogens. Interestingly, we found that defense gene markers such as PR-1, PR-2 and PR-3 are still activated by the pathogens. This observation provided one of the first indications that cell death and defense gene activation may be uncoupled in this type of interaction. By switching the incubation atmosphere between oxygenic and anoxygenic conditions under our experimental protocol, we examined the time at which ROS may be required to signal or execute HR cell death. With a bacterial pathogen, *Pseudomonas syringae pv. phaseolicola* 3121, that induces non-host resistance we found that ambient oxygen concentration is needed only within the first 3 hours of incubation after the inoculation of the bacteria into the tobacco leaf. Switching to nitrogen atmosphere after 3 hours does not inhibit HR cell death, which is manifested morphologically only until about 15 hours post-inoculation. This observation indicated that ROS may be required only as a signal for cell death in this experimental system. In contrast, TMV (tobacco mosaic virus) induced HR apparently requires oxygen continuously to sustain cell death progression after virus inoculation, which suggests that ROS may be required for the execution process in addition to functioning as a cell death signal (Mittler *et al* 1996).

2.2 Effects of Peptide Inhibitors of Proteases on HR Cell Death

To determine whether caspase-like proteases (CLPs) may be involved in HR cell death, we have utilized synthetic peptide inhibitors of caspases that have been used widely in animal apoptosis research (Villa *et al* 1997). Two different tetrapeptide inhibitors, YVAD-CMK and DEVD-CHO, were found to have dramatic effects on cell death induction by *P. syringae pv. phaseolicola* 3121. HR cell death induction by inoculation of these bacteria on tobacco can be effectively blocked by coinfiltration with these peptide inhibitors. In contrast, protease inhibitors such as PMSF, TLCK or TPCK were without significant effects at similar concentrations (del Pozo and Lam 1998). Two early HR marker genes in tobacco, Hin1 and Hsr203J, which are induced to a maximum level within 8 hours of bacteria challenge have been cloned by differential screening (Gopalan *et al* 1996, Pontier *et al* 1994). We found that the induction of these genes are suppressed coordinately with the inhibition of HR cell death by the caspase inhibitors. In contrast, the induction of the late genes such as PR-1 or PR-2, the expression of which peaks at about 20 hrs. post-inoculation, was not affected dramatically. These results are consistent with a model in which the pathways for cell death induction and disease resistance may bifurcate at some point after the initial signaling step between the plant host and the pathogen. Early induced genes such as Hin1 and Hsr203J may be associated with the PCD branch of this model while late genes such as PR-1 and PR-2 may be more closely correlated with the resistance pathway.

More recently, indirect evidence for the presence of CLPs in plant cell death have emerged in studies with plant cell cultures. In one study, Zhao *et al* (1999) found that the cytosol of carrot cells can provide the necessary cofactor(s) to activate apoptosis of isolated mouse liver nuclei in a cytochrome c-dependent fashion. This presumed plant cytosolic activity can be inhibited by YVAD-CHO and DEVD-CHO, suggestive of the involvement of a CLP that may be analogous to our early findings. The same group of researchers also found that menadione-induced oxidative stress in tobacco protoplasts can lead to DNA fragmentation and poly-ADP-ribose-polymerase (PARP) cleavage which can be inhibited by DEVD-CHO (Sun *et al* 1999). This is the first report that PARP, a well-characterized caspase-3 substrate in animal apoptosis, is also cleaved in some instances of plant cell death. This study also found evidence that cytochrome c is relocated to the cytosol from mitochondria after menadione treatment, concomitant with the appearance of DNA laddering. As mentioned above, cytochrome c repartitioning is correlated with the activation of caspase-9 through interaction with Apaf-1 and dATP (Hu *et al* 1999). In addition to cell death induction by menadione, two other recent reports on plant cell death also observed cytochrome c redistribution to the cytosol. One study involves the induction of *Arabidopsis* and maize cell death by D-mannose (Stein and Hansen 1999) while heat stress of cucumber plants was the experimental system in the other work (Balk *et al* 1999). In the latter report, it was shown that fumarase, an enzyme of 50 kDa which is present in the matrix of mitochondria, is retained within the mitochondria after heat stess while greater than 70% of the cytochrome c was lost from this organelle to the cytosolic fraction. Although this observation is consistent with the membrane of the mitochondria being relatively intact and that the partitioning of cytochrome c is specific, it should be pointed out that the latter protein is much smaller than 50 kDa and thus may be more easily leaked upon small perturbation of the mitochondrial outer membrane. Nevertheless, these studies raise the possibility that active cytochrome c relocalization to the cytosol may be one common phenomenon associated with PCD in plants. Furthermore, it provided additional evidence for the involvement of CLPs in plant cell death since the only function at present attributed to cytosolic cytochrome c is the activation of caspases.

2.3 P35 Overexpression Delays TMV Induced HR Cell Death in Tobacco

To test for CLP involvement in HR cell death, we have generated tobacco plants that express the baculovirus P35 gene under the CaMV 35S promoter (del Pozo and Lam, unpublished results). P35 has been shown to be critical for the efficient suppression of host cell apoptosis upon infection by the insect virus. Its action essentially targets cellular caspases by physical association with these proteases after cleavage. The preferential caspase cleavage site in P35 is aspartate 87 (D87), which is situated in an accessible loop that is well-suited to fit into the active sites of caspases (Fisher et al 1999). This structural feature explains the fact that P35 is an effective inhibitor for all known caspases examined thus far but does not appear to inhibit other known proteases significantly (Zhou et al 1998). Given the specificity and efficacy of P35 as a broad range caspase inhibitor, it is ideally suited to test for possible involvement of CLPs in a heterologous system. In tobacco plants expressing P35, we found that the rate of cell death as measured by membrane leakage is significantly delayed when HR is induced by either TMV or P. syringae inoculation. However, cell death is morphologically evident and defense gene markers such as PR-1 are activated as in untransformed plants. This incomplete inhibition of cell death may be due to very low level of expression of the P35 protein as detected by western blot analysis, in spite of high levels of P35 transcript in the transgenic plants. The low P35 protein levels may thus be due to poor translation of the RNA resulting from incompatible codon bias in the P35 coding sequence or instability of the P35 protein in plants after translation. Aside from the delay in cell death that can be measured by membrane leakage, a dramatic phenotype in TMV inoculated transgenic plants expressing P35 is the systemic spread of the virus in spite of the HR in the primary infected leaf. Under normal circumstances, systemic movement of TMV is suppressed completely upon HR induction and virus is restricted to the infected leaf. The cell death delay in the P35-expressing plants apparently allowed the escape of the virus. This observation provides evidence for HR cell death as an important mechanism to efficiently restrict certain invading viral pathogens to the infection site.

To correlate the effects of P35 in plants to its inhibitory action on CLPs, we constructed mutants of P35 in which the D87 residue or together with its neighboring aspartate residues are converted to alanines. In vitro, these mutant variants of P35 protein are not cleaved efficiently or at all by purified caspase-1 and thus not expected to inhibit caspases. In transgenic tobacco plants, these P35 variants do not retard HR cell death and TMV is effectively restricted to the infected leaf. Thus, these results provide critical evidence that the effects of P35 *in planta* are due to its inhibition of CLPs.

2.4 Detection of Caspase-like Protease Activities in Tobacco Leaf Tissues

To biochemically detect CLP activity in plants, we took advantage of the temperature-dependent nature of TMV-induced HR in tobacco to synchronize cell death in the infected tissues. At $30^\circ C$ or higher temperature, TMV infection fails to induce HR and disease resistance in spite of the presence of the resistance gene N. After the virus has systemically infected the plant in about 1 to 2 weeks, massive cell death can be activated by transferring the infected plant to $23^\circ C$. Using this protocol and the fluorogenic substrate YVAD-AMC, we detected CLP activity that is induced upon activation of HR cell death by TMV (del Pozo and Lam 1998). This inducible protease activity is inhibited by synthetic caspase inhibitors and peaks at about 20 to 30 hours after shifting from $30^\circ C$ to $23^\circ C$. One interesting observation is that although this CLP activity can be inhibited by DEVD-CHO, it is unable to cleave DEVD-AMC. This may suggest that the active site structure of the plant CLP is somewhat different from its animal counterparts. Another interesting feature of the HR-induced CLP activity is its transient induction profile. It drastically declines after peaking at about 30 hours of the shift from $30^\circ C$ to $23^\circ C$. One possible explanation may be the induction of HR cell death could result in more rapid turnover of cellular proteins in general. In any case, the low specific activity combined with the unstable induction of the plant CLP make its biochemical purification rather difficult. Nevertheless, the direct detection of CLP activity that can be induced during the HR provided strong support for the inhibitor studies.

3. CONCLUSIONS

The evidence for ROS and CLP involvement in HR cell death signaling remains mostly indirect with the identities of the major players still elusive.

Future experiments involving the genetic manipulation of enzymes that regulate ROS metabolism and the plant CLPs should be critical in advancing our understanding of these phenomena. The cloning and characterization of plant CLPs will be an important step toward this goal. In addition to ROS and CLPs, salicylic acid is also a likely potentiator of HR cell death while the function of the alternative oxidase as a modulator of ROS generation from the mitochondria has emerged in recent years (Lam *et al* 1999). Given the expanding repertoire of molecular and genomic tools for plant research, our understanding of cell death signaling in plant systems should improve rapidly in the near future.

ACKNOWLEDGEMENTS

The work on PCD in our laboratory is supported by the New Jersey Commission on Science and Technology and a competitive grant from the USDA.

REFERENCES

Balk, J., Leaver, C.J., and McCabe, 1999, Translocation of cytochrome c from the mitochondria to the cytosol occurs during heat-induced programmed cell death in cucumber plants. *FEBS Lett.* **463**: 151-154.

Cryns, V., and Yuan, J., 1998, Proteases to die for. *Genes Dev.* **12**: 1551-1570.

del Pozo, O., and Lam, E., 1998, Caspases and programmed cell death in the hypersensitive response of plants to pathogens. *Curr. Biol.* **8**:1129-1132.

Fisher, A.J., de la Cruz, W., Zoog, S.J., Schneider, C., and Friesen, P.D., 1999, Crystal structure of baculovirus P35: role of a novel reactive site loop in apoptotic caspase inhibition. *EMBO J.* **18**: 2031-2039.

Glazener, J.A., Orlandi, E.W., and Baker, J.C., 1996, The active oxygen response of cell suspensions to incompatible bacteria is not sufficient to cause hypersensitive cell death. *Plant Physiol.* **110**: 759-763.

Gopalan, S., Wei, W. and He, S.Y., 1996, *hrp* gene-dependent induction of hin1: a plant gene activated rapidly by both harpins and the avrPto gene mediated signal. *Plant J.* **10**: 591-600.

Hu, Y., Benedict, M.A., Ding, L., and Nunez, G., 1999, Role of cytochrome c and dATP/ATP hydrolysis in Apaf-1-mediated caspase-9 activation and apoptosis. *EMBO J.* **18**: 3586-3595.

Kawasaki, T., Henmi, K., Ono, E., Hatakeyama, S., Iwano, M., Satoh, H., and Shimamoto, K., 1999, The small GTP-binding protein Rac is a regulator of cell death in plants. *Proc. Natl. Acad. Sci. USA* **96**: 10922-10926.

Lam, E., Pontier, D., and del Pozo, O., 1999, Die and let live-programmed cell death in plants. *Curr. Opin. Plant Biol.* **2**: 502-507.

Lamb, C., and Dixon, R.A., 1997, The oxidative burst in plant disease resistance. *Annu. Rev. Plant Physiol. Plant Mol. Biol.* **48**: 251-275.

Mehmet, H., 2000, Caspases find a new place to hide. *Nature* **403**: 29-30.

Mittler, R., and Lam, E., 1996, Sacrifice in the face of foes: pathogen-induced programmed cell death in plants. *Trends Microbiol.* **4**: 10-15.

Mittler, R., Shulaev, V., Seskar, M., and Lam, E., 1996, Inhibition of programmed cell death in tobacco plants during pathogen-induced hypersensitive response at low oxygen pressure. *Plant Cell* **8**: 1991-2001.

Pontier, D., Godiard, L., Marco, Y., and Roby, D., 1994, *hsr203J*, a tobacco gene whose ativation is rapid, highly localized and specific for incompatible plant pathogen interactions. *Plant J.* **5**: 507-521.

Sakahira, H., Enari, M., Ohsawa, Y., Uchiyama, Y., and Nagata, S., 1999, Apoptotic nuclear morphological change without DNA fragmentation. *Curr. Biol.* **9**: 543-546.

Stein, J.C. and Hansen, G., 1999, Mannose induces an endonuclease responsible for DNA laddering in plants cells. *Plant Physiol.* **121**: 71-79.

Sun, Y.L., Zhao, Y., Hong, X., and Zhai, Z.H., 1999, Cytochrome c release and caspase activation during menadione-induced apoptosis in plants. *FEBS Lett.* **462**: 317-321.

Villa, P., Kaufmann, S.H., and Earnshaw, W.C., 1997, Caspases and caspase inhibitors. *Trends Biochem. Sci.* **22**: 388-393.

Zhao, Y., Jiang, Z.F., Sun, Y.L., and Zhai, Z.H., 1999, Apoptosis of mouse liver nuclei induced in the cytosol of carrot cells. *FEBS Lett.* **448**: 197-200.

Zhou, Q., Krebs, J.F., Snipas, S.J., Price, A., Alnemri, E.S., Tomaselli, K.J., and Slavesen, G.S., 1998, Interaction of the baculovirus anti-apoptotic protein p35 with caspases. Specificity, kinetics and characterization of the caspase/p35 complex. *Biochemistry* **37**: 10757-10765.

The Role of HSF in Heat Shock Signal Transduction and Heat Shock Response in Plants

RALF PRÄNDL, CHRISTIAN LOHMANN, STEFANIE DÖHR, AND FRITZ SCHÖFFL
Zentrum für Molekekularbiologie der Pflanzen (ZMBP), Allgemeine Genetik, Universität Tübingen, Auf der Morgenstelle 28, D-72076 Tübingen, Germany

1. INTRODUCTION

All organisms are exposed to different stress conditions in their natural environment. Particularly in plants, survival and yield are strongly influenced by environmental conditions. The response to heat stress is a model for acquisition of thermotolerance and it may also apply for the development of common stress tolerance of cells and organisms. The knowledge gained in recent years about stress signal transfer and regulation of gene expression led also to the first successes in changing stress tolerance traits in transgenic plants. This paper deals with the central role of the heat shock transcription factors in the signal transfer, the regulation of the heat shock response and the development of thermotolerance.

1.1 Acquisition of Stress Tolerance and Stress Injuries

The sessile life style of plants requires a great deal of "flexibility" in order to secure survival under temporarily unfavourable environmental conditions. This flexibilitity is genetically determined by genes coding for the potential to generate temporarily enhanced levels of stress tolerance, a basic feature of all organisms. Environmental stresses are heat, cold, drought, salt, heavy metal, *etc.* Each stress induces an enhanced tolerance against the inducing stress, some also induce a broader effect, common stress tolerance. The response to heat stress and the acquisition of thermotolerance is a well studied phenomenon, cross protection against heavy metal stress suggests that the heat shock response is also responsible for common stress tolerance. Therefore, the heat shock response has been a model for the different aspects of stress and tolerance including sensing, signaling, gene expression and protective mechanism (for an overview see Schöffl *et al* 1998).

Acquisition of thermotolerance is induced by a relatively mild, sublethal heat stress. Conditioned cells and plants are protected from lethal effects of a subsequently applied severe heat stress. Primary and major targets for protection from heat injury are proteins and protein function. Heat causes denaturation by unfolding of the secondary structure and leads to exposures of otherwise internal hydrophobic domains on the surface of the polypeptides. This may result in intermolecular hydrophobic interactions, formation of aggregates and precipitation. As a consequence, cells become deplenished of essential proteins and enzymes and proteotoxic and oxidative damages increase. Hence, protection of proteins from denaturation is a key process in the mechanism leading to acquired thermotolerance. The induction of thermotolerance is paralleled by a general reprogramming of gene expression: shut off of normal protein synthesis and switch on of *de novo* synthesis of heat shock proteins (HSP). These changes in gene expression are transient in most organisms, reverting to normal, even during extended periods of heat stress. Thus HSP expression shows the features of a feedback regulation-excess HSP repress HSP synthesis. The model of feedback control suggests that the final products of the heat shock response counteract their own synthesis. In this model, sensors for measuring both, the dosage of heat stress and the level of available HSP must be present in the cell. Furthermore, these data have to be integrated and linked to a response system that is able to induce the expression of HSP and to adjust their levels exactly to the demand for protective function (for an overview see Schöffl *et al* 1998).

1.2 Heat Shock Proteins and Chaperone Function

The heat shock response is, in its principles of regulation and protection, evolutionil conserved from bacteria to man. The strong conservation of HSP suggests that they have very important, possibly essential functions in the cell. All known HSPs (Tab. 1) tested to date act as molecular chaperones *in vitro* (for overview see Beissinger and Buchner 1998). It is believed that HSP protect cells from detrimental effects of protein denaturation and aggregation by exerting this property. Molecular chaperones have the capacity to interact temporarily with other proteins within the cell in order to assist proper folding and to promote interaction with other proteins and formation of functional protein complexes. It is not surprising that the importance of HSP is paralleled by genetic redundancy. Each class of HSP contains a number of isoforms, some of them are expressed under non-stress conditions. Certain isoforms of HSP70 are involved in inter-organellar transport (Voisinen *et al* 1999), HSP90 in regulating the activity of other important proteins (transcription factors, hormone receptors, protein kinases) in the cell (for an overview see Mayer and Bukau 1999).

Table 1. Ular and functional properties of hsp/hsc* proteins

Family	Molecular and functional properties
HSP20	small HSP, belonging to several families, form larger complexes, ATP-independent binding of denatured proteins, isoforms present in different cellular compartments
HSP/HSC60	GroEL-homologue; multimeric barrel structure; association with HSP10 (GroES); ATP-dependend folding of proteins
HSP/HSC70	ATP-dependend folding of proteins (co- and post-translationally); participation in organellar trans-membrane protein transport and in chaperone complexes with diverse cellular regulators including HSF; isoforms present in different cellular compartments
HSP/HSC90	ATP-dependend reversible binding of protein kinases, steroid hormone receptors, transcription factors, HSF and others regulatory proteins; interaction is essential for the functional activation of associated proteins
HSP/HSC100	ATP-dependend dissociation of protein aggregates

*) HSC are constitutively expressed isoforms, so-called heat shock cognate genes/proteins

2. STRESS SENSING, SIGNALING AND FEEDBACK REGULATION

The current model for the regulation of the heat shock response suggests that the transcription factor HSF plays an important role. HSF is constitutively expressed and maintained in an inactive conformation, probably by an interaction and complex formation with chaperones. Upon heat stress denaturation of cellular proteins causes a higher demand for chaperone activity. Dissociation of the HSF-chaperone complex takes place, free HSF becomes trimerized, and binds to heat shock promoter elements (HSE). Binding of HSF is prerequisite to the transcriptional stimulation and subsequently to the expression of HSP (Fig. 1). Massive synthesis of HSP meets the request for chaperones and enables cells to cope with the detrimental effects of protein denaturation. Excess of free HSP leads eventually to an interaction with HSF, induces its inactivation and shuts off the heat shock response. Components of HSF recycling complex are possibly HSP70 (Shi *et al* 1998), HSP90 (*Zou et al* 1998) and probably also other proteins. This model of chaperone titration and complex formation can explain both, the transient nature of the heat shock response and the properties of stress sensing and signaling. HSF and HSP, both types of molecules fulfil multiple functions in the heat shock response. They are components of the cellular "thermometer", the HSF-chaperone complex. Perception, transfer and conversion of the stress signal requires probably only one component of the thermometer, the HSF. This pathway must be very short, HSP mRNA is detected within minutes after heat stress (Schöffl and Key 1982). The dual function of HSP is the "execution" of the heat shock response (generation of thermotolerance) and the co-suppression (regulation) of HSF activity.

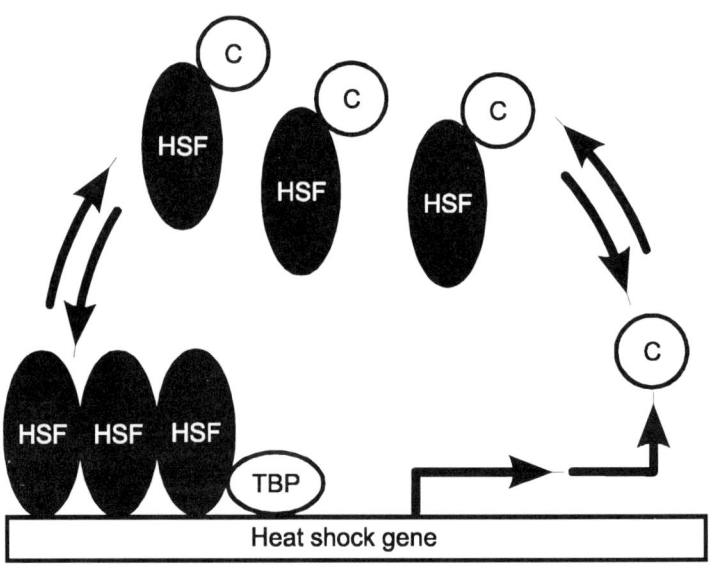

Figure 1. Model of HSF regulation. In the absence of heat stress, HSF is kept in a repressed status by interaction with the chaperones HSP70 and HSP90 (C). Upon heat shock, the chaperones associate preferentially with denatured proteins (not shown). HSF released from chaperone interaction, trimerizes, and gains DNA binding and results in transcriptional competence. A prerequisite for transcription of heat shock genes is the interaction of HSF with the TATA box binding protein (TBP). The increased biosynthesis of chaperones produces a surplus of HSP70 and HSP90 that inactivates HSF.

This model integrates many but not yet all available data on HSF regulation. The regulatory circuit is complicated by the fact, that several different HSF are co-expressed in plant tissues. In *Arabidopsis*, possibly more than 15 different HSF genes are present in the genome and a great number of them was found to be expressed (Schöffl *et al* 1997, Schöffl and Prändl 1999, Lohmann *et al* 2000). Based on sequence similarities within the highly conserved DNA-binding and hydrophobic repeat regions (multimerization), HSF are grouped in two distinct classes A and B (Fig. 2). HSF of other plant species fall also devided into these classes (Nover *et al* 1996). The evolutionary conservation of differences in structure are probably of functional importance but functional implications of these differences are as yet unknown. Members of both groups of HSF are able to recognize the same target sequence (HSE), but there may be differences in binding affinities *in vivo*. However, in transgenic assays only members of class A have been found to perform as transcription factors, e.g. prompting expression of native HSP or heat shock promoter-driven reporter genes (Hübel *et al* 1995, Lee *et al* 1995, Prändl *et al* 1998). Class B HSF seem to be inert with respect to transcriptional activation of known heat shock target genes. Their capacity to bind HSE sequences has lead to the speculation that HSF of class B have a negative (repressor) effect on HSP expression (Czarnecka-Verner *et al* 1997). HSF-like repressors may be involved in turning off the heat shock response or in maintaining the "off" state at higher temperature. Interestingly, by contrast to the constitutive expression of most *Arabidopsis* HSF, expression of AtHSF7 (class B) is strongly induced by heat stress (Lohmann *et al* 2000). The heat inducibility of this HSF suggests that its function is required later in or after the heat shock response.

3. HSF "DOWNSTREAM EFFECTS"

In higher eukaryotes the transcription of HSF target genes (HSP) is dependent on HSF bound to HSE in the promoter region. The trans-activation domains of HSF, required for stimulation of transcription, are usually located near the C-terminus of the protein. The primary amino acid sequences are not well conserved in this region, a so-called AHA motif, containing aromatic, bulky hydrophobic and acidic amino acids, seems to be important for transcriptional activation by tomato HSF (for an overview see Nover *et al* 1996). It is generally accepted that transcriptional activators promote transcription *via* interaction with components of the basal transcription machinery located at the TATA box. In *Arabidopsis*, for the first time in a plant species, it was shown that the TATA box binding protein AtTBP is a target protein for an interaction with the transcription activator AtHSF1 (Reindl and Schöffl 1998). Interaction between these proteins is consistent with their putative role in transcription. However, the interaction of a truncated HSF protein (lacking the activation domain at the C-terminus) with TBP (Reindl and Schöffl 1998) would be compatible with a model that suggests recruitment of HSF to HSE is governed by AtTBP. Constitutive attachment of TBP and assembly of a basal complex is consistent with a pausing of RNA polymerase at the *hsp70* promoter in the absence of heat stress in *Drosophila* (for an overview see Lis 1998). This view is further supported TBP-dependent access of HSF to reconstituted chromatin *in vitro* (Taylor *et al* 1991).

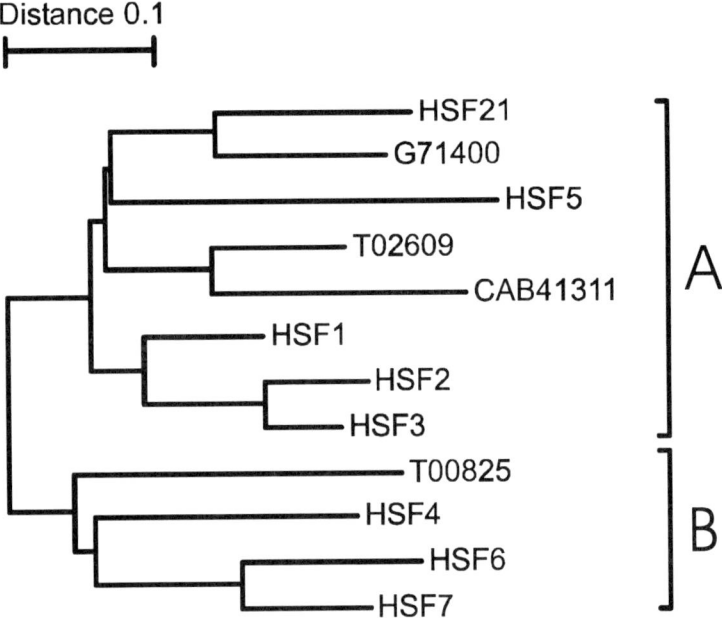

Figure 2. Phylogenetic relationship of *Arabidopsis* HSFs revealed by comparison of the amino acid sequences of the DNA binding domains. *Arabidopsis* HSFs group into the two major classes A and B. Phylogenetic trees were constructed using the Treecon software (version 1.3b 1997). Distances were calculated using the Poisson correction, and the tree topology was inferred using the neighbor-joining method. HSFs are designated according to the published numbering (Hübel and Schöffl 1994, Nover *et al* 1996, Prändl *et al* 1998; EMBL/GenBank/DDBJ submissions) or by the respective accession numbers (ORFs revealed in the *Arabidopsis* genome project). Criterion for the classification as HSF was the presence of domains with homology to previously published DNA binding domains and hydrophobic regions.

4. HSF FUNCTION IN GROWTH AND DEVELOPMENT

The activity of HSF may not only depend on interaction with other proteins, covalent modifications may also play an important role in modulation of HSF activity (for an overview see Wu 1995). Using plant cell extracts of *Arabidopsis*, serine phosphorylation of recombinant AtHSF1 by the cyclin-dependent kinase CDC2a correlates with a decreased capacity of this HSF in HSE binding (Reindl et al 1997). A cell cycle-dependent interference with HSF activity suggests that HSF function is not only required for stress sensing and signaling, HSF function may be also involved in processes of cell growth and development. In human cells, serine phosphorylation of HSF is stimulated by kinases of the mitogen activated Raf/ERK pathway (Knauf et al 1996, Chu et al 1996). It is conceivable that HSF activity is modulated during certain stages of the cell cycle under non-stress conditions but it is unknown, whether its function is mediated through expression of HSP or other proteins. Experimental data indicate an involvement of HSF in both.

In plants, the requirement of HSE sequences for developmental expression of heat shock genes (Prändl and Schöffl 1996) and the expression and presence of active HSF in seeds (Lohmann et al 2000) indicate an involvement of HSF in HSP expression during seed maturation in *Arabidopsis*. In addition, seed specific expression of most HSPs is regulated by ABI3, a central regulator of seed maturation (Wehmeyer et al 1996).

In *Drosophila*, mutations in the unique HSF gene have been isolated (Jedlicka et al 1998). The analysis showed, not surprisingly, that lack of HSF activity results in loss of the capacity to express HSP and to develop thermotolerance in adult flies. However, HSF activity seemed to be essential during early stages of larval development but this requirement of HSF was not related to HSP synthesis. HSF mutants have been also isolated in *Arabidopsis*, but owing to the genetic redundancy of HSF in plants, phenotypic effects are less obvious (Lohmann et al 2000). In summary, HSF seem to have also important functions in development, probably by regulating the activities of other "non-heat shock genes" and proteins.

5. MOLECULAR AND APPLIED ASPECTS OF HSF MANIPULATION

Owing to its central role in stress signaling and HSP expression, HSF is an ideal target for genetic manipulation of the heat shock response and for an improvement of stress tolerance in plants. In *Arabidopsis*, transgenic overexpression of AtHSF3 resulted in a co-ordinate expression of all HSP at normal temperature (Prändl et al 1998). This de-repression of the heat shock response is attributed to a stress-independent activation of HSF3. The exact molecular mechanism of de-repression is unknown, however, excess of HSF3 may, in accordance to the chaperone titration model neutralize the negative regulator (see Fig. 1) and surplus of HSF3 is in an active (trimeric) conformation. Similar effects on the heat shock response have been observed in transgenic *Arabidopsis* plants expressing AtHSF1-GUS (glucuronidase) fusion protein (Lee *et al* 1995). The phenomenon of HSF-GUS de-repression has been further investigated in tobacco. Different HSF-GUSΔ deletions and HSF-NPTII (neomycin phosphotransferase) constructs were tested in transgenic plants. Their analysis led to the conclusion, that alternatively to chaperone titration, HSF may become de-repressed *via* structural properties of fused proteins (Döhr *et al* 2000). Both, GUS and NPTII enzymes are active as homo-multimeric (di- or tetrameric) complexes and it seems possible that HSF monomers are converted to active complexes by "external" multimerization through GUS or NPTII proteins.

Common features of HSF-transgenic lines (overexpressing HSF in its active conformation) are: the synthesis of HSPs at normal, non-heat shock temperatures and a higher levels of basal thermotolerance. The potential for the acquisition of enhanced thermotolerance seems to be unaffected. In summary, results of transgenic expression and manipulation of HSF have important implications on plant breeding and biotechnology:

- Heat shock transcription factors can be functionally expressed in heterologous plant systems
- The functional activities can be retetained in both parts of fusion proteins (HSF and GUS/NPTII); GUS can be used as a reporter, NPTII as a selection marker for transgenic expression
- HSF, although genetically redundant, can be used for genetic engineering of stress tolerance
- Overexpression and/or genetic manipulations of HSF are sufficient to switch on HSP synthesis and consequently increase basal thermotolerance in plants.

ACKNOWLEDGEMENTS

The research work in our laboratory has been supported by grants of the Deutsche Forschungsgemeinschaft and the EC to F. S.

REFERENCES

Beissinger, M., and Buchner, J., 1998, How chaperones fold proteins. *Biol. Chem.* **379**: 245-259.

Czarnecka-Verner, E., Yuan, C. X., Nover, L., Scharf, K.-D., English, G. and Gurley, W. B.,1997, Plant heat shock transcription factors: positive and negative aspects of regulation. *Acta Physiol. Plant.* **19**: 529-537.

Chu, B., Soncin, F., Price, B.D., Stevenson, M.A. and Calderwood, S.T., 1996, Sequential phosphorylation by mitogen-activated protein kinase and glycogen synthase kinase 3 represses transcriptional activation by heat shock factor 1. *J. Biol. Chem.* **271**: 30847-30857.

Döhr, S., Wunderlich, M. and Schöffl, F., 2000, Derepression of the heat shock response in transgenic tobacco expressing *Arabidopsis* HSF1 fusion proteins (submitted).

Hübel, A. and Schöffl, F., 1994, *Arabidopsis* heat shock factor: characterization of the gene and the recombinant protein. *Plant Mol. Biol.* **26**: 353-362.

Hübel, A., Lee, J. H., Wu, C. and Schöffl, F., 1995, *Arabidopsis* heat shock factor is constitutively active in *Drosophila* and human cells. *Mol. Gen. Genet.* **24**: 136-141.

Jedlicka, P. Mortin, M. A. and Wu, C., 1998, Multiple functions of *Drosophila* heat shock factor *in vivo*. *EMBO J.* **16**: 2452-2462.

Knauf, U., Newton, E. M., Kyriakis, J. and Kingston, R. E., 1996, Repression of human heat shock factor 1 activity at control temperature by phosphorylation. *Genes Dev.* **10**:2782-2793.

Lee, J. H., Hübel, A. and Schöffl, F., 1995, Derepression of the activity of genetically engineerd heat shock factor causes constitutive synthesis of heat shock proteins and increased thermal tolerance in transgenic *Arabidopsis*. *Plant J.* **8**: 603-612.

Lis, J., 1999, Promoter-associated pausing in promoter architecture and postinitiation transcriptional regulation. In: *Coldspring harbor Symposia on Quantitative Biology,* Vo. **LXIII**, pp. 347-356, Cold Spring Harbor Laboratory Press.

Lohmann, C., Prändl, R., Eggers-Schumacher, G., Ward, J. and Schöffl, F., 2000, Determination of the activity profiles of heats hock factors (HSF) in *Arabidopsis* and identification of a HSF3 knock-out mutant (submitted).

Mason, P.B. Jr., and Lis J.T., 1997, Cooperative and competitive protein interactions at the HSP70 promoter. *J. Biol. Chem.* **272**: 33227-33233.

Mayer, M.P. and Bukau, B., 1999, Molecular chaperones: the busy life of HSP90. *Current Biology* **9**: R322-R325.

Nover, L., Scharf, K.D., Gagliardi, D., Vergne, P., Czarnecka-Verner, E., and Gurley, W.B., 1996, The HSF world: classification and properties of plant heat stress transcription factors. *Cell Stress Chaper.* **1**: 215-223.

Owens-Grillo, J.K., Stancato, L.F., Hoffmann, K., Pratt, W.B., and Krishna, P., 1996, Binding of immuophilins to the 90 kDa heat shock protein (hsp90) via a tetratricopeptide repeat domain is a conserved protein interaction in plants. *Biochemistry* **35**: 15249-15255.

Prändl, R. and Schöffl, F., 1996, Heat shock elements are involved in heat shock promoter activation during tobacco seed maturation. *Plant Mol. Biol.* **31**: 157-162.

Prändl, R., Hinderhofer, K., Eggers-Schumacher, G. and Schöffl, F., 1998, HSF3, a new heat shock factor from *Arabidopsis thaliana*, derepresses the heat shock response and confers thermotolerance when overexpressed in transgenic plants. *Mol. Gen. Genet.* **258**: 269-278.

Reindl, A., Schöffl, F., Schell, J., Koncz, C. and Bako, L., 1997, Phosporylation by a cycline-dependent kinase modulates DNA-binding of the *Arabidopsis* heat shock transcription factor HSF1 *in vitro*. *Plant Physiol.* **115**: 93-100.

Reindl, A. and Schöffl, F, 1998, Interaction between the *Arabidopsis thaliana* heat shock transcription factor HSF1 and the TATA-binding protein TBP. *FEBS Lett.* **436**: 318-322.

Stepanova, L., Leng, X., Parker, S.B., and Harper, J.W., 1996, Mammalian p50^{Cdc37} is a protein kinase-targeting subunit of HSP90 that binds and stabilizes Cdk4. *Genes Dev.* **10**: 1491-1502.

Shi, Y., Mosser, D.D., Morimoto, R.I., 1998, Molecular Chaperones as HSF1-specific transcriptional repressors. *Genes Dev.* **12**: 654-666.

Schöffl, F. and Key, J.L., 1982, An analysis of mRNAs for a group of heat shock proteins of soybean using cloned cDNAs. *J. Mol. Appl. Gen.* **1**: 301-314.

Schöffl, F., Prändl, R. and Reindl, A., 1998, Regulation of the heat shock response. *Plant Physiol.* **117**: 1135-1141.

Schöffl, F., Prändl, R. and Reindl, A., 1998, Molecular responses to heat stress. In *Molecular Responses to Cold, Drought, Heat and Salt Stress in Higher Plants*. (eds. K Shinozaki, K. Yamaguchi-Shinozaki), pp. 81-98, Austin Tx: R.G. Landes Company.

Schöffl, F. and Prändl, R., 1999, Derepression of the heat shock protein synthesis in transgenic plants. In *Plant Responses to Environmental Stress*. (eds. M. Smallwood, C. Calvert, D. Bowles), pp. 65-73. Oxford (UK): BIOS Scientific Publishers Limited.

Taylor, I.A.C., Workman, J.L., Schuetz, T. J. and Kingston, R.E., 1991, Facilitated binding of GAL4 and heat shock factor to nucleosomal templates: differential function of DNA-binding proteins. *Genes Dev.* **5**:1285-1298.

Voisine, C., Craig, E.A., Zufall, N, von Ahsen, O., Pfanner, N., and Voos W., 1999, The protein import motor of mitochondria: Unfolding and trapping of preproteins are distinct and separable functions of matrix Hsp70. *Cell* **97**: 565-574.

Wehmeyer, N., Hernandez, L. D., Finkelstein, R. R. and Vierling, E., 1996, Synthesis of small heat shock proteins is part of the developmental program of late seed maturation. *Plant Physiol.* **112**: 747-757.

Wu, C., 1995, Heat shock transcription factors: structure and regulation. *Annu. Rev. Cell Dev. Biol.* **11**: 441-469.

Zou, J., Guo, Y., Guettouche, T., Smith, D.F., and Voellmy, R., 1998, Repression of heat shock transcription factor HSF1 activation by HSP90 (HSP90 complex) that forms a stress-sensitive complex with HSF1. *Cell* **94**: 471-480.

Towards Understanding the Recognition and Signal Transduction Processes in the Soybean-Phytophthora Sojae Interaction

MADAN K. BHATTACHARYYA[1], BONNIE G. ESPINOSA[1], TAKAO KASUGA[1], YONGQING LIU[1], SHANMUKHASWAMI S. SALIMATH[1], MARK GIJZEN[2], VAINO POISA[3] AND RICHARD BUZZELL[3]
[1]Noble Foundation, USA; [2]Agriculture & Agri-Food Canada, London, Canada; [3]Agriculture & Agri-Food Canada, Harrow, Canada.

1. INTRODUCTION

Soybean [*Glycine max* L. (Merrill)] is a major oil seed crop and is grown throughout much of the world. North America is a major producer of soybean. The United States alone produces over half of the world output. Soybean seed contains 40% protein and 20% oil and is used, in addition to human consumption, for livestock feed and industrial purposes. In North America, soybean suffers yield loss from a fungal disease root and stem rot caused by *Phytophthora sojae*. It was estimated that in 1994 the yield loss of soybean in the United States from this disease was worth 0.12 billion dollars (Wrather *et al* 1994) Monogenic resistance, encoded by *Rps* genes, introduced into the commercially grown cultivars through a series of backcrosses, has been providing a reasonable protection to the soybean crop against this pathogen over the last three decades. It is generally believed that this fungus originated in the United States, although soybean was only recently introduced into the United States from China (Förster *et al* 1994). There are several physiological races of this fungal pathogen. The number of races is increasing rapidly. For example, in 1994 there were 37 recorded races of the fungus (Förster *et al* 1994). Now this number has been increased to 45 (Abney *et al* 1997, Schmitthenner *et al* 1994*)* concluded that *P. sojae* is a highly variable pathogen and exists in soil in a wide variety of virulence phenotypes to which most *Rps* genes are ineffective. They also concluded that unless new *Rps* genes are identified or existing *Rps* genes are pyramided in single cultivars, resistance available in the present day cultivars might not be effective in controlling the disease in future.

The major source of genetic variation in this pathogen is the outcome of rare outcrosses and spontaneous mutation at the avirulence genes (Förster *et al* 1994). At present, there are 14 *Rps* genes that confer race-specific resistance of soybean to different physiological races of *P. sojae*. These genes were obtained from different *Glycine max* lines, and mapped in seven different loci (Anderson and Buzzell 1992, Polzin *et al* 1994, Schmitthenner 1989). The *Rps1* and *Rps7* loci are linked at a 12.5 cM distance, while *Rps2, Rps3, Rps4, Rps5* and *Rps6* are located in individual linkage groups (Anderson and Buzzell 1992, Diers *et al* 1992). The genetics of resistance conferred by *Rps* genes is well established. Recently, genetics of most of the avirulence genes (*Avr*) that correspond to specific *Rps* genes have also been reported (Gijzen *et al* 1996, Tyler *et al* 1995, Whisson *et al* 1994). The interaction between these 14 genes with the corresponding genes for avirulence in *P. sojae* follows the 'gene-for-gene' hypothesis of Flor (Flor 1995).

The soybean-*P. sojae* interaction is probably the most extensively studied host-pathogen interaction (Graham 1995, Keen and Yoshikawa 1990, Paxton 1995, Ward 1990). Biochemical studies on this classical model system contributed towards understanding the mechanisms of phytoalexin accumulation (Bhattacharyya and Ward 1987, Moesta and Grisebach 1981, Yoshikawa *et al* 1979), role of phytoalexins in disease resistance (Yoshikawa *et al* 1978), and the concept of elicitors in defense responses (Keen 1975). Elicitors derived from *P. sojae* have been used in studying receptors from soybean (Cheong and Hahn 1991, Ebel *et al* 1993, Umemoto 1997) and also from non-host plants (Nürnberger *et al* 1994). Recently, a soybean cDNA encoding D-glucan-elicitor-binding protein has been isolated (Umemoto *et al* 1997). The phytoalexin glyceollin, the major active defense compound with antimicrobial activity, accumulates to a significant level 8 h following inoculation in the incompatible interaction but only 12 h following inoculation in the compatible interaction (Bhattacharyya and Ward 1986). The rapid induction of glyceollin results from the transcriptional activation of genes in the phenylpropanoid pathway in an incompatible interaction. Some of the key enzymes and genes of this pathway have been shown to be activated 3 h following inoculation in the resistant but not in the susceptible response (Bhattacharyya and Ward 1988, Ebel and Grisebach 1988, Esnault *et al* 1987). Inoculation of etiolated hypocotyls or roots with zoospore suspensions of *P. sojae* revealed that the extent of colonization by the fungus was comparable immediately following infection in both near-isogenic lines that differ in a single *Rps* gene. However, host cells associated with the penetrated hyphae were alive and healthy in the compatible interaction, while they were all dead by 3-4 h following inoculation in the incompatible interaction (Enkerli *et al* 1997, Ward *et al* 1989). The major differences between the two interactions appeared to relate to timing of the host responses. Both physiological and genetic studies firmly established that a recognition process does occur between a specific soybean cultivar and a specific *P. sojae* race, and this determines the outcome of a soybean-*P. sojae* interaction. In this recognition process, most likely a direct association or interaction between Rps and Avr proteins takes place.

Several plant disease resistance genes that follow the classical gene-for-gene hypothesis (Flor 1955) have now been cloned. Cloned resistance genes that follow Flor's hypothesis can be classified into four groups based on their protein structures: i) proteins with serine/threonine kinase activity, e.g., *Pto* (Martin *et al* 1993); ii) proteins with nucleotide binding sites (NBS) and leucine rich repeat regions, e.g. *RPS2, N, L6, RPM1, Prf, M, I2* and *RPP5* (Anderson *et al* 1977, Bent *et al* 1994, Grant *et al* 1995, Lawrence *et al* 1995, Mindrinos *et al* 1994, Oro *et al* 1997, Parker *et al* 1997, Salmeron *et al* 1996, Whitham *et al* 1994); iii) proteins with leucine rich repeat regions and transmembrane domain, e.g. *Cf2, Cf4, Cf5, Cf9,* and *Hs1^{pro-1}* (Cai *et al* 1977, Dixon *et al* 1998, Jones *et al* 1994, Thomas *et al* 1997) and iv) proteins with leucine rich repeat regions, transmembrane and serine/threonine kinase domains, e.g. *Xa21* (Song *et al* 1995). A tomato gene, *Pti1*, encoding a serine/threonine kinase was cloned (Zhou *et al* 1995). Pti1 is phosphorylated by Pto, and is involved in the hypersensitive response caused by infection of the bacterial pathogen *Pseudomonas solanocearum* pv. *tomato*. *In vivo* interaction between resistance gene and avirulence gene products has also been shown (Leister *et al* 1996, Scofield *et al* 1996, Tang *et al* 1996). Sequencing and genetic analyses of several complex disease resistance loci have shed light on evolution of disease resistance loci (Anderson *et al* 1977, Caicedo *et al* 1999, Dixon *et al* 1998, Ellis *et al* 1997, Grant *et al* 1998, Meyers *et al* 1998, Parniske *et al* 1997, Song *et al* 1997).

We are interested in investigating two important aspects of the soybean-*P. sojae* interaction. They are 1) recognition and 2) signal transduction. The recognition process, determined by Rps from soybean and Avr from *P. sojae*, presumably activates a signal pathway involved in the expression of defense genes and resistance. We have applied a map-based cloning strategy to isolate the soybean disease resistance gene *Rps1*-k that determines this recognition process. This gene may allow us to isolate the downstream Rps1-k-interacting factor(s) that regulates the signal transduction process involved in the expression of defense responses. Thus, it may also assist us in investigating how the downstream defense responses are regulated by the recognition event. Towards understanding the signal transduction pathway a chemical mutagenesis approach has been also applied in isolating candidate signal transduction mutants.

2. RESULTS AND DISCUSSION

We have isolated molecular markers and mapped these markers using a very large segregating population. From the high density and high resolution mapping of the *Rps1* region we have observed that *Rps1*-k mapped to one end of the introgressed region that was transferred from the cultivar Kingawa to Williams 82 through several generations of backcrossing. We observed that *Rps1*-k is flanked by two AFLP markers TC1 and CG1 in a 0.13 cM distance (Kasuga *et al* 1997). Physical mapping data indicate that these two markers are most likely located in a ~145 kb DNA fragment (Fig. 1). We have constructed two BAC libraries and isolated BAC clones for the *Rps1* region (Salimath and Bhattacharyya 1999, unpublished data). A BAC library constructed in Dr. R. C. Shoemaker's laboratory was also used in isolating BACs for this region. We observed that BAC libraries are highly under-represented in the *Rps1*-k region. Molecular markers mapped at 0.5-0.64 cM distances from *Rps1* are well represented in the BAC library (Fig. 2). Most likely the *Rps1* locus is composed of repetitive sequences that are unstable in *Escherichia coli*.

Sequences of a BAC end, close to *Rps1*-k, showed homology to leucine-rich repeat regions (LRR) of disease resistance genes. We applied this LRR sequence to identify a BAC clone (from a different BAC library constructed in GenomeSystems, Inc., St. Louis) that mapped to the gap of the contigs in the *Rps1* region. This BAC is referred to as "Island BAC" (Fig. 3). This BAC carries two copies of the LRR like sequences. One copy has been sub-cloned and sequenced. Sequence data indicate that this particular copy has high identity to NBS-LRR type disease resistance genes such as *I2C-1* gene, a homologue of which confers resistance of tomato to *Fusarium oxysporum* f. sp. *lycopersici*. This LRR copy that we have sequenced could be either the *Rps1*-k gene or a homologue of *Rps1*-k. Current experiments include filling gaps of the contigs in the *Rps1* region (Fig. 3) and identification of *Rps1*-k. The candidate *Rps1*-k gene showing similarity to NBS-LRR-type disease resistance genes will be used in the complementation analysis. To rule out the possibility that another open reading frame could be *Rps1*-k, sequencing of the region between TC1 and CG1 will be carried out in parallel to complementation analysis of the candidate *Rps1*-k gene.

Toward understanding the signal transduction process involved in the expression of disease resistance we have initiated a chemical (ethyl methane sulphonate) mutagenesis experiment. Mutagenised seeds of HARO1572 carrying two *Rps* genes, *Rps1*-k and *Rps7*, were inoculated with two *P. sojae* races, race 1 and 35. HARO1572 is resistant to race 1 that carries *Avr1*-k but not *Avr7*. This line is also resistant to race 35 carrying *Avr7* but not *Avr1*-k. We hypothesized that there could be two alternative possible signal pathways involved in the expression of *Rps*-specific resistance. Common signal pathway could be involved for both *Rps1*-k and *Rps7* encoded race-specific resistance.

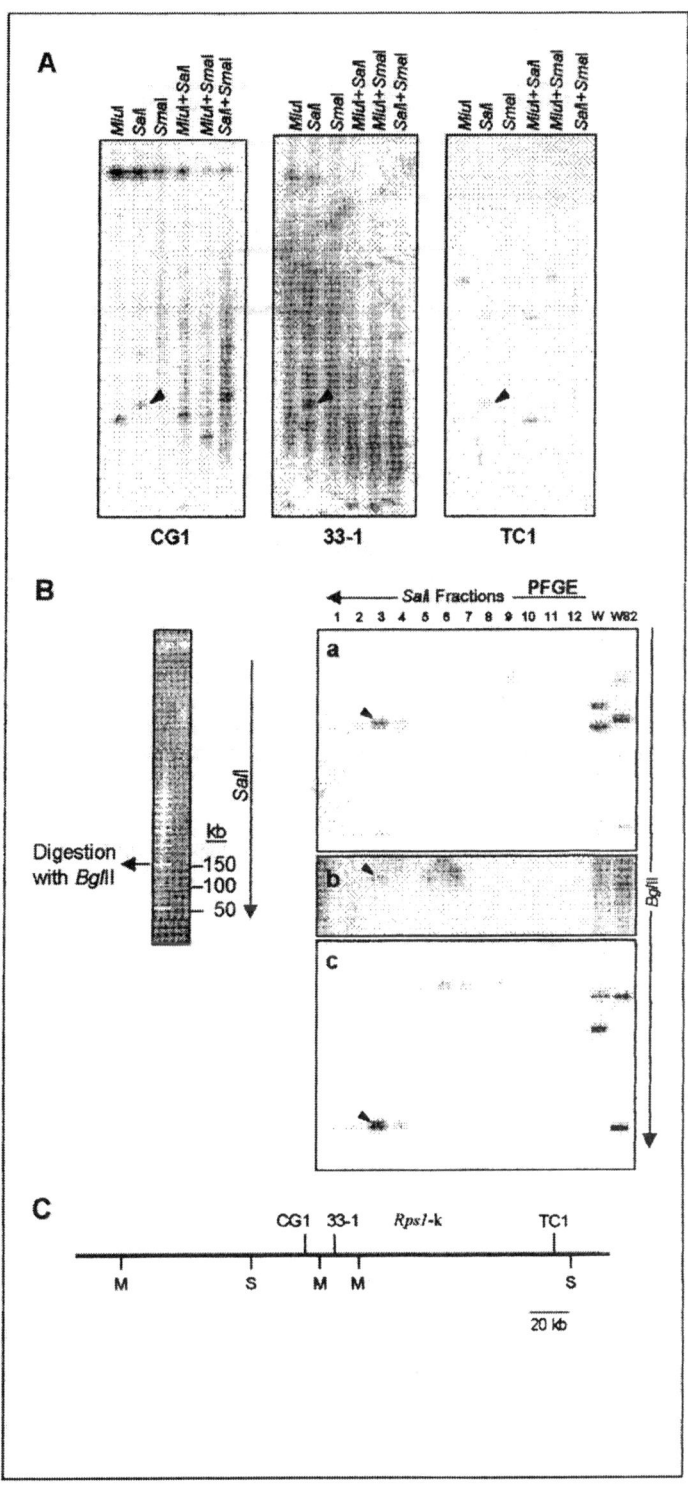

Figure 1. Physical mapping of the *Rps1*-k region.

(A) A contour-clamped homogeneous electrophoresis (CHEF) of soybean DNA.
High molecular weight (HMW) DNA from the cv. Williams 82, digested with different enzyme combinations as shown in Figure A (top of the panel), was separated by CHEF in a CHEF mapper apparatus (BioRad, CA), and then transferred to nylon membrane. The same membrane was hybridized independently to three different probes TC1, 33-1, and CG1. 33-1 is a dominant RFLP marker representing one end of a BAC clone that carries CG1. Southern blot data indicated that an ~145 kb *Sal*I-fragment (shown by arrowheads) hybridized to TC1, 33-1 and CG1.

(B) Two-dimensional electrophoresis of soybean DNA.
Since both markers TC1 and CG1 hybridized to duplicated sequences, a two-dimensional electrophoresis experiment was carried out to investigate if the same 145 kb *Sal*I-fragment carries TC1, 33-1 and CG1. Gel slices were cut from the bottom of a CHEF gel, in which *Sal*I digested HMW DNA is separated (as shown in the left side of B), and 12 gel fraction were obtained. Gel fractions 1- 12, representing approximately 100 kb to 550 kb, respectively, were digested with *Bgl*II and Southern analysis was carried out for three independent probes TC1, 33-1 and CG1, and results are presented in B: a, TC1; b, 33-1; c, CG1. This analysis confirmed that PFGE gel fraction No. 3 carrying the 145 kb *Sal*I fragment (arrow) hybridized to these three molecular markers.

(C) Physical map of the *Rps1*-k region.
The physical map was constructed based on the gel electrophoresis of genomic DNA followed by Southern analysis (Fig. 1A and 1B). The BAC33 end 33-2 overlaps with BAC120, which carries a *Sal*I site. The BAC120 end 120-1 is a single copy sequence and did not hybridize to the same 145 *Sal*I fragment. Therefore, 33-1 is located in between CG1 and TC1.

Alternatively, two independent pathways may be involved in the expression of *Rps1*-k and *Rps7*-specific resistance. Mutation in either *Rps1*-k or any other genes necessary for the expression of resistance will cause susceptibility against race 1. To distinguish the *rps1*-k mutants from downstream signal transduction or defense-related mutants, the mutants were re-screened with a second race 35. The *rps1*-k mutants will be resistant against this race because of the *Rps7*-specific resistance. If the mutants susceptible to race 1 are also susceptible to race 35, then the mutations are not in *Rps1*-k, but in genes that function downstream from the recognition event governed by either *Rps1*-k or *Rps7*. These mutants can be considered as the candidate signal transduction mutants, essential in the expression of resistance against *P. sojae*.

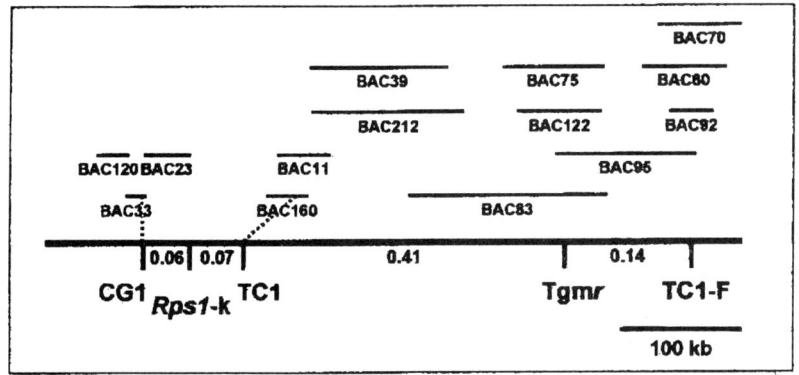

Figure 2. The *Rps1* region is highly underrepresented in BAC library.
BAC libraries carrying 14 genome equivalent and a cosmid library carrying 8 genome equivalent clones were screened for AFLP markers CG1 and TC1. No cosmid clones were obtained for these markers. BAC33, BAC120 and BAC160 were obtained from a BAC library carrying 5 genome equivalent clones of average size ~45 kb (Salimath and Bhattacharyya, unpublished). BAC23, BAC11, BAC39 and BAC212 were isolated from a separate BAC library (Marek and Shomaker 1997) that carries 5 genome equivalent clones of average size ~150 kb. BAC33 and BAC160 carry the CG1- and TC1-specific RFLPs, respectively. The library carrying four genome equivalent BAC clones (Salimath and Bhattacharyya 1999), was screened for Tgm*r* and TC1-F. Sequences away from *Rps1*-k are well represented in these BAC libraries, while those next to *Rps1* are highly underrepresented in BAC libraries. We have estimated from physical mapping data and BAC contigs that the gap in the *Rps1* locus is ~70 kb.

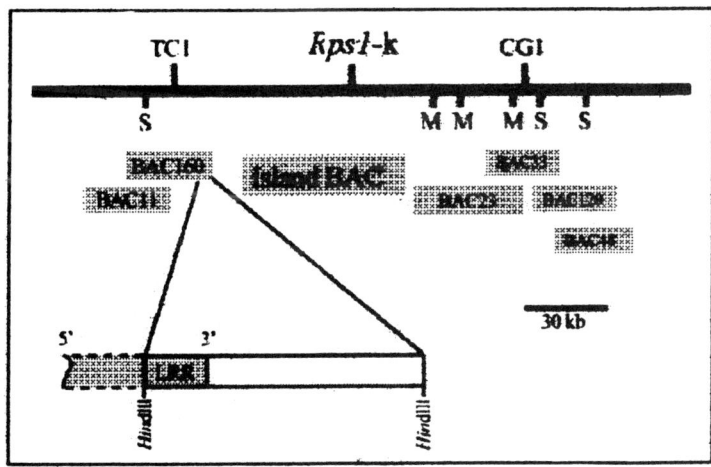

Figure 3. Identification of a BAC that cosegregates with *Rps1*-k and carries disease resistance gene-like sequences.

In an effort to identify simple sequences from the BAC160-end close to *Rps1*-k, we have isolated a *Hind*III DNA restriction fragment that carries a leucine- rich repeat region (LRR) showing high identity to LRRs of cloned resistance genes. RFLP analysis of the recombinants identified an 1.6 kb *Eco*RI fragment that co-segregates with *Rps1*-k. This LRR sequence was used to screen a four genome equivalent BAC library (constructed in GenomeSytems, Inc.) and isolated 20 clones. One clone carries this 1.6 kb *Eco*RI fragment and is refered to as "Island BAC," because neither ends of this clone hybridize to the BAC contigs of the *Rps1* region. The figure is not to the scale for the gap region and "Island BAC." Based on pulse field gel electrophoresis data we have estimated that the size of the *Sal*I fragment carrying both TC1 and CG1 (shown by "S" for *Sal*I; "M" for *Mlu*I in the map) is ~145 kb. Therefore, the size of the gap region including Island BAC could be ~70 kb.

Etiolated seedlings of M_2 generation were first screened with race1. We had screened over 3456 M_2 families obtained from 1533 M_1 plants. Approximately 20 etiolated seedlings from each M_2 family were inoculated. We identified 61 putative mutants (putants). Progeny testing of these putants identified 21 mutants that are susceptible to race 1. Some of them were re-screened with race 35. Three mutants showed susceptibility to both races. Mutant 631 segregates for the mutation (Fig. 4). Etiolated hypocotyls and leaves of mutants 531and 668 showed loss of resistance to both races. These three mutants can be considered as candidate signal transduction mutants. These results indicate a common signal pathway that may be operative in mediating both *Rps1*-k and *Rps7*-specific resistance of soybean against *P. sojae*. We are interested in studying if the resistance of these EMS mutants is compromised against pathogens other than *P. sojae*.

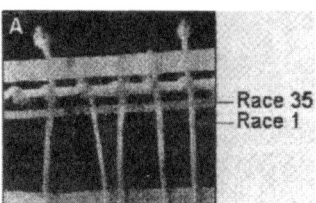

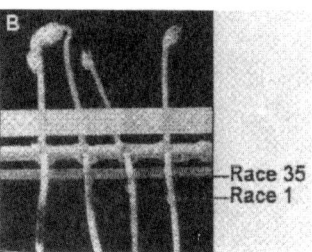

Figure 4. Identification of putative signal transduction mutants.

The EMS mutants were developed from the culitivar HARO1572 that carries *Rps1*-k and *Rps7* conferring resistance against race 1 and 35, respectively. We identified several mutants that have lost resistance against both races 1 and 35. It is very unlikely that mutations occurred simultaneously at *Rps1*-k and *Rps7* in these mutants. Therefore, we conclude that single mutations affecting *Rps1*-k and *Rps7*-specific functions might have occurred in genes that could be members of the pathway involved in the expression of resistance against *P. sojae*.
(A) Progeny of mutant 531-F have shown susceptibility to *P. sojae* race 1 and race 35.
(B) Progeny of mutant 763-1E segregated for host responses. Seedlings 3 and 4 from the left are susceptible to both races; while seedlings 1 and 2 show wild-type resistance responses.

Isolation of the *Rps1*-k and signal transduction pathway genes that are essential for the expression of *Rps*-specific resistance may assist us not only in understanding the recognition and signal transduction processes, but also in increasing the resistance of soybean against *P. sojae* and other pathogens. It is possible that over-expression of the *Rps1*-k gene could enhance the resistance of soybean against not only *P. sojae* but also other soybean pathogens. Over-expression of disease resistance genes *Pto* and *Prf* has been shown to increase resistance of tomato against several pathogens (Oldroyd and Staskawicz 1998, Tang *et al* 1999). Similarly over-expression of signal transduction pathway genes may enhance the resistance of soybean against many pathogens. It has been shown that over-expression of *NPR1*, a key regulator of the salicylic acid-mediated systemic acquired resistance, enhances broad-spectrum disease resistance in *Arabidopsis* (Cao and Dong 1998). Likewise, expression of the tomato signal transduction gene *Pti* in tobacco enhances the hypersensitive response against the *Pseudomonas syringae* pv. *tabaci* strain that carries *avrPto* (Zhou *et al* 1995).

ACKNOWLEDGEMENT

We thank Ann R. Harris and Angela D. Scott for assistance with oligonucleotide synthesis and DNA sequencing, Darla Boydston and Cuc Ly for photographs and illustrations. We thank Randy C. Shoemaker and Laura F. Marek of the Iowa State University, and Catherine Baublite of GenomeSystems, Inc., St. Louis, for providing the access to their BAC libraries. The work was funded by the Samuel Roberts Noble Foundation.

REFERENCES

Abney, T.S., Melgar, J.C., Richards, T.L., Scott, D.H., Grogan, J., and Young J:, 1997, New races of *Phytophthora sojae* with *Rps*1-d virulence. *Plant Dis.* **81**: 653-655.

Anderson, P.A., Lawrence, G.J., Morrish, B.C., Ayliffe, M.A., Finnegan, E.J., and Ellis, J.G., 1977, Inactivation of the flax rust resistance gene *M* associated with loss of a repeated unit within the leucine-rich repeat coding region. *Plant Cell* **9**: 641-651.

Anderson, T.R., and Buzzell, R.I., 1992, Inheritance and linkage of the *Rps*7 gene for resistance to *Phytophthora* rot of soybean. *Plant Dis.* **76**: 958-959.

Bent, A.F., Kunkel, B.N., Dahlbeck, D., Brown, K.L., Schmidt, R., Giraudat, J., Leung, J., and Staskawicz, B.J., 1994, RPS2 of *Arabidopsis thaliana*: A leucine-rich repeat class of plant disease resistance genes. *Science* **265**: 1856-1860.

Bhattacharyya, M.K., and Ward, E.W.B., 1986, Resistance, susceptibility and accumulation of glyceollins I-III in soybean organs inoculated with *Phytophthora megasperma* f.sp. *glycinea*. *Physiol. Mol. Plant Pathol.* **29**: 227-237.

Bhattacharyya, M.K., and Ward, E.W.B., 1987, Biosynthesis and metabolism of glyceollin I in soybean hypocotyls following wounding or inoculation with *Phytophthora megasperma* f.sp. *glycinea*. *Physiol. Mol. Plant Pathol.* **31**: 387-405.

Bhattacharyya, M.K., and Ward, E.W.B., 1988, Phenylalanine ammonia-lyase activity in soybean hypocotyls and leaves following infection with *Phytophthora megasperma* f.sp. *glycinea*. *Can. J. Bot.* **66**: 18-23.

Cai, D., Kleine, M., Kifle, S., Harloff, H.-J., Sandal, N.N., Marcker, K.A., Klein-Lankhorst, R.M., Salentijn, E.M.J., Lange, W., Stiekema, W.J., Wyss, U., Grundler, F.M.W., and Jung, C., 1977, Positional cloning of a gene for nematode resistance in sugar beet. *Science* **275**: 832-834.

Caicedo, A.L., Schaal, B.A., and Kunkel, B.N., 1999, Diversity and molecular evolution of the RPS2 resistance gene in *Arabidopsis thaliana*. *Proc. Natl. Acad. Sci. USA* **96**: 302-306.

Cao, H., Li, X., and Dong, X., 1998, Generation of broad-spectrum disease resistance by overexpression of an essential regulatory gene in systemic acquired resistance. *Proc. Natl. Acad. Sci. USA* **95**: 6531-6536.

Cheong, J.J., and Hahn, M.G., 1991, A specific, high-affinity binding site for the hepta-b-glucoside elicitor exists in soybean membranes. *Plant Cell Rep.* **3**: 137-147.

Diers, B.W., Mansur, L., Imsande, J., and Shoemaker, R.C., 1992, Mapping Phytophthora resistance loci in soybean with restriction fragment length polymorphism markers. *Crop. Sci.* **32**: 377-383.

Dixon, M.S., Hatzixanthis, K., Jones, D.A., Harrison, K., and Jones, J.D., 1998, The tomato Cf-5 disease resistance gene and six homologs show pronounced allelic variation in leucine-rich repeat copy number. *Plant Cell* **10**: 1915-1925.

Ebel, J., Cosio, E.G., Feger, M., Frey, T., Kissel, U., Reinold, S., and Waldmüller, T., 1993, Glucan elicitor-binding proteins and signal transduction in the activation of plant defence. In: Nester EW, Verma DPS (eds) Advances in Molecular Genetics of Plant-Microbe Interactions, pp. 477-484. Kluwer Academic Publishers, Dordrecht, Netherlands.

Ebel, J., and Grisebach, H., 1988, Defense strategies of soybean against the fungus *Phytophthora megasperma* f.sp. *glycinea*: a molecular analysis. *Trends Biochem. Sci.* **13**: 23-27.

Ellis, J., Lawrence, G., Ayliffe, M., Anderson, P., Collins, N., Finnegan, J., Frost, D., Luck, J., and Pryor, T., 1997, Advances in the molecular genetic analysis of the flax-flax rust interaction. *Annu. Rev. Phytopathol.* **35**: 271-291.

Enkerli, K., Hahn, M.G., and Mims, C.W., 1997, Ultrastructure of compatible and incompatible interactions of soybean roots infected with the plant pathogenic oomycete *Phytophthora sojae*. *Can. J. Bot.* **75**: 1493-1508.

Esnault, R., Chibbar, R.N., Lee, D., Van Huystee, R.B., and Ward, E.W.B., 1987, Early differences in production of mRNAs for phenylalanine ammonia-lyase and chalcone synthase in resistant and susceptible cultivars of soybean inoculated with *Phytophthora megasperma* f.sp. *glycinea*. *Physiol. Mol. Plant Pathol.* **30**: 293-297.

Flor. H.H., 1955, Host-parasite interaction in flax rust - its genetics and other implications. *Phytopathology* **45**: 680-685.

Förster, H., Tyler, B.M., and Coffey, M.D., 1994, *Phytophthora sojae* races have arisen by clonal evolution and by rare outcrosses. *Mol. Plant-Microbe Interact.* **7**: 780-791.

Gijzen, M., Förster, H., Coffey, M.D., and Tyler, B., 1996, Cosegregation of *Avr4* and *Avr6* in *Phytophthora sojae*. *Can. J. Bot.* **74**: 800-802.

Graham, T.L., 1995, Cellular biochemistry of phenylpropanoid responses of soybean to infection by *Phytophthora sojae*. In: Daniel M, Purkayastha RP (eds.) Handbook of phytoalexin metabolism and action, pp. 85-116. *Marcel Dekker, New York.*

Grant, M.R., Godiard, L., Straube, E., Ashfield, T., Lewald, J., Sattler, A., Innes, R.W., and Dangl, J.L., 1995, Structure of the *Arabidopsis RPM1* gene enabling dual specificity disease resistance. *Science* **269**: 843-846.

Grant, M.R., McDowell, J.M., Sharpe, A.G., de Torres Zabala, M., Lydiate, D.J., and Dangl, J.L., 1998, Independent deletions of a pathogen-resistance gene in *Brassica* and *Arabidopsis*. *Proc. Natl. Acad. Sci. USA* **95**: 15843-15848.

Jones, D.A., Thomas, C.M., Hammond-Kosack, K.E., Balint-Kurti, P.J., and Jones, J.D.G., 1994, Isolation of the tomato Cf-9 gene for resistance to *Cladosporium fulvum* by transposon tagging. *Science* **266**: 789-793.

Kasuga, T., Salimath, S.S., Shi, J., Gijzen, M., Buzzell, R.I., and Bhattacharyya, M.K., 1997, High resolution genetic and physical mapping of molecular markers linked to the *Phytophthora* resistance gene *Rps1-k* in soybean. *Mol. Plant-Microbe Interact.* **10**: 1035-1044.

Keen, N.T., 1975, Specific elicitors of plant phytoalexin production: determinants of race specificity in pathogens? *Science* **187**: 74-75.

Keen, N.T., and Yoshikawa, M., 1990, The expression of resistance in soya beans to *Phytophthora megasperma* f.sp. *glycinea*. In: Hornby D, Cook RJ, Henis Y, Ko WH, Rovira AD, Schippers B, Scott PR (eds) Biological Control of Soil-Borne Plant Pathogens, pp. 329-344. *CAB International, Wallingford, Oxon, United Kingdom.*

Lawrence. G.J., Finnegan, E.J., Ayliffe, M.A., and Ellis, J.G., 1995, The L6 gene for flax rust resistance is related to the *Arabidopsis* bacterial resistance gene *RPS2* and the tobacco viral resistance gene N. *Plant Cell* **7**: 1195-1206.

Leister, R.T., Ausubel, F.M., and Katagiri, F., 1996, Molecular recognition of pathogen attack occurs inside of plant cells in plant disease resistance specified by the *Arabidopsis* genes RPS2 and RPM1. *Proc. Natl. Acad. Sci. USA* **93**: 15497-15502.

Marek, L.F., and Shoemaker, R.C., 1997, BAC contig development by fingerprint analysis in soybean. *Genome* **40**: 420-427.

Martin, G.B., Brommonschenkel, S.H., Chunqongse, J., Frary, A., Ganal, M.W., Spivey, R., Wu, T., Earle, E.D., and Tanksley, SD., 1993, Map-based cloning of a protein kinase gene conferring disease resistance in tomato. *Science* **262**: 1432-1436.

Meyers, B.C., Shen, K.A., Rohani, P., Gaut, B.S., and Michelmore, R.W., 1998, Receptor-like genes in the major resistance locus of lettuce are subject to divergent selection. *Plant Cell* **10**: 1833-1846.

Mindrinos, M., Katagiri, F., Yu, G.-L., and Ausubel, F.M., 1994, The A. thaliana disease resistance gene *RPS2* encodes a protein containing a nucleotide-binding site and leucine-rich repeats. *Cell* **78**: 1089-1099.

Moesta, P., and Grisebach, H., 1981, Investigation of the mechanism of glyceollin accumulation in soybean infected by *Phytophthora megasperma* f.sp. *glycinea*. *Arch. Biochem. Biophys.* **212**: 462-467.

Nürnberger. T., Nennstiel, D., Jabs, T., Sacks, W.R., Hahlbrock, K., and Scheel, D., 1994, High affinity binding of a fungal oligopeptide elicitors to parsley plasma membranes triggers multiple defense responses. *Cell* **78**: 449-460.

Oldroyd, G.E.D., and Staskawicz, B.., 1998,: Genetically engineered broad-spectrum disease resistance in tomato. *Proc. Natl. Acad. Sci. USA* **95**: 10300-10305.

Ori, N., Eshed, Y., Paran, I., Presting, G., Aviv, D., Tanksley, S., Zamir, D., and Fluhr, R., 1997, The *I2C*family from the wilt disease resistance locus *I2* belongs to the nucleotide binding, leucine-rich repeat superfamily of plant resistance genes. *Plant Cell* **9**: 521-532.

Parker, J.E., Coleman, M.J., Szabó, V., Frost, L.N., Schmidt, R., van der Biezen, E.A., Moores, T., Dean, C., Daniels, M.J., and Jones, J.D.G., 1997, The *Arabidopsis* downy mildew resistance gene RPP5 shares similarity to the toll and interleukin-1 receptors with *N* and *L6*. *Plant Cell* **9**: 879-894.

Parniske, M., Hammond-Kosack, K.E., Golstein, C., Thomas,C.M., Jones, D.A., Harrison, K., Wulff, B.B.H., and Jones, J.D.G.,1997, Novel disease resistance specificities result from sequence exchange between tandemly repeated genes at the *Cf-4/9* locus of tomato. *Cell* **91**: 821-832.

Paxton, J., 1995, Soybean phytoalexins: Elicitation, nature, mode of action, and role. In: Daniel M, Purkayastha RP (eds) Handbook of phytoalexin metabolism and action, pp. 69-83. Marcel Dekker, New York.

Polzin, K.M., Lorenzen, L.L., Olson, T.C., and Shoemaker, R.C., 1994, An unusual polymorphic locus useful for tagging *Rps*1 resistance alleles in soybean. *Theor. Appl. Genet.* **89**: 226-232.

Salimath, S.S., and Bhattacharyya, M.K., 1999, Generation of a soybean BAC library, and identification of DNA sequences tightly linked to the *Rps1*-k disease resistance gene. *Theor. Appl. Genet.* **98**: 712-720.

Salmeron. J.M., Oldroyd. G.E.D., Rommens, C.M.T., Scofield, S.R., Kim, H.-S., Lavelle, D.T., Dahlbeck, D., and Staskawicz, B.J., 1996, Tomato *Prf* is a member of the leucine-rich repeat class of plant disease resistance genes and lies embedded within the *Pto* kinase gene cluster. *Cell* **86**: 123-133.

Schmitthenner, A.F., 1989, Phytophthora rot. In: Sinclair JB, Backman PA (eds.) Compendium of soybean diseases, pp. 35-38. *APS Press*, St. Paul, MN.

Schmitthenner, A.F., 1994, *Phytophthora sojae* races in Ohio over a 10-year interval. *Plant Dis.* **78**: 269-276.

Scofield, S.R., Tobias, C.M., Rathjen, J.P., Chang, J.H., Lavelle, D.T., Michelmore, R.W., and Staskawicz, B.J., 1996, Molecular basis of gene-for-gene specificity in bacterial speck disease of tomato. *Science* **274**: 2063-2065.

Song, W.-Y., Pi, L.-Y., Wang, G.-L., Gardner, J., Holsten, T., and Ronald, P.C., 1997, Evolution of the rice *Xa21* disease resistance gene family. *Plant Cell* **9**: 1279-1287.

Song, W.-Y., Wang, G.-L., Kim, H.-S., Pi, L.-Y., Holsten, T., Gardner, J., Wang, B., Zhai, W.-X., Zhu,.L.-H., Fauquet, C., and Ronald, P., 1995, A receptor kinase-like protein encoded by the rice disease resistance gene, Xa21. *Science* **270**: 1804-1806.

Tang, X., Frederick. R.D., Zhou, J., Halterman, D.A., Jia,Y., and Martin, G.B., 1996, Initiation of plant disease resistance by physical interaction of AvrPto and Pto kinase. *Science* **274**: 2060-2063.

Tang, X., Xie, M., Kim, Y.J., Zhou, J., Klessig, D.F., and Martin, G.B., 1999, Overexpression of P to activates defense responses and confers broad resistance. *Plant Cell* **11**: 15-30.

Thomas, C.M., Jones, D.A., Parniske, M., Harrison, K., Balint-Kurti, P.J., Hatzixanthis, K., and Jones, J.D.G., 1997, Characterization of the tomato *Cf-4* gene for resistance to *Cladosporium fulvum* identifies sequences that determine recognitional specificity in Cf-4 and Cf-9. *Plant Cell* **9**: 2209-2224.

Tyler, B.M., Förster, H., and Coffey, M.D., 1995, Inheritance of avirulence factors and restriction fragment length polymorphism markers in outcrosses of the oomycete *Phytophthora sojae*. *Mol. Plant-Microbe Interact.* **8**: 515-523.

Umemoto, N., Kakitani, M., Iwamatsu, A., Yoshikawa, M., Yamaoka, N., and Ishida, I., 1997, The structure and function of a soybean b-glucan-elicitor-binding protein. *Proc. Natl. Acad. Sci. USA* **94**: 1029-1034.

Ward, E.W.B., 1990, The interaction of soya beans with *Phytophthora megasperma* f.sp. *glycinea*: Pathogenicity. In: Hornby D (ed) Biological Control of Soil-Borne Plant Pathogens, pp. 311-327. *C.A.B. International, Wallingford, United Kingdom*.

Ward, E.W.B., Cahill, D.M., and Bhattacharyya, M.K., 1989, Early cytological differences between compatible and incompatible interactions of soybeans with *Phytophthora megasperma* f.sp. *glycinea*. *Physiol. Mol. Plant Pathol.* **34**: 267-283.

Whisson, S.C., Drenth, A., Maclean, D.J., and Irwin, J.A.G., 1994, Evidence for outcrossing in *Phytophthora sojae* and linking of a DNA marker to two avirulence genes. *Curr. Gen.* **27**: 77-82.

Whitham, S., Dinesh-Kumar, S.P., Choi, D., Hehl, R., Corr, C., and Baker, B., 1994, The product of the tobacco mosaic virus resistance gene N: Similarity to toll and the interleukin-1 receptor. *Cell* **78**: 1101-1115.

Wrather, J.A., Anderson, T.R., Arsyad, D.M., Gai, J., Ploper, L.D., Porta-Puglia, A., Ram, H.H., and Yorinori, .J.T., 1997, Soybean disease loss estimates for the top 10 soybean producing countries in 1994. *Plant Dis.* **81**: 107-110.

Yoshikawa, M., Yamauchi, K., and Masago, H., 1978, Glyceollin: its role in restricting fungal growth in resistant sobyean hypocotyls infected with *Phytophthora megasperma* var. *sojae*. *Physiol. Plant Pathol.* **12**: 73-82.

Yoshikwa, M., Yamauchi, K., and Masago, H., 1979, Biosynthesis and biodegradation of glyceollin by soybean hypocotyls infected with *Phytophthora megasperma* var. *sojae*. *Physiol. Plant Pathol.* **14**: 157-169.

Zhou, J., Loh, Y.-T., Bressan, R.A., and Martin, G.B., 1995, The tomato gene *Pti1* encodes a serine/threonine kinase that is phosphorylated by Pto and is involved in the hypersensitive response. *Cell* **83**: 925-935.

Elements of Signal Transduction Involved in Thylakoid Membrane Dynamics

[1]PETRA WEBER, [1]ANNA SOKOLENKO, [4]SAID ESHAGHI, [2]HRVOJE FULGOSI, [3]ALEXANDER V. VENER, [4]BERTIL ANDERSSON, [5]ITZHAK OHAD, AND [1]REINHOLD G. HERRMANN

[1]*Botanisches Institut der Ludwig-Maximilians-Universität, Menzinger Str. 67, D-80638 München, Germany,* [2]*Botanisches Institut der Christian-Albrechts-Universität, Olshausenstr. 40, D-24098 Kiel, Germany,* [3]*Department of Horticulture, University of Wisconsin, Madison, WI 53706-1590, USA,* [4]*Department of Biochemistry, Linköping University, S-58183 Linköping, Sweden, and* [5]*Department of Biological Chemistry, The Hebrew University, Jerusalem 91904, Israel*

1. INTRODUCTION

Photosynthesis, the most fundamental biochemical process for maintenance of life and the only renewable energy source of our planet, is a highly regulated, but also one of the most light- and heat-sensitive processes in plants and cyanobacteria (Berry and Björkman 1980, Weis and Berry 1988, Havaux and Tardy 1996). Light, obviously, is not only useful, but can also cause problems, and even exert harm to a plant. In their natural habitat, plants are exposed to light intensity and quality changes that can lead to an imbalance between the excitation rates of the two photosystems, and thus diminish photosynthetic efficiency. Light can also induce photoinactivation of electron transport and damage both photosystems (Prasil *et al* 1992, Aro *et al* 1993, Sonoike 1995). Comparably, high temperatures dramatically inhibit carbon dioxide fixation (Berry and Björkman 1980, Feller *et al* 1998), and the heat tolerance limits of leaves, which possess surprisingly little heat capacity, can depend on the thermal sensitivity of photochemical reactions in thylakoid membranes (Berry and Björkman 1980, Weis and Berry 1988, Havaux and Tardy 1996). Even short exposure to increasing temperatures in the range of $40°C$ can cause a progressive destacking of stacked thylakoids (Gounaris *et al* 1984), a lateral redistribution of the membrane protein complexes (Sundby *et al* 1986) that constitute this membrane (Herrmann *et al* 1991), an increase in the ionic permeability and fluidity of the membrane (Havaux *et al* 1996, Tardy and Havaux 1997), rapid dephosphorylation and degradation of the photosystem II core protein D1 (Rokka *et al* 2000), and a release of extrinsic proteins of the photosystem II oxygen-evolving complex (Enami *et al* 1994).

To avoid serious consequences for their life due to impairment of photosynthesis, plants have evolved specific and sophisticated protective and acclimative mechanisms against unfavorable environmental conditions to adapt and protect their photosynthetic system. The ways how this is managed are different from those of animals, last not least because of the inability of a plant to escape unfavorable milieus (Weis and Berry 1988, Schöffl *et al* 1998). However, the molecular details of plant signal transduction pathways, and which of the elements involved resemble or differ from those in animals and prokaryotic cells, are not yet clearly understood. Thylakoid membranes are highly dynamic structures. The mechanisms involved operate in different time scales and can be based on a transient reprogramming of cellular activities. Much of the sensing of external signals and signal response occurs in the photosynthetic membrane itself. Various lines of evidence suggest that light-induced thylakoid protein phosphorylation/dephosphorylation and energy dissipation *via* the xanthophyll cycle play an important role predominantly in short-term adaptation, the former also in the recovery of photosynthetic activity following stress conditions (Allen 1992, Gal *et al* 1997, Vener *et al* 1998, Rintamäki *et al* 1997). Proteolytic activities, in turn, can be involved in short-term processes (see below), but are frequently operating in long-term acclimation (Adam 1996, Andersson and Aro 1997).

The reversible phosphorylation of proteins is now recognized to be a major mechanism for the control of intracellular events in eukaryotic cells and a universal molecular mechanism for adaptation. The phosphorylation or dephosphorylation of serine, threonine, histidine, and tyrosine residues generally triggers conformational changes in target proteins that alter their biological properties. Protein phosphorylation involved in adaptation and repair processes in photosynthesis and its regulation become most obvious after relatively drastic changes in environment. In the eightieth, heat shock, and later light stress, were found to trigger rapid phosphorylation/dephosphorylation processes of a set of thylakoid proteins which both suggested that such phosphorylation plays a pivotal role in sensing and transducing stress in plants (Krishnan and Pueppke 1987). The primary site of light as well as of thermal damage is thought to be associated with photosystem II, light can cause oxidative damage to its reaction center (Berry and Björkman 1980, Weis and Berry 1988, Havaux and Tardy 1996, Keren *et al* 1997, Keren and Ohad 1998).

In photosynthesis, protein phosphorylation/dephosphorylation appears to be involved in at least three major processes. First, thylakoids contain an intriguing redox-controlled membrane protein kinase (Allen 1992, Vener *et al* 1995, 1997, 1998, Gal *et al* 1997) which is activated by light and leads to phosphorylation of approximately 20 thylakoid membrane proteins (Silverstein *et al* 1993, Vener *et al* 1995, Gal *et al* 1997). This protein kinase is thought to regulate the balance of the light-driven electron transfer between the two photosystems through reversible association of the light-harvesting chlorophyll *a/b* antenna (LHCII) with photosystems II or I due to the reversible phosphorylation of LHCII. This process modulates the relative light absorption cross section of the two photosystems (State I/State II transitions, Allen 1992, Gal *et al* 1997). Increasing evidence suggests that the cytochrome b_6/f complex is involved as a redox sensor *via* the PQH_2/PQ ratio (Vener *et al* 1995, 1997, 1998, Zito *et al* 1999). The major phosphorylated thylakoid proteins are those of the light-harvesting complex of LHCII at the outer face of the thylakoid membrane and several polypeptides of photosystem II core complex itself, besides CP43 and a 9 kDa (PsbH) polypeptide notably the two proteins D1 and D2 of the innermost reaction center.

Second, reversible phosphorylation/dephosphorylation of photosystem II proteins has also been inferred to regulate the stability, degradation and turnover of this photosystem, i.e. the turnover of damaged photosystem II reaction center subunits (Andersson and Aro 1997, Kruse *et al* 1997, Vener *et al* 1998). The reaction center protein D1 has the highest turnover rate of all thylakoid proteins due to light stress-induced damage (Aro *et al* 1993). Reversible phosphorylation of photosystem II appears to be a crucial regulatory element of its functionality. In particular, the dephosphorylation of D1 protein seems to be crucial in the control of photosystem II repair, required to resist light, and probably heat stress. Repair of photosynthetic function requires proteolytic degradation of the damaged protein, synthesis of a new D1 molecule and its integration into photosystem II (Andersson and Aro 1997, Keren and Ohad 1998). The phosphorylation status appears to be a key regulatory event in coordinating D1 protein degradation with the integration and assembly of newly synthesised D1 protein, since the photodamaged D1 protein can only be proteolytically degraded, and hence replaced, after dephosphorylation (Andersson and Aro 1997, Rintamäki et al 1996, 1997). Under conditions of light stress, the D1 protein is highly phosphorylated. This appears to increase its stability against proteolysis. Moreover, phosphorylation stabilises the dimeric structure of photosystem II (Barber et al 1997) which can be converted into the monomeric form only when photosystem II proteins are dephosphorylated (Kruse et al 1997).

A third process involving reversible protein phosphorylation has been proposed for redox controlled expression of photosynthetic genes based on "classical" two-component signal transduction systems known from prokaryotes (Allen 1992), but proof for this is lacking (cf. Race et al 1999).

The enzymology and molecular biology of thylakoid protein phosphorylation and dephosphorylation, the identity of various phosphorylated proteins as well as the role and interaction between distinct signalling regulatory components are not yet well understood. Inspite of substantial efforts, thylakoid protein kinases and phosphatases have largely resisted identification, and very little is known about the regulation of the latter (Carlberg and Andersson 1996). Recent work has uncovered that there are several intrinsic membrane protein kinases (Weber et al 1999, Snyders and Kohorn 1999, Zer et al 1999) and phosphatase(s) (Vener et al 1999), that one of the kinase activities is under redox control (Allen 1992, Race and Hind 1996, Andersson and Aro 1997, Gal et al 1997, Vener et al 1998) involving plastoquinol binding at the quinol oxidation site of the cytochrome b_6/f complex (Vener et al 1995, 1997, 1999, Zito et al 1999), and that light-induced conformational changes of the LHCII substrate exposes the phosphorylation site to the protein kinase (Zer et al 1999). Several lines of evidence suggest as well that dephosphorylation of thylakoid phosphoproteins is not only catalysed by membrane phosphatase activity(ies) (Bennett 1980) but probably by membrane-attached phosphatases (Sun et al 1989, Kieleczawa et al 1992, Hammer et al 1997) and phosphatases of the chloroplast stroma as well (Hammer et al 1997).

We summarise here work on the isolation, characterization and localization of several new components involved in thylakoid protein phosphorylation/dephosphorylation processes in an attempt to study and understand their role in the dynamics of the photosynthetic membrane.

2. RESULTS AND DISCUSSION

2.1 Thylakoid Kinases and their cDNAs

Fractions from spinach thylakoids exhibiting approximately three orders of magnitude enrichment in kinase activity were obtained in our laboratories based on ammonium sulfate (AMS) fractionation of partially lysed thylakoid membranes (Gal *et al* 1997). Employing blot renaturation assays with histone as a substrate (Sokolenko *et al* 1995) as many as four kinase-active bands with molecular masses of 33, 56 (designated AMS9), 64 (AMS6, cf. also Zer *et al* 1999) and 87 kDa could be identified in such fractions (Fig. 1; Weber *et al* 1999). Using a refined strategy we have recently been able to identify an additional activity migrating with an apparent mass of 45 kDa. None of these five kinase activities corresponds to the recently described intrinsic thylakoid kinase of 55.5 kDa from *Arabidopsis thaliana*, designated TAK-1 (Snyders and Kohorn 1999).

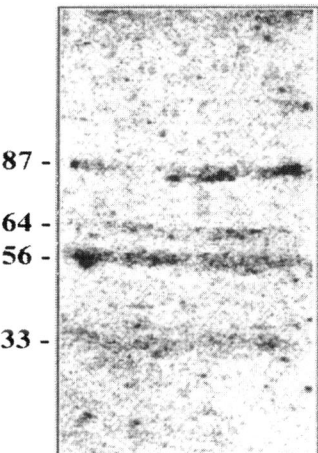

Figure 1. Autoradiograph illustrating protein kinase activities after renaturation of polypeptides electrotransferred onto a PVDF membrane and incubation with histones in the presence of [γ^{32}P]-ATP (Sokolenko *et al* 1995). Note that the 45 kDa component is not visible in the gel shown. Primary separation was on a 10% SDS/polyacrylamide gel. Left: molecular weight standards.

In an attempt to isolate the cDNAs/genes for these kinases, we have begun to purify the kinase-active bands and to determine their N-terminal as well as internal peptide sequences by gas phase sequencing. A database search with sequences obtained from the 56 and 64 kDa compounds using the BLASTP program at the NCBI www-service gave no match in the latter case. However, a moderate homology score to the gene *slr0311* from *Synechocystis* was found in the former case. This gene encodes a potential histidine kinase of 85 kDa.

The serum generated against the overexpressed cyanobacterial 85 kDa component was found to crossreact with two chloroplast thylakoid components of 56 and 33 kDa (Weber *et al* 1999). This serum was used to isolate the corresponding cDNA(s) from recombinant λgt11 phage cDNA libraries from spinach. In fact, phages with two different kinds of inserts could be selected. One of these contained a full-length cDNA which encoded a polypeptide of 349 amino acid residues corresponding to a molecular mass of 39.46 kDa (accession no. Y14198). Hydropathy analysis suggested a mono- or bitopic protein with a potential C-terminal membrane-buried segment of 18 residues that may serve as a membrane anchor (Fig. 2). The organelle location of the polypeptide was checked by *in organello* assays according to Clausmeyer *et al* (1993). Transcripts made *in vitro* from TTP30 cDNA could be decoded into a protein of the expected size in cell-free translation systems derived either from wheat germ or rabbit reticulocytes. This protein represents a precursor molecule, since it could be imported into isolated spinach or pea chloroplasts. It was found to be correctly processed and to integrate stably into thylakoid membranes. It could not or only partially be removed by chaotropic salts, such as 2 M NaSCN consistent with its stable integration as an integral protein (data not shown). The imported polypeptide could be immunoprecipitated with the antiserum elicited against the cyanobacterial component (see above). It comigrated with the authentic immunoreactive 33 kDa protein, and was sensitive to protease treatment indicating that its large hydrophilic moiety is stroma-exposed.

Structural evaluation of the deduced amino acid sequence revealed two notable features of the polypeptide that are known from phosphate-transferring enzymes and their regulators, (i) a highly charged N-terminal segment with 29 acidic and 16 basic residues, and (ii), as the most conspicuous and intriguing feature, three tandemly arranged, almost perfect tetratricopeptide repeats (TPR) in the central region (Fig. 2). The polypeptide was therefore designated TTP30 (tetratricopeptide-containing thylakoid protein of approximately 30 kDa). TPR motifs provide a protein fold for protein/protein interaction (Sikorski *et al* 1990, Das *et al* 1998). Each TPR motif is comprised of a pair of amphipathic, antiparallel α-helices which are punctuated by proline-induced turns. The subhelices are stereochemically complementary and fit into a knob/hole structure. The individual repeats are generally sequentially organized into a parallel arrangement. TPR motives were first identified in proteins involved in regulating the cell division cycle (e.g. Sikorski *et al* 1990) and are known from various eukaryotes. It is relevant to note that they can interact with phosphoproteins in a defined way (Das *et al* 1998). They can also act as molecular links in various biological processes, frequently with immunophilins, as protein kinase inhibitors, or protein phosphatase activators. TPR motifs may even be elements of these enzymes (Das *et al* 1998).

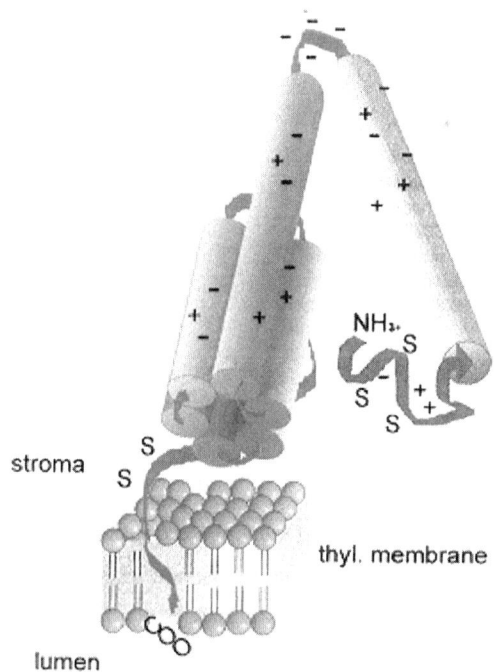

Figure 2. Hypothetical secondary structure of TTP30. Cylinders depict six TPR amphipatic helices forming a bulky moiety with exposed charged residues. Larger cylinders represent two helices forming a putative helix-loop-helix domain. The separating loop is highly negatively charged. Regions rich in serine are indicated with S; charged residues are presented by "+" or "-". The protein is presumed to be

anchored mono- or bitopically in the thylakoid membrane with its carboxy-terminal hydrophobic stretch (Fulgosi 1999).

The TTP30 cDNA was expressed in *E. coli,* and the resulting protein as well as oligopeptides synthesized from the 21 residue N-terminal sequence of the 62 kDa component AMS6 were used to generate antisera in rabbits. Both polyclonal antisera obtained were monospecific and found to react with corresponding bands from thylakoids (see Section 2.4).

2.2 Thylakoid Phosphatases

Thylakoid protein phosphorylation is reversible implying protein phosphatases as another class of central members of thylakoid signal transduction chains. Integral, peripheral and extrinsic protein phosphatases have been claimed to be involved to catalyse dephosphorylation of the thylakoid phosphoproteins (Hammer *et al* 1995, 1997, Vener *et al* 1997, 1999). Several thylakoid-associated phosphatases have been identified and partially characterised (Hammer *et al* 1995, Vener *et al* 1999). These enzymes were generally found to be difficult to assay because of a high substrate specificity towards thylakoid phosphoproteins or phosphothreonine-containing peptides mimicking the N-termini of these proteins (Sun *et al* 1993, Cheng *et al* 1994).

In general, the major eukaryotic serine/threonine phosphatases are classified into two families, designated PPM and PPP, according to their structure, substrate specificity, requirement for divalent cations and susceptibility to inhibitors (Barford 1996, Cohen 1997, Zolnierowicz and Bollen 2000). The protein phosphatases of plant thylakoid membranes are considered to be unrelated to the conventional families of serine/threonine phosphatases, mainly because of the failure to obtain inhibition by group-specific inhibitors during *in membrano* dephosphorylation experiments (MacKintosh *et al* 1991, Sun and Markwell 1992). However, we could recently characterise and purify an intrinsic thylakoid protein phosphatase of 39 kDa molecular mass (Vener *et al* 1999). This phosphatase has been shown to be related to conventional eukaryotic protein phosphatases of the PPP family, and probably belongs to the PP2A phosphatase subfamily as judged from its susceptibility to inhibitors specific to PP2A, PP4, PP5 and PP6. This is a first case for the existence of such an enzyme in plants.

Furthermore, this phosphatase appears to be regulated by a lumenal immunophilin-like protein (see below).

2.3 Other Compounds of Thylakoid Signal Transduction Processes

Protein kinases and phosphatases are controlled in a variety of ways. Often, they consist of a constant catalytic subunit and one or more regulatory subunits that control their activity, substrate specificity and subcellular location. For instance, the catalytic subunits of PPP phosphatases are under sophisticated control by binding regulatory and targeting subunits (reviewed in Barford 1996, Cohen 1997, Zolnierowicz and Bollen 2000). Regarding the PP2A family approximately twenty distinct regulatory and targeting subunits have been characterised. These proteins include compounds that are collectively designated immunophilins which comprise two subfamilies: cyclophilins and FKBPs (Liu et al 1991, Walsh et al 1992). Immunophilins are peptidyl-prolyl *cis-trans* isomerases (PPIases) that operate as protein folding catalysts (Fruman et al 1994, Marks 1996).

The recently found intrinsic thylakoid phosphatase activity mentioned above appears to be regulated *via* a 40 kDa cyclophilin-like protein, designated TLP40, in a way that is reminiscent to the regulation of the heterodimeric protein phosphatase calcineurin. This PP2B phosphatase has been shown to be the target of cytoplasmic immunophilins that are receptors of immunosuppressive drugs. Inhibition of PP2B by complexes between immunophilins and the corresponding immunosuppressive drugs, such as cyclosporin A, is recognised as a key step in the arrest of T cell signal transduction causing immunosuppression in mammals (Liu et al 1991, Walsh et al 1992). An analogous process appears to be a key step in regulating the photosynthetic process and to be probably involved in D1 dephosphorylation and degradation (Fulgosi et al 1998, Vener et al 1999).

TLP40 is a member of the relatively rare category of complex immunophilins which are generally constituents of supramolecular machineries. As those it possesses domain structure with an N-terminal leucin zipper, potential phosphatase binding sites, and a PPIase moiety (Fulgosi et al 1998, Vener et al 1999). It is relevant to mention that TLP40 co-purified with the intrinsic phosphatase and that it is located in the thylakoid lumen of stroma lamellae but appears to be involved in protein dephosphorylation, although the catalytic domain of the membrane phosphatase resides at the opposite, the outer face, of the membrane.

Nevertheless, binding of the TLP40 catalytic site by prolyl-containing peptides clearly regulates phosphatase activity across the membrane (Vener et al 1999). TLP40 can occur in three conformations, as a monomer, a dimer, and membrane attached (Fulgosi et al 1998). It is conceivable therefore that the transmembrane protein phosphatase is regulated *via* reversible association of TLP40 at the lumenal face of the membrane, i.e. the phosphatase is activated when TLP40 is released into the lumen (*Vener et al* 1999). The findings that, after heat stress, grana destack, TLP40 is rapidly detached from thylakoids, and D1 dephosphorylated and degraded within a few minutes is consistent with this notion (Rokka et al 2000). Collectively, the outlined data suggest an unanticipated and unique cyclophilin-controlled, *reverse* (i.e. inside-out-directed) signal transduction process in thylakoid membranes. It is conceivable that these reactions and compounds are part of a pathway for the coordination of protein folding and protein turnover, with potential significance for the regulation of protein turnover in and repair of photoinhibitory damage to photosystem II under stress. It is not known whether this phosphatase is also involved in the regulation of LHCII mobility, nor are its topographical location and perhaps positional changes in thylakoid membranes, except that it can be associated with the cytochrome complex (see below).

2.4 Subthylakoid Location of Regulatory Compounds

The distribution within the thylakoid system of the minor components described was checked serologically by Western analysis after partial lysis of membranes and fractionation of the lysate by sucrose gradient centrifugation into the various supramolecular thylakoid complexes (Eshaghi et al 1999). The positions of thylakoid kinases/phosphatase and regulatory components was monitored relative to the position of the major thylakoid assemblies, photosystem II supercomplex, dimer and monomer, photosystem I, cytochrome b_6/f, ATP synthase, and LHCII using antisera against subunits of these assemblies and against the catalytic subunit of human PP2A phosphatase, TLP40, AMS6, TTP30 and 56 kDa kinase. This analysis uncovered that the putative regulatory components occupy different and distinct topographical locations within thylakoids. The results obtained are illustrated by Fig. 3.

One of the most remarkable findings of this study is that the five components under study, namely TTP30, the intrinsic phosphatase, the 56 and 64 kDa kinases as well as the thylakoid-associated fraction of TLP40, were all found to comigrate with the cytochrome b_6/f complex and photosystem assemblies. These data fit nicely with accumulating evidence that these molecular machines are "major players" in thylakoid dynamics, the cytochrome complex being a principal regulatory element in redox sensing and signal transduction. Expectedly, a major amount of TLP40 was also present in the soluble fraction. Refined analysis using fractions of grana thylakoids, stroma lamellae and grana margins (Andersson and Anderson 1980) has shown that TLP40 was exclusively located in stroma lamellae into which newly synthesised polypeptides integrate. On the other hand, TTP30 is not found in the cytochrome b_6/f complex. This indicates that it may be involved in different processes than TLP40.

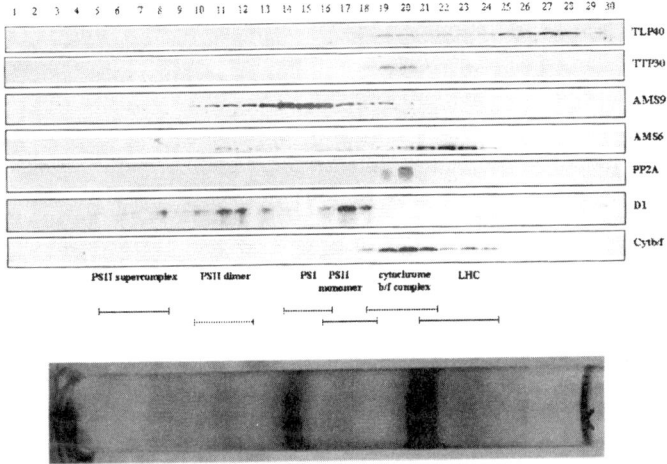

Figure 3. The distribution of phosphatase/kinase and regulatory components within thylakoid membrane complexes. The major photosynthetic complexes were prepared after sucrose gradient centrifugation as described by Eshaghi *et al* (1999), and 1.0 ml fractions were collected from the bottom of the gradient using an ISCO 640 Gradient Fraction Collector. The fractions were separated by SDS-PAGE, transferred onto nitrocellulose membrane and immunodecorated with antisera against TLP40, TTP30, 56 kDa protein (AMS9), AMS6 (N-terminal peptide), human PP2A phosphatase), D1 protein of photosystem II, and cytochrome b_6 of the cytochrome b_6/f complex.

AMS9 (56 kDa protein) represents a peripheral protein kinase. It can be extracted from thylakoids by salt treatment and recovered from the soluble fraction. It was found to be associated with photosystem I. Its substrate is not known presently.

AMS6 is a novel thylakoid component as well, associated with the trimeric LHCII complex, but does not represent the redox-controlled protein kinase as anticipated initially (Weber *et al* 1999). It is closely migrating with this kinase even in high-resolution gels but not identical with it, since it could be separated recently from the 64 kDa kinase by perfusion chromatography.

3. CONCLUSIONS AND PERSPECTIVES

Signals such as light, heat, gravity or hormones, control diverse physiological processes in plants including the photosynthetic process. Thylakoid membranes are highly dynamic structures that can cope high levels of environmental stresses, especially light and heat, that plants can encounter. In addition to the management of conversion of solar energy the membrane system can receive, record and transduce signals in various ways and directions, including messages from the thylakoid lumen to the stroma face. The outlined findings indicate that phosphorylation/dephosphorylation processes are important elements of thylakoid-based signal transduction. Moreover, the relatively high number of the components involved in these processes at or within the membrane identified up to now suggests that thylakoid signal transduction chains and their regulation are substantially more complex than presently assumed. The various components and processes that have been identified to be likely members of thylakoid-associated signal transduction chains include half-a-dozen protein kinases including the redox-controlled one with a molecular mass of approx. 64 kDa involved in StateI/StateII transitions, light-controlled substrate (LHCII) induction for that kinase, the cytochrome b_6/f complex as redox sensor *via* the PQH_2/PQ ratio, an intrinsic protein phosphatase, and an immunophilin of the complex type located in the thylakoid lumen that appears to interact with that phosphatase. These compounds are associated with different membrane structures and reside in different regions of thylakoid membranes, although several of them are attached to the cytochrome complex and photosystem II assembly. Some of them, notably TTP30 and the 56 kDa component, display characteristics of two-component signal transduction elements. Apart from deciphering the role, function and molecular mechanisms for each of these components crucial questions to be answered are their topography within the heterogeneous thylakoid system, and the mechanisms relating signal perception and signal transduction in thylakoids with processes at the organelle envelope or the nucleo/cytosolic compartment. Eukaryotic thylakoid membranes are of dual genetic origin. The compartmentalised integrated genetic system of the plant cell requires the regulation of plastid and nuclear genes to be coordinated in quantity, time, in multicellular plants also in space which complicates their biogenesis and maintenance when compared to analogous processes in prokaryotes (Herrmann 1997). Almost all of the regulatory components are of nuclear origin and need to be regulated through information from the organelle.

ACKNOWLEDGMENTS

This work was supported by the Deutsche Forschungsgemeinschaft (SFB 184), the Human Frontier Science Program (HFSP), the Royal Swedish Academy of Sciences, and the Swedish Council for Agriculture and Forestry.

REFERENCES

Adam, Z., 1996, Protein stability and degradation in chloroplasts. *Plant Mol. Biol.* **32**: 773-783.

Allen, J.F., 1992, Protein phosphorylation in regulation of photosynthesis. *Biochim. Biophys. Acta* **1098**: 275-335.

Andersson, B. and Anderson, J. M., 1980, Lateral heterogeneity in the distribution of chlorophyll-protein complexes of the thylakoid membrane of spinach chloroplasts. *Biochim. Biophys. Acta* **593**: 427 – 440.

Andersson, B., and Aro, E.-M., 1997, Proteolytic activities and proteases of plant chloroplasts. *Physiol. Plant.* **100**: 780-793.

Aro, E.-M., Virgin, I., and Andersson, B., 1993, Photoinhibition of photosystem II. Inactivation, protein damage and turnover. *Biochem Biophys. Acta* **1143**: 113-134.

Barber, J., Nield, J., Morris, E.P., Zheleva, D., and Hankamer, B., 1997, The structure, function and dynamics of photosystem two. *Physiol. Plant.* **100**: 817-827.

Barford, D., 1996, Molecular mechanisms of the protein serine/threonine phosphatases. *Trends Biochem. Sci.* **21**: 407-412.

Bennett, J., 1980, Chloroplast phosphoproteins: Evidence for a thylakoid-bound phosphoprotein phosphatase. *Eur. J. Biochem.* **104**: 85-89.

Berry, J., and Björkman, O., 1980, Photosynthetic response and adaptation to temperature in higher plants. *Annu. Rev. Plant Physiol.* **31**: 491-543.

Carlberg, I., and Andersson, B., 1996, Phosphatase activities in spinach thylakoid membranes–Effectors, regulation and location. *Photosynth. Res.* **47**: 145-156.

Clausmeyer, S., Klösgen, R.B., and Herrmann, R.G., 1993, Protein import into chloroplasts. The hydrophilic lumenal proteins exhibit unexpected import and sorting specificities in spite of structurally conserved transit peptides. *J. Biol. Chem.* **268**: 13869-13876.

Cheng, L., Spangfort, M.D., and Allen, J.F., 1994, Substrate specificity and kinetics of thylakoid phosphoprotein phosphatase reactions. *Biochim. Biophys. Acta* **1188**: 151-157.

Cohen, P., 1997, Novel protein serine/threonine phosphatases: variety is the spice of life. *Trends Biochem. Sci.* **22**: 245-251.

Das, A.K., Cohen, P.T.W., and Barford, D., 1998, The structure of the tetratricopeptiderepeats of protein phosphatase 5: implications for TPR-mediated protein-protein interactions. *EMBO J.* **17**: 1192–1199.

Enami I, Kitamura M, Tomo T, Isokawa Y, Ohta H and Katoh S., 1994. Is the primary cause of thermal inactivation of oxygen evolution in spinach PSII membranes release of the extrinsic 33 kDa protein or of Mn? *Biochem. Biophys. Acta* **1186**: 52-58

Eshaghi, S., Andersson, B., and Barber, J., 1999, Isolation of a highly active PSII-LHCII supercomplex from thylakoid membranes by a direct method. *FEBS Lett.* **446**: 23 –26.

Feller, U., Crafts-Brandner, S.J., and Salvucci, M.E., 1998, Moderately high temperatures inhibit ribulose-1,5-bisphosphate carboxylase/oxygenase (Rubisco) activase-mediated activation of Rubisco. *Plant Physiol.* **116**: 539-546.

Fruman, D.A., Burakoff, S.J., and Bierer, B.E., 1994, Immunophilins in protein folding and immunosuppression. *FASEB J.* **8**: 391-400.

Fulgosi, H., 1999, Molecular characterization of auxiliary thylakoid components involved in regulation of photosynthesis. *PhD thesis*, Universität München

Fulgosi, H., Vener, A.V., Altschmied, L., Herrmann, R.G., and Andersson B., 1998, A novel multi-functional chloroplast protein: identification of a 40 kDa immunophilin-like protein located in the thylakoid lumen. *EMBO J.* **17**: 1577-1587.

Gal, A., Zer., H., and Ohad, I., 1997, Redox-controlled thylakoid protein phosphorylation. News and views. *Physiol. Plant.* **100**: 869-885.

Gal, A., Zer, H. Roobol-Boza, M., Fulgosi, H., Herrmann, R.G., Ohad, I., and Andersson, B., (1995), Use of perfusion chromatography for the rapid isolation of thylakoid kinase enriched preparations. In *Photosynthesis; from Light to Biosphere* (Mathis, P., ed.), Kluwer Academic Publishers, Dordrecht, The Netherlands, vol. III: 341-344.

Gounaris, K., Brain, A.R.R., Quinn, P.J., and Williams, W.P., 1984, Structural reorganisation of chloroplast thylakoid membranes in response to heat-stress. *Biochim. Biophys. Acta* **766**: 198-208.

Hammer, M.F., Sarath, G., Osterman, J.C., and Markwell, J., 1995, Assessing modulation of stromal and thylakoid light-harvesting complex-II phosphatase activities with phosphopeptide substrates. *Photosynth. Res.* **44**: 107–115.

Hammer, M.F., Markwell, J., and Sarath, G., 1997, Purification of a protein phosphatase from chloroplast stroma capable of dephosphorylating the light-harvesting complex-II. *Plant Physiol.* **113**: 227-233.

Havaux, M., and Tardy, F., 1996,Temperature-dependent adjustment of the thermal stability of photosystem II in vivo: Possible involvement of xanthophyll-cycle pigments. *Planta* **198**: 324-333.

Havaux, M., Tardy, F., Ravenel, J., Chanu, D., and Parot, P., 1996. Thylakoid membrane stability to heat stress studied by flash spectroscopic measurements of the electrochimic shift in intact potato leaves: influence of xanthophyll content. *Plant Cell Environ.* **19**: 1359-1368.

Herrmann, R.G., Oelmüller, R., Bichler, J., Schneiderbauer, A., Steppuhn, J., Wedel, N., Tyagi, A.K., and Westhoff, P., 1991, The thylakoid membrane of higher plants: genes, their expression and interaction. In *Plant Molecular Biology 2* (R.G. Herrmann and B.A. Larkins, eds.), NATO ASI Series A: Life Sciences, vol. 212, Plenum Publ. Corp., New York, pp.411-427.

Herrmann, R.G., 1997, Eukaryotism, towards a new interpretation. In *Eukaryotism and Symbiosis* (H.E.A. Schenk, R.G. Herrmann, K.W. Jeon, N.E. Müller and W. Schwemmler, eds.), Springer, Heidelberg, New York, 73-118.

Keren, N., Berg, A., van Kan, P.J.M., Levanon, H., and Ohad, I., 1997, Mechanism of photosystem II photoinactivation and D1 protein degradation at low light: the role of back electron flow. *Proc. Natl. Acad. Sci. USA* **94**: 1579-1584.

Keren, N., and Ohad, I., 1998, State transmission and photoinhibition. In *Advances in Photosynthesis series „Molecular biology of* Chlamydomonas*: chloroplast and mitochondria"* (J.-D. Rochaix, M. Goldscmidt-Clermont and S. Merchant eds.), Kluwer Academic Publishers, vol **7**: pp.569-596.

Kieleczawa, J., Coughlan, S.J., and Hind, G., 1992, Isolation and characterization of an alkaline phosphatase from pea thylakoids. *Plant Physiol.* **99**: 1029-1936.

Krishnan, H.B., and Pueppke, S.G., 1987, Heat shock triggers rapid protein phosphorylation in soybean seedlings. *Biochem. Biophys. Res. Commun.* **148**: 762-767.

Kruse, O., Zheleva, D., and Barber, J., 1997, Stabilization of photosystem II dimers by phosphorylation: implication for the regulation of the turnover of D1 protein. *FEBS Lett.* **408**: 276-280.

Liu, J., Farmer, J.D., Lane, W.S., Friedman, J., Weissman, I., and Schreiber, S.L., 1991, Calcineurin is a common target of cyclophilin-cyclosporin a and FKBP-FK506 complexes. *Cell* **66**: 807-815.

Marks A. R., 1996, Cellular functions of immunophilins. *Physiol. Rev.* **76**: 631– 649.

MacKintosh, C., Coggins, J., and Cohen, P., 1991, Plant protein phosphatases. Subcellular distribution, detection of protein phosphatase 2C and identification of protein phosphatase 2A as the major quinate dehydrogenase phosphatase. *Biochem. J.* **273**: 733-738.

Prasil, O., Adir, N., and Ohad, I., 1992, Dynamics of photosystem II: Mechanism of photoinhibition and recovery processes. In *The Photosystems: Structure, Function and Molecular Biology* (J. Barber, ed.), Amsterdam: Elsevier, pp.295-348.

Race, L.H., Eaton-Rye, J.J., and Hind, G., 1995, A 64-kDa protein is a substrate for phosphorylation by a distinct thylakoid protein kinase. *Photosynth. Res.* **43**: 231-239

Race, H. L. and Hind G., 1996, A protein kinase in the core of photosystem II. *Biochem.* **35**: 13006 - 13010

Race, H.L., Herrmann, R.G., and Martin W., 1999, Why have organelles retained genomes? *Trends Genet.* **15**: 364-370.

Rintamäki, E., Kettunen, R., and Aro, E.-M., 1996, Differential D1 dephosphorylation in functional and photodameged photosystem II centres. Dephosphorylation is a prerequisite for degradation of damaged D1. *J. Biol. Chem.* **271**: 14870-14875.

Rintamäki, E., Salonen, M., Suoranta, U.M., Carlberg, I., Andersson, B., and Aro, E.-M., 1997, Phosphorylation of light-harvesting complex II and photosystem II core proteins shows different irradiance-dependent regulation in vivo. Application of phosphothreonine antibodies to analysis of thylakoid phosphoproteins. *J. Biol. Chem.* **272**: 30476-30482.

Schöffl, F., Prandl, R., and Reindl, A., 1998, Regulation of the heat-shock response. *Plant Physiol.* **117**: 1135-1141.

Sikorski, R.S., Bogunski, M.S., Goebl, M., and Hieter, P., 1990, A repeating amino acid motif in CDC23 defines a family of proteins and a new relationship among genes required for mitosis and RNA synthesis. *Cell* **60**: 307-317.

Silverstein, T., Cheng, L., and Allen, J.F., 1993, Redox titration of multiple protein phosphorylations in pea chloroplast thylakoids. *Biochim. Biophys. Acta* **1183**: 215-220.

Snyders, S., and Kohorn, B.D., 1999, TAKs, thylakoid membrane protein kinases associated with energy transduction. *J. Biol. Chem.* **274**: 9137-91405.

Sokolenko, A., Fulgosi, H., Gal, A., Altschmied, L., Ohad, I., and Herrmann, R.G., 1995, The 64 kDa polypeptide of spinach may not be the LHCII kinase, but a lumen-located polyphenol oxidase. *FEBS Lett.* **371**: 176-180.

Sonoike, K., Terashima, I., Iwaki, M., and Itoh, S., 1995, Destruction of photosystem I iron-sulfur centres in leaves of *Cucumis sativus* L. by weak illumination at chilling temperatures. *FEBS Lett.* **362**: 235-238.

Sun, G., Bailey, D., Jones, M.W., and Markwell, .J, 1989, Chloroplast thylakoid protein phosphatase is a membrane surface-associated activity. *Plant Physiol.* **89**: 238–243.

Sun, G., and Markwell, J., 1992, Lack of types 1 and 2A protein serine(P)/threonine(P) phosphatase activities in chloroplasts. *Plant Physiol.* **100**: 620-624.

Sun, G., Sarath, G., and Markwell, J., 1993, Phosphopeptides as substrates for thylakoid protein phosphatase activity. *Arch. Biochem. Biophys.* **304**: 490-495.

Sundby, C., Melis, A., Mäenpää, P., and Andersson, B., 1986, Temperature-dependent changes in antenna size of photosystem II. Reversible conversion of photosystem II$^\alpha$ to photosystem II$^\beta$. *Biochem. Biophys. Acta* **851**: 475-483.

Tardy, F., and Havaux, M., 1997, Thylakoid membrane fluidity and thermostability during the operation of the xanthophyll cycle in higher-plant chloroplasts. *Biochem. Biophys. Acta* **1330**: 179-193.

Vener, A.V., van Kan, P.J., Gal, A., Andersson, B., and Ohad, I., 1995, Activation/deactivation cycle of redox-controlled thylakoid protein phosphorylation. Role of plastoquinol bound to the reduced cytochrome *bf* complex. *J. Biol. Chem.* **270**: 25225-25232.

Vener, A.V., van Kan, P.J.M., Ohad, I., and Andersson, B., 1997, Plastochinol at the Qo-site of reduced cytochrome *bf* mediates signal transduction between light and protein phosphorylation: Thylakoid protein kinase deactivation by a single turnover flash. *Proc. Natl. Acad. Sci. USA* **94**: 1585-1590.

Vener, A., Ohad, I., and Andersson, B., 1998, Protein phosphorylation and redox sensing in chloroplast thylakoids. *Curr. Opin. Plant Biol.* **1**: 217-223.

Vener, A.V., Rokka, A., Fulgosi, H., Andersson, B., and Herrmann R.G., 1999, A cyclophilin-regulated PP2A-like protein phosphatase in thylakoid membranes of plant chloroplasts. *Biochemistry* **38**, 14955 – 14965.

Walsh, C.T., Zydowski, L.D., and McKeon, F.D., 1992, Cyclosporin A, the cyclophylin class of peptidyl prolyl isomerases, and blockade of T cell signal transduction. *J. Biol. Chem.* **267**: 13115-13118.

Weber, P., Fulgosi, H., Sokolenko, A., Karnauchov, I., Andersson, B., Ohad, I., and Herrmann, R.G., 1998, Evidence for four thylakoid-located protein kinases. In: *Photosynthesis: Mechanisms and Effects* (G. Garab, ed.), Kluwer Academic Publ., Dordrecht, The Netherlands, pp.1883-1886.

Weis, E., and Berry, J.A., 1988, Plants and high temperature stress. *Symp. Soc. Exp. Biol.* **42**: 329-346.

Zer, H., Vink, M., Keren, N., Dilly-Hartwig, H.G., Paulsen, H., Herrmann, R.G., Andersson, B., and Ohad, I., 1999, Regulation of thylakoid protein phosphorylation at the substrate level: Reversible light-induced conformational changes expose the phosphorylation site of the light-harvesting complex II. *Proc. Natl. Acad. Sci. USA* **96**: 8277-8282.

Zito, F., Finazzi, G., Delosme, R., Nitschke, W., Picot, D., and Wollman, F.-A., 1999, The Qo site of cytochrome *b6f* complexes controls the activation of the LHCII kinase. *EMBO J.* **18**: 2961-2969.

Zolnierowicz, S., Bollen, M., 2000, Protein phosphorylation and protein phosphatases. De Panne, Belgium, September 19–24, 1999. *EMBO J.* **19**: 483 – 488.

Novel Aspects in Photosynthesis Gene Regulation

[1]RALF OELMÜLLER, [1]TATJANA PESKAN, [2]MARTIN WESTERMANN, [1]IRENA SHERAMETI, [1]MEENA CHANDOK, [3]SUDHIR K. SOPORY, [1]ANKE WÖSTEMEYER, [4]VICTOR KUSNETSOV, [1]STAVER BEZHANI AND [1]THOMAS PFANNSCHMIDT

[1]*Friedrich-Schiller-Universität Jena, Pflanzenphysiologie, Dornburger Str. 159, D-07743 Jena, Germany:* [2]*Institut für Ultrastrukturforschung des Klinikums der Friedrich-Schiller-Universität Jena, Ziegelmühlenweg 1, D-07740 Jena, Germany:* [3]*International Centre for Genetic Engineering and Biotechnology, P.O. Box-10504, Aruna Asaf Ali Marg, New Dehli-110067, India:* [4] *Institut of Plant Physiology, Russian Academy of Science, Botanicheskaya 35, 127276 Moscow, Russia.*

1. INTRODUCTION

Probably more than 2000 nuclear genes code for proteins located in the plastids (Abdallah *et al* 2000). A common feature of many of them is that their expression is regulated by a variety of cell-internal and -external signals, such as light, phytohormones, biotic and abiotic factors, photosynthesis, metabolic conditions within the cell (organelle and cytoplasm) or the developmental stage of the plastids. All these signals initiate transduction pathways which ultimately operate at defined regulatory elements. This causes changes in transcription rates of the responsive genes, but effects also posttransscriptional processes. In this overview, we focus on the following aspects:

(1)   How can a signal be perceived at the plasma membrane?

(2) How do plastids control nuclear gene expression and do higher plants posses chloroplast-localized response regulators?

(3) We present three genes (*AtpC*, *PsaF*, *PsaD*) with unusual regulatory mechanisms.

2. RESULTS AND DISCUSSION

2.1 HOW CAN A SIGNAL BE PERCEIVED AT THE PLASMA MEMBRANE?

The major question in signal transduction today is how a specific signaling is achieved and maintained between the different signaling pathways. Simons and Ikonen (1997) described distinct lipid microdomains in the plasma membrane ("rafts") which selectively attach specific proteins and exclude others. The lipid rafts are enriched in sphingolipids and cholesterol and are insoluble in the detergent Triton X-100 at 4°C. Due to the high lipid content, those detergent-insoluble glycolipid-enriched domains (DIGs) can be isolated as a low density fraction during gradient centrifugation, enabling the identification of the raft proteins.

DIGs also exist in caveolae, nonclathrin-coated plasma membrane pits with a proposed role in endocytosis and signal transduction. The major component of caveolae is caveolin, a 21-kDa structural protein, which appears to be essential for caveolar formation (Harder and Simons 1997) in mammalian membranes. A variety of signal transduction components are enriched in caveolae, including G-protein-mediated signaling components, tyrosine kinase-mitogen-activated protein kinase pathway components and calcium-mediated signaling molecules (Shaul and Anderson 1998). Although the caveolae and caveolins are not present in all mammalian cells, membrane domains with the characteristics of DIGs were isolated from different cell types and also from yeasts.

We isolated plasma membrane fractions of higher plants with the characteristics of DIGs. Triton X-100 insoluble membranes from tobacco leaves had slightly higher bouyant density than mammalian DIGs and appeared as vesicles with diameters of 200-500 nm, which tend to aggregate protein complexes (Fig. 1a). Immunolocalization (Fig. 1b) shows that heterotrimeric G-proteins are present in plant Triton X-100 insoluble membranes suggesting their role in G-protein-coupled signaling.

Identification of other components present in this membrane microdomains could enlighten our understanding of plant signaling pathways.

Heterotrimeric G-proteins are proposed to be involved in different plant signaling pathways, however, the supporting evidence relay mostly on the use of G-protein activators and inhibitors, like cholera and pertussis toxin or GTP and GDP analogs (Yang 1996, Hooley 1998). We employed the antisense RNA strategy in order to investigate the function of heterotrimeric G-proteins in chloroplast development.

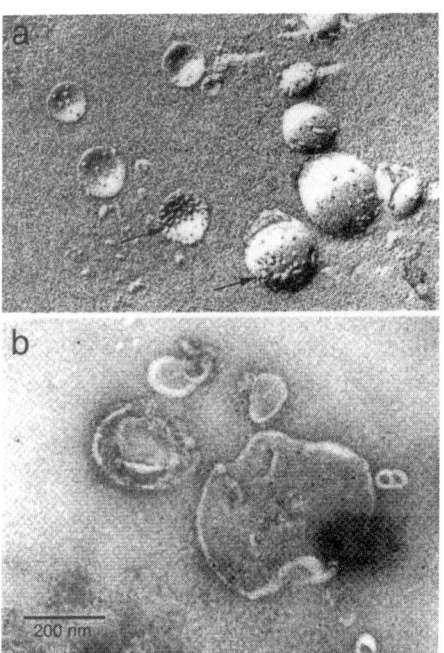

Figure 1. Freeze-fracture electron microscopy of the Triton X-100-insoluble plasma membrane subfraction (a) and immunolocalization of the heterotrimeric G-protein β-subunit (Gβ) (b). Gβ was immunolabeled with 10-nm gold particles on negatively stained Triton X-100-insoluble vesicles.

Transgenic tobacco lines with up to 75% reduced level of the heterotrimeric G-protein β-subunit (Gβ) were analysed (A15-6, Fig. 2). The transgenic line did not differ from the wild type plants under the long day conditions, but when the dark-grown seedlings were incubated in the light for several hours, the antisense line showed slower accumulation of chlorophyll and less efficient photosynthesis when compared to wild type seedlings. The photosynthesis yield of the antisense seedlings was reduced by 37% after 6 hours incubation in light and by 20% after 16 or 24 hours of illumination (Tab. 1). The components of the photosynthetic apparatus are obviously accumulating slower in the antisense line, suggesting a function of the heterotrimeric Gβ in the regulation and / or stability of the chloroplast proteins.

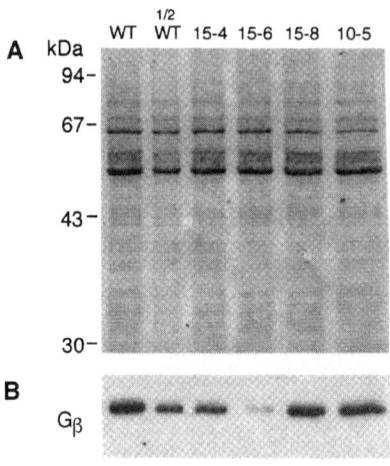

Figure 2. Gβ content in the leaf microsomal fraction of the antisense lines: 15-4, 15-6, 15-8 and 10-5, compared to wild type tobacco (WT). The same amount of proteins was separated on a 12.5% SDS- polyacrylamidgel (A) and analysed with the anti-Gβ antibody (B).

2.2 HOW DO PLASTIDS CONTROL NUCLEAR GENE EXPRESSION AND DO HIGHER PLANTS POSSES CHLOROPLAST-LOCALIZED RESPONSE REGULATORS?

More than 15 years ago it was discovered that plastids control the expression of those genes in the nucleus which code for plastid proteins or for proteins with functions related to plastids. In the meantime a number of factors have been identified and postulated which might be involved in the crosstalk between both organelles, such as chlorophyll precursors, sugars, proteins synthesized on plastid ribosomes, or components related to the redox-state of the plastids, *etc*. None of these factors alone is sufficient to explain the whole scenario. The crosstalk is more likely the result of the sum of all of these factors, which might influence the metabolic state in the cytoplasm and - as a consequence - influences regulatory factors or pathways involved in the expression of nuclear genes for plastid proteins. It is still enigmatic whether protein factors are directly involved in the crosstalk. We are currently investigating one putative class of proteins, which might be involved in these processes, the so-called ‚two-component systems' (reviewed in Chang and Stewart 1998, D´Agostino and Kieber 1999, Urao *et al* 2000, Sakakibara *et al* 2000).

Table 1. Chlorophyll *a*-fluorescence measurement with the antisense lines 15-6 and 10-5 compared to wild type. The parameters were according to Genty *et al* (1989): Φ_{II}, effective quantum yield; F_v/F_m, maximal quantum yield; qP, photochemical quenching; qN, non-photochemical quenching.

6 h light	Φ_{II}	F_v/F_m	qP	qN
WT	0,22	0,294	0,8	0,4
10-5	0,27	0,374	0,9	0,42
15-6	0,14	0,212	0,5	0,44
11 h light	Φ_{II}	F_v/F_m	qP	qN
WT	0,47	0,598	0,8	0,27
10-5	0,46	0,628	0,9	0,3
15-6	0,38	0,510	0,8	0,3
16 h light	Φ_{II}	F_v/F_m	qP	qN
WT	0,58	0,720	0,9	0,23
10-5	0,56	0,746	0,85	0,34
15-6	0,46	0,617	0,8	0,25

"Two component systems" operate in a new class of plant signaling pathways and were first discovered in the prokaryotic kingdom. This signaling system is wide spread among all classes of bacteria and consists of a sensor histidine kinase (mostly membrane-associated) and an interacting response regulator, which acts commonly as a regulator of transcription. In response to an environmental signal which is detected by a N-terminal ‚sensor domaine' the kinase undergoes autophosphorylation at a highly conserved histidine residue in the C-terminal region. This phosphoryl group is then transferred to a highly conserved aspartate residue within the N-terminal ‚receiver domaine' of the response regulator. The phosphorylation event activates the regulator which then initiates a physiological response (for example transcription) *via* its C-terminal ‚output domaine'. The available sequence information of the *Arabidopsis* genome has provided us with a new tool for the identification of such new genes. Databank analyses revealed that *Arabidopsis* contains a gene family of at least 14 different response regulators (Imamura *et al* 1999). Detailed structural analysis of the *Arabidopsis* response regulator (ARR) gene family revealed that the identified genes can be divided into two distinct groups, 7 small genes (encoding proteins with 184-234 amino acid residues; type A ARR) and 7 large genes (encoding proteins with 382-669 amino acid residues; type B ARR). Both types show high homologies to each other within the conserved N-terminal ‚receiver domaine' but differ in their C-terminal ‚output domaines' (data not shown). Those of type B ARRs show long gene-specific extensions and include nuclear localisation sequences (NLSs) as well as a motif related to the nuclear transcription factor Myb (Imamura *et al* 1999). For the type B ARR11 it was shown that it is indeed nuclear-localised and that the C-terminal domain acts as a transcription activating region (Lohrmann *et al* 1999). Type A ARRs have very short ‚output domains' lacking any recognisable NLS (Fig. 3). Nothing is known about their cellular localisation and physiological role, except that their transcripts accumulate in response to cytokinine or nitrate. Since type A ARRs are similar in size to those of photosynthetic bacteria, since photosynthetic reaction centre genes in higher plant chloroplasts are redox-controlled (Pfannschmidt *et al* 1999), and since it was hypothesised that redox controlled gene transcription in higher plant chloroplasts is regulated *via* bacteria-like ‚two-component' systems (Allen 1993, Race *et al* 1999), we were interested to determine whether some members of this subfamily are targeted into the chloroplasts.

Initially, we isolated all known ARR genes from our *Arabidopsis* cDNA libraries. During this procedure we identified a new member of the gene family, named ARR15, according to the nomenclature proposed by Imamura *et al* (1999). Its structural design clearly identifies ARR15 as a type A ARR (Fig. 3). The ChloroP-programme (Emanuelson *et al* 1999) did not predict any chloroplast transit peptide within type A ARRs, while the Psort Prediction programme (HYPERLINK http://psort.nibb.ac.jp/ form.html; http://psort.nibb.ac.jp/form.html) suggested the existance of such sequences for a few ARRs (cf. also Abdallah *et al* 2000). More detailed sequence comparisons of type A ARRs with bacterial and organelle-encoded RRs involved in photosynthesis or redox signaling revealed that ARR 3,4,5,6 and 15 contains N-terminal extensions with similarities to transit peptides (protein start = MA; characteristic S residues). The bacterial prrA (*Rhodobacter sphaeroides*, Eraso and Kaplan 2000) and rppA (*Synechocystis* PCC 6803, Li and Sherman 2000, cf. Fig. 3) are known to be involved in photosynthesis gene expression, while *ycf27* and *ycf29*, both putative response regulator genes located on the plastome of *Cyanophora paradoxa* appear to be part of an redox signalling pathway. The non-photosynthesis-related cheY response regulator from *E. coli* involved in chemotaxis, which was included as a control, does not contain such an extension.

ARR8 and 9 do not contain those N-terminal extensions, however they contain an additional short segment within the receiver domain which is not (or only to a small extent) present in the sequences of ARRs 3, 4, 5, 6 and 15. ARR 8 and 9 therefore seem to represent a different class of type A ARRs and we propose to distinguish between type A1 and type A2. The availability of all of these cDNAs in appropriate vectors will allow us to perfom chloroplast and mitochondrial import experiments and to answer the question, whether higher plant organelles contain ARRs.

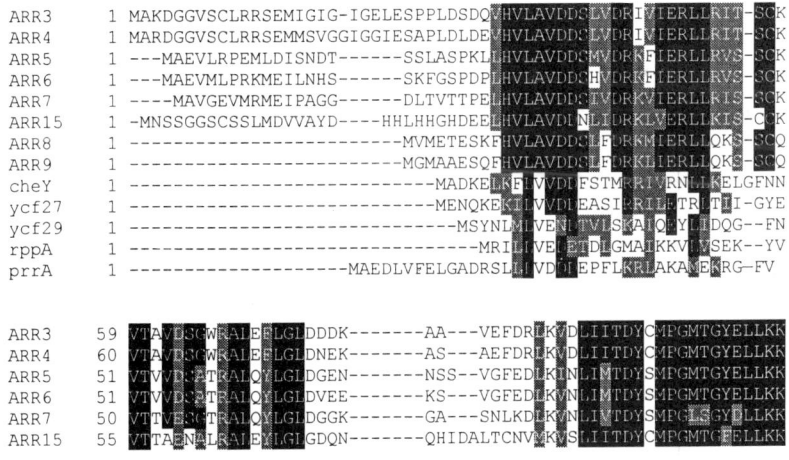

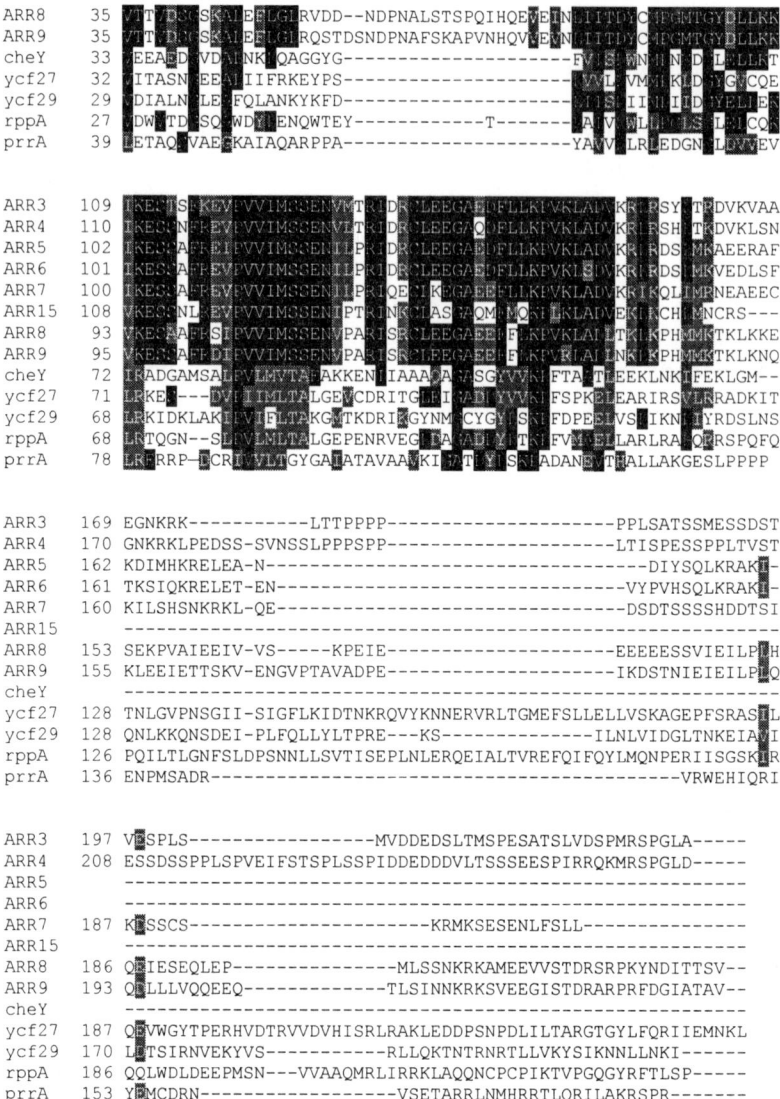

Figure 3. The alignment has been performed with the ClustalW multiple alignment application (HYPERLINK http://dot.imgen.bcm.tmc.edu:9331/multi-align/multi-align.html) in the global alignment modus. Shadowing was introduced by the boxshade link from the same site. Black amino acid residues are identical, grey are homologues. Acc. no. of ARR15: AJ279079.

2.3 THE PHOTOSYNTHESIS GENES *ATPC*, *PSAF* AND *PSAD* CONTAIN UNUSUAL REGULATORY *CIS*-ELEMENTS AT THE DNA AND RNA LEVEL. THE INVOLVEMENT OF PHOSPHOLIPIDS IN PLANT SIGNALING.

Studies during the last decade lead to three mayor conclusions: (i) Whenever gene expression is regulated at the transcriptional level, signals from different regulators converge and operate *via* a small number of defined *trans*-acting factors, which altimately interact with *cis*-regulatory elements in the respective promoters to initiate transcription. (ii) Many *cis*-elements in different gene promoters are conserved and might have evolved from a common ancestor, while others followed their own strategies: although they respond to the same stimuli, gene activation occurs *via* different *cis*- and *trans*-acting factors and *via* different modes of action. (iii) Apparently, regulated expression of nuclear-encoded genes for photosynthesis proteins are also controlled posttranscriptionally. Components and mechanisms involved in the transduction of these signals as well as their interaction are poorly investigated in plants.

(a) *AtpC* - one of three nuclear-encoded genes for the plastid ATP synthase.

The nuclear-encoded genes for the plastid ATP synthase subunit γ exhibits unusual features, because crucial *cis*-regulatory elements which are involved in the regulated expression are located in close vicinity of the transcription start site. More detailed studies uncovered that a single nucleotide immediately upstream of the CAAT box is crucial for the light- and cytokinin-regulated, as well as plastid-dependent expression (Kusnetsov *et al* 1999). If this nucleotide is exchanged by site-directed mutagenesis, chimeric *AtpC* promoter::*uidA* constructs in transgenic tobacco exhibit a constitutive high level of expression. This suggests (i) that different signals which regulate *AtpC* gene expression converge prior to gene regulation, (ii) that different regulators operate *via* the same *cis*-element and (iii) that the high level of expression in the mutant promoter is caused by the release of a repressor which normally prevents *AtpC* gene expression under conditions when transcription is low. Two genes which encode polypeptides involved in this scenario were cloned by Southwestern screens. The first gene encodes subunit C of a CAAT box binding complex previously described for human and yeast (Kusnetsov *et al* 1999). The second gene, AtpC-2, encodes a novel class of protein with similarities to DNA helicases (Bezhani *et al* 2001),

(Fig. 4). A phylogenetic tree of proteins with similarities to other sequences available in the Databanks uncovered that AtpC-2 has a procaryotic origin (Fig. 5), and that there are also homologues in eucaryots. Surpringly the last domain which is conserved in all DNA helicases of this type and which is required for helicase function is lost in AtpC-2. Thus we suggest that the protein has still DNA-binding capability, but lost its helicase activity. Comparison of DNA binding studies, competition experiments with both polypeptides and *in vivo* promoter activities suggest that either the repressor complex or the CAAT-box complex can be formed at the *AtpC* promoter segment. Formation of both complexes simultaneously is not possible for sterical reasons. Whenever the promoter activity is high, the CAAT box complex is assembled and the repressor is not bound, and *vice versa*. The above described mutation in the promoter prevents repressor binding, which results in a constitutive assembly of the CAAT-box-binding complex and thus in a high promoter activity. A model is presented which describes this novel type of gene regulation in plants (Fig. 6).

```
D86988      1   MSVEAYGPSSQTLTFLDTEEAELLGADTQGSEFEFTDFTLPSQTQTPPGGPGGPGGGGAG

AB026643    1                                              MESWKESESWKLLKEI
D86988     61   GPGGAGAGSAAGQLDAQVGPEGILQNGAVDDSVAKTSQLLAELNFEEDEEDTYYTKDLPI

B75105      1                                        MNITSFINRLKELVEIEREAEIEAMR
E71080      1                                        MNIKSFINRLKELVEIEKEAEIEAMR
G72429      1                                        MTVQQFIKKLVRLVELERNAEINAML
D69085      1                                                     MEREAEISAMM
C69423      1                                              MKEFIEGLIRLTEVERDAQISAMM
P40694      1                                     MASSTVESFVAQQLQLLELERDAEVEERR
L24544      1                                     MASAAVESFVTKQLDLLELERDAEVEERR
AJ300306    1                                          MSLEAFVSTMAPLIDMEKEAEISMSL
Q57568      1                              MQNWFKSGDSLNLVDLYVKKFMDLIEIERRCEMDFHK
T41580      1                            MSSSNEQICLDEDWIQKFGDREIEFVDEAQKSEVDETE
P34243      1                                             MNKELASKFLSSIKHEREQDIQTTS
AB026643   17   ANSAQHREVARKAAQAKPVQGVLGMDSEKVKAIQERIDEFTSQMSQLLQVERDTELEVTQ
D86988    121   HACSYCGIHDPACVVYCNTSKKWFCNGRGNTSGSHIVNHLVRAKCKEVTLHKDGPLGETV

B75105     27   LEMRRLSG----------------------------------------------------
E71080     27   LEMKRLSG----------------------------------------------------
G72429     27   DEMKRLSG----------------------------------------------------
D69085     12   NEIRRLSP----------------------------------------------------
C69423     25   DEIRRLSG----------------------------------------------------
P40694     30   SWQEHSSL----------------------------------------------------
L24544     30   SWQENISL----------------------------------------------------
AJ300306   27   TSGASRNI----------------------------------------------------
Q57568     38   NEIIKLG-----------------------------------------------------
T41580     39   KSIKRFPL----------------------------------------------------
P34243     26   RLLTTLSI----------------------------------------------------
AB026643   77   EELDVVPTPDESSD--------------------------------------SSKPI
D86988    181   LECYNCGCRNVFLLGFIPAKADSVVVLLCRQPCASQSSLKDINWDSSQWQPLIQDRCFLS

B75105     35   -------IERERLGRAILNLNG--------------------------------------
E71080     35   -------VERERLGRAILNLNG--------------------------------------
G72429     35   -------EEREKKGRAVLGLTG--------------------------------------
D69085     20   -------RQRERARRAVNGLNG--------------------------------------
C69423     33   -------EKRERKGRAVLGLRG--------------------------------------
P40694     38   -------RELQSRGVCLLKLQVS-------------------------------------
L24544     38   -------KELQSRGVCLLKLQVS-------------------------------------
AJ300306   35   -------ETAQKKGTTILNLKCV-------------------------------------
Q57568     45   -------KKRENVGRAILNLKG--------------------------------------
T41580     47   -------SVLQRKGLALINLRIG-------------------------------------
P34243     34   -------QQLVQNGLAINNIHLE-------------------------------------
AB026643   96   EFLVRHGDAPQELCDTICNLYAV-------------------------------------
D86988    241   WLVKIPSEQEQLRARQITAQQINKLEELWKENPSATLEDLEKPGVDEEPQHVLLRYEDAY
```

```
B75105     50  --------------------KIIGEELGYFLVKYGRNRE-------IKTEISVGDLVVI
E71080     50  --------------------KIAGEELGYFLVKYGRNRE-------IKTEISVGDLVVI
G72429     50  --------------------KFIGEELGYFLVRFGRRKK-------IDTEIGVGDLVLI
D69085     35  --------------------KITGRELGFHLVKYGRRDP-------IDTQISVGDLVLI
C69423     48  --------------------KVVGEELGFKLVRYGRRKA-------IETEISVGDEVLI
P40694     54  --------------------SQRTGLYGQRLVTFEPRKFGP-AVVLPSNSFTSGDIVGL
L24544     54  --------------------SQRTGLYGRLLVTFEPRRYGS-AAALPSNSFTSGDIVGL
AJ300306   51  --------------------DVQTGLMGKSLIEFQSNKG----DALPAHKFGNHDVVVL
Q57568     60  --------------------KFLGESLGCTIVRFGRKKP-------FKTEISPGDVVLV
T41580     63  --------------------VVKTGFGGKTIIDFEKDPAFSNGEELPANSFSPGDVVSI
P34243     50  --------------------NIRSGLIGKLYMELGPNLAVN--DKIQRGDIKVGDIVLV
ABO26643  119  --------------------STSTGLGGMHLVLFKVGGN----HRLPPTTLSPGDMVCI
D86988    301  QYQNIFGPLVKLEADYDKKLKESQTQDNITVRWDLGLNKKRIAYFTLPKTDSGNEDLVII

B75105     82  SKRD------------------PLKSDLLGTVVEKGKRFIVVALETVPEWALRDVR---
E71080     82  SKRD------------------PLKSDLLGTVVEKGKRFIVVALETVPEWALRDVR---
G72429     82  SKGN------------------PLKSDYTGTVVVEKGERFITVAVDRLPSWKLKNVR---
D69085     67  SRGN------------------PLRSDLTGTVAMKGKRFLVVALEHVPGWALKNVR---
C69423     80  SRGD------------------PLKSDLRGVVVEKGSRYLTVSLESVPEWALRDVR---
P40694     92  YDTN------------------ENSQLATGVLTRITQKSVTVAFDESHDLQLNLDREN-
L24544     92  YDAAN-----------------EGSQLATGILTRVTQKSVTVAFDESHDFQLSLDREN-
AJ300306   86  KLNKSDL---------------GSSPLAQGVVYRLKDSSITVVFDEVPEEGLNTS----
Q57568     92  SKEN------------------PLQSDLYANVIYVGKNFIDVAFDVDVPKWVYKER---
T41580    102  RQDFQSS-----------KKKRPNETDISVEGVVTRVHERHISVALKSEEDIPSSVTR--
P34243     87  RPATKTVNTKTKPKVKKVSEDSNGEQAECSGVVYKMSDTQITIALEESQDVIATTFYSYS
ABO26643  154  RVCDSRG---------------AGATACTQGFVHNLGEDGCS1GVALESRHGDPTFSKLF
D86988    361  WLRDMRLMQGDEICLRYKG-DLAPLWKGIGHVIKVPDNYGDEIAIELRSSVGAPVEVTHN

B75105    120  --------IDLYANDITFRRWLENLNTIKK------AGKRALRFYLGLDEPSQ-------
E71080    120  --------IDLYANDITFRRWLENLDRVKK------AGKRALEFYLGLDEPSQ-------
G72429    120  --------IDLFASDITFRRQIENLMTLSS------EGKKALEFLLGKRKPEE-------
D69085    105  --------IDLYANDVTFQRMIDNLKSPTR------NVLRVLGFLCGTEKPSDG------
C69423    118  --------IDLYASDLTFKRWIENLENLTE------NGKRALKFALGLEEPSK-------
P40694    132  -----TYRLLKLANDVTYKRLKKALMTLKK---YHSGPASSLIDILLGSSTPSP------
L24544    133  -----SYRLLKLANDVTYRRLKKALIALKK---YHSGPASSLIEVLFGRSAPSP------
AJ300306  126  ------LRLEKLANEVTYRRMKDTLIQLSKG--VLRGPASDLVPVLFGERQPSVS-----
Q57568    130  -----VRVDLYVNDITFKRMKEALREFAR-------KRDKLAYIILGIEHPEKP------
T41580    149  ------LSVVKLVNRVTYERMRHTMLEFKRS--IPEYRNSLFYTLIGRKKADVSID----
P34243    147  KLYILKTTNVVTYNRMESMTMRKLSEI--S-SPIQDKIIQYLVNERPFIPNT---
ABO26643  199  GKSVRIDRIHGLADALTYERNCEALMLLQKNGLQKKNPSISVVATLFGDGEDITWLEQND
D86988    420  ------FQVDFVWKSTSFDRMQSALKTFAVD--ETSVSGYIYHKLLGHEVEDVIIK----

B75105    159  -----GEEVNFVPFDKSLNRSQRKAISKALGS-EDFFLVHGPFGTGKTRTLVELIRQEVK
E71080    159  -----GEEVSFEPFDKSLNSSQRKAIAKGLGS-SDFFLIHGPFGTGKTRTLVELIRQEVK
G72429    159  -----SFEEEFTPFDEGLNESQREAVSLALGS-SDFFLIHGPFGTGKTRTLVEYIRQEVA
D69085    145  -----VDVVDFQAVDPELNESQRDAIRMALGS-EDFFLIHGPFGTGKTRTLHELIRQEVM
C69423    157  -----TECEDFKPFDSSLNRAQLKAVGCAVST-DDFFLIHGPFGTGKTRTVVEVVRQLVK
P40694    178  -----AMEIPPLSFYNTTLDLSQKEAVSFALAQ-KELAIIHGPPGTGKTTTVVEIILQAVK
L24544    179  -----ASEIHPLTFFNTCLDTSQKEAVLFALSQ-KELAIIHGPPGTGKTTTVVEIILQAVK
AJ300306  173  -----KKDVKSFTPFNKNLDQSQKDAITKALSS-KDVFLLHGPPGTGKTTTVVEIVLQEVK
Q57568    172  -----LREDIKLEFYDKNLNESQKLAVKKAVLS-RDLYLLHGPPGTGKTRTITEVIVQEVK
T41580    197  -----QKLIGDIKYFNKELNASQKKAVKFSIAV-KELSLLHGPPGTGKTHTLVEIIQQLVL
P34243    195  -----NSFQNIKSFLNPNLNDSQKTAINFAIN--NDLTIIHGPPGTGKTFTLIELIQQLLI
ABO26643  259  YVDWSEAELSDEPVSKLFDSSQRRAIALGVNKKRPVMIVQGPPGTGKTGMLKEVITLAVQ
D86988    468  ---CQLPKRFTAQGLPDLNHSQVYAVKTVLQR--PLSLIQGPPGTGKTVTSATIVYHLAR
                                                                *****I******

B75105    213  R--GNKVLATAESNVAVDNLVERLSRSG--IKIVRIGHPSR--VSKHLHETTLAYLITQH
E71080    213  R--GNKVLATAESNVAVDNLVERLAKDG--VKIVRVGHPSR--VSRHLHETTLAYLITQH
G72429    213  R--GKKILVTAESNLAVDNLVERLWG-K--VSLVRIGHPSR--VSSHLKESTLAHQIETS
D69085    199  R--GSRVLVTAESNAAVDNLLEGIAG-H--VKCVRLGHPDP--VSRENLRETLAYKIENH
C69423    211  R--GERVLVTAESNTAVDNLVELLSD----MKIVRLGHPSR--VEKRLKEHTLASLVLNH
P40694    233  Q--GLKVLCCAPSNIAVDNLVERLALCK--KRILRLGHPAR--LLESVQHHSLDAVLARS
L24544    234  Q--GLKVLCCAPSNIAVDNLVERLALCK--QRILRLGHPAR--LLESIQQHSLDAVLARS
AJ300306  228  R--GSKILACAASNIAVDNIVERLVPHK--VKLVRVGHPAR--LLPQVLDSALDAQVLKG
Q57568    227  FN-KHKVLATADSNIAADNIIEYLIKKYPDLKVVRVGHPTR--ISKDLIQHSLPYLIENH
T41580    252  R--NKRILVCGASNLAVDNIVDRLSSSG--IPMVRLGHPAR--LLPSILDHSLDVLSRTG
P34243    249  KNPEERILICGPSNISVDTILERLTPLVPNNLLLRIGHPAR--LLDSNKRHSLDILSKK-
ABO26643  319  Q--GERVLVTAPTNAAVDNMVEKLLHLG--LNIVRVGNPAR--ISSAVASKSLSGEIVNSK
D86988    523  QG-NGPVLVCAPSNIAVDQLTEKIHQTG--LKVVRLCAKSREAIDSPVSFLALHNQIRNM
                ********I********

B75105    267  ELYGELRELRVIGQSLAEKRDTYTKPTPKFRRGLSDEEIIKLAERKRGARGLSARLIMEM
E71080    267  ELYGELRELRVIGQSLAEKRDTYTKPTPKFRRGLSDEEIIKLAERKRGIRGLSARLIKEM
G72429    266  SEYEKVKKMKEELAKLIKKRDSFTKPSPQWRRGLSDKKILEYAEKNWSARGVSKEKIKEM
D69085    252  PEYSKVREYQEKIDELIEERERHQKPTPQIRRGLSDTQILINATKRRGARGISPNVMISM
C69423    263  PDYKRIEEIKGKIEEIERRMERLTKPSPQLRRGLSDEEILRLARSNRGARGVAAKKIRSM
P40694    287  DNAQIVADIRRDIDQVFGK--N---------------------------KKTQDK
L24544    288  DSAQIVADIRKDIDQVFVK--N---------------------------KKTQDK
AJ300306  282  DNSGLANDIRKEMKALNGK--L---------------------------LKAKDK
Q57568    284  EKYQEILALREKIKEIKEQRDKFLKPSPRWWRRGMSDEQILKVAKRKKSYRGIPKEKIVSM
T41580    306  DNGDVIRGISEDIDVCLSK--------------------------------ITKTKNG
P34243    306  N--TIVKDISQEIDKLIQE--------------------------------NKKLKNY
ABO26643  373  LASFRAELERKKSDLRKDLR--------------------------------QCLRDD
D86988    580  DSMPELQKLQQLKDETGELS

B75105    327  AEWIKLNRQVQKAFDDARKLEERIARDIIREADVVLTTNSSAALEVVD--------YDTY
E71080    327  AEWIKLNRQVQKAFEDARKLEERIARDIIREADVLTTNSSAALEVVD---------ATDY
G72429    326  AEWIKLNSQIQDIRDLIERKEEIIASRIVREAQVVLSTNSSAALEILS--------GIVF
```

```
D69085     312  ARWIETNQRIDDLHSKLQEAEMRIADRILRESQVVLSTNSSAALEYID--------GLRF
C69423     323  AEWIEARKALDQLYTEMKEEEERIVKEIIEESDVVISTNSSAFLLEES----------F
P40694     313  REKGNFRSEIKLLRKELKEREEAAIVQSLTAADVVLATNTGASSDGPLKLL---PE-DYF
L24544     314  REKSNFRNEIKLLRKELKEREEAAMLESLTSANVVLATNTGASADGPLKLL---PE-SYF
AJ300306   308  NTRRLIQKELRTLGKEERKRQQLAVSDVIKNADVILTTLTGALTR----KL---DN-RTF
Q57568     344  AEWIIRNKKIKRIINNLDEITEKIMNEILAEADVIVATNSMAGSEILKG--------WEF
T41580     332  RERREIYKNIRELRKDYRKYEAKTVANIVSASKVVFCTLHGAGSRQLK--------GQRF
P34243     330  KQRENWNEIKLLRKDLKKREFKTIKDLIIQSRIVVTTLHGSSSRELCSLYRDDPNFQLF
ABO26643   399  VLAAGIRQLLKQLGKTLKKKEKETVKEILSNAQVVFATNIGAADPLIRR-------LETF
D86988     600  ----------SADEKRYRALKRTAERELLMNADVICCTCVGAGDPRLAK--------MQF

B75105     379  DVAIIDEATQSTIPSILIPLNKVE----RFVLAGDHKQLPPTILSL----EAQELSRTLF
E71080     379  DVAIIDEATQATIPSILIPLNKVD----RFILAGDHKQLPPTILSL----EAQELSHTLF
G72429     378  DVVVVDEASQATIPSILIPISKGK----KFVLAGDHKQLPPTILSE----DAKDLSRTLF
D69085     364  DVAIVDEASQATIPSILIPLARAP----RFILAGDHKQLPPTILSR----DASELERTLF
C69423     372  DTAVIDEASQATIPSVLIPINRAR----KFILAGDHRQLPPTVM------KAEKLSETLF
P40694     369  DVVVVDECAQALEASCWIPLLKAP----KCILAGDHKQLPPTTVSHR--AALAGLSRSLM
L24544     370  DVVVVIDECAQALEASCWIPLLKAR----KCILAGDHKQLPPTTVSHK--AALAGLSLSLM
AJ300306   360  DLVIIDEGAQALEVACWIALLKGS----RCILAGDHLQLPPTIQSAE--AERKGLGRTLF
Q57568     396  DVIVVIDEGSQAMEPSCLIPIVKGR----KLIMAGDHKQLPPTVLS------ENEELKKTLF
T41580     384  DAVIIDEASQALEPQCWIPLLGMN----KVILAGDHMQLSPNVQS-----KRPYIS--MF
P34243     390  DTLIIDEVSQAMEPQCWIPLIAHQNQFHKLVLAGDNKQLPPTIKTEDDKNVIHNLETTLF
ABO26643   452  DLVVIDEAGQSIEPSCWIPILQGK----RCILSGDPCQLAPVVLSR--KALEGGLGVSLL
D86988     642  RSILIDESTQATEPECMVPVVLGAK---QLILVGDHCQLGPVVMCKK--AAKAGLSQSLF
                **II***                    ***III***

B75105     431  EGLIERYPWKSE--MLVVQYRMNERIMEFPSKEFYGGKIIADESVRGITLRDLVEYQ-SP
E71080     431  EGLIERYPWKSE--MLTIQYRMNERIMEFPSKEFYDGKIVADERVKNITLGDLGIKV-NA
G72429     430  EELITRYPEKSS--LLDTQYRMNELLMEFPSEEFYDGKLKAAEKVRNITLFDLGVEIPNF
D69085     416  EELIKRHPGRSR--MLNCQYRMNEPAIMEFPNREFYDGRIRAHPSLEDISIRDIIEDV--P
C69423     422  EKLIELYPEKSQ--LLNVQYRMNEKLMEFPSREFYGGRIVAHESCTAIALSQIAKRE---
P40694     423  ERLAEKHGAGVVR-MLTVQYRMHQAIMCWASEAMYHGQFTSHPSVAGHLLKDLPGVTD--
L24544     424  ERLAEEYGARVVR-TLTVQYRMHQAIMRWASDTMYLGQLTAHSSVARHLLRDLPGVAA--
AJ300306   414  ERLADLYGDEIKS-MLTVQYRMHELIMNWSSKELYDNKITAHSSVASHMLFDLENVTK--
Q57568     447  ERLIKKYPEFSS--LIEIQYRMNEKIMEFPNKMFYNNKLKADESVKNITLLDLVKEEE--
T41580     433  ERLVKSQGDLVKC-FLNIQYRMHELISKFPSDTFYDSKLVPAEEVVKKRLLMDLENVEET
P34243     450  DRIIKIFPKRDMVKFLNVQYRMNQKIMEFPSHMSHMYMLLADATVANRLLIDLPTVDATP
ABO26643   506  ERAASLHDGVLAT-KLTTQYRMNDVIAGWASKEMYGGWLKSAPSVASHLLIDSPFVKATW
D86988     697  ERLVVLGIRPIR--LQVQYRMHPALSAFPSNIFYEGSLQNGVTAADRVKKGFDFQWP--
                              ****IV****

B75105     488  NDSWGKILNPENVIVFIDTSKAENK-----WERQRRGSESRENPLEAEIVAKIVDKLLSI
E71080     488  TGIWRDILDPSNVLVFIDTCMLDNR-----FERQRRGSESRENPLEAKIVSKIVEKLLES
G72429     488  GKFWDVVLSPKNVLVFIDTKNRSDR-----FERQRKDSPSRENPLEAQIVKEVVEKLLSM
D69085     472  DSDICQKLADPDPVLFIDTSGLDG------CERRLKGSTSIQNPLEADLAVIISRSLMRM
C69423     477  -AEKLREILGDEPLVFIDTSKCKNR-----WEGKLADSTSRYNRLEAEIVTEIVTELLKM
P40694     480  -------TEETRVPLLLIDTAGCGL-----FELEEEDSQSKGNPGEVRLVTLHIQALVDA
L24544     481  -------TEETGVPLLLVDTAGCGL-----FELEEEDEQSKGNPGEVRLVSLHIQALVDA
AJ300306   471  -------SSSTEATLLLVDTAGCDM-----EEKKDE-EESTYNEGEAEVAMAHAKRLMES
Q57568     503  ---IDEVDRDIINEIPVQFINVG-------IERKDKESPSYYNIEEAEKVLEIVKKLVKY
T41580     491  --------ELTDSPIYFYDTLGNYQEDDRSEDMQNFYQDSKSNHWEAQIVSYHISGLLEA
P34243     510  ----SEDDDDTKIPLIWYDTQGDE--FQETADEATILGSKYNEGEIAIVKEHIENLRSF
ABO26643   565  ITQCPLVLLDTRMPYGSLSVGCEER------LDPAGTGSLYNEGEADIVVNHVISLIYA
D86988     752  --------Q-PDKPMFFYVTQGQE--------EIASSGTSYLNRTEAANVEKITTKLLKA

B75105     543  GVKPEWIGVITPYDDQRDLISMKV----------PEDVEVKTVDGYQGREKEVIILSLVR
E71080     543  GVKAEMIGVITPYDDQRDLISLM-----------PEEVEVKTVDGYQGREKEVIILSFVR
G72429     543  GVKEDWIGIITPYDDQVNLIRELI----------EAKVEVHSVDGFQGREKEVIIISFVR
D69085     526  GVKPEEIGIITPYDDQVDLISS------------MIDVEVNSVDGFQGREKEDVIIISMVR
C69423     531  GLKKEQIGVITPYDDQVDLLREK-----------VDVEVSSVDGFQGREKEVIIISFVR
P40694     528  GVQAGDIAVIAPYNLQVDLLRQS--LS-----NKHPELEIKSVDGFQGREKEAVLLTFVR
L24544     529  GVPARDIAVVSPYNLQVDLLRQS--LV-----HRHPELEIKSVDGFQGREKEAVLSFVR
AJ300306   518  GVQPSDIGIITPYAAQVMLLRILRGKE-----EKLKDMEISTVDGFQGREKEAIIISMVR
Q57568     554  KIPTN---VITPYDAQVRYLRRLFEEH----N--IDIEVMTVDGFQGRENEAIVISFVR
T41580     543  GLEAKDIAVVFPYNAQVALIRQLLK-------EKGIEVEMGSVDKVQGREKEAIIFSLVR
P34243     563  NVPENSIGVISPYNAQVSHLKKLIHDE-----LKLTDIEISTVDGFQGREKDVIILSLVR
ABO26643   618  GVSPMAIAVQSPYVAQVQLLRERLDDF-----PVADGVEVATIDSFQGREADAVIISMVR
D86988     795  GAKPDQIGIITPYEGQRSYLVQYMQFSGSLHTKLYQEVEIASVDAFQGREKDFIILSCVR
                                                    >----------------CON-

B75105     593  SNKLGEIGFLKDLRRLNVSLTRAKRKLIVVGDSSTLSSH--ETYKKMIEFFKEKGCFIEF
E71080     593  SNKVGEIGFLKDLRRLNVSLTRAKRKLIMIGDSSTLSSH--ETYKRLIEHVREKGLYVVL
G72429     593  SNKNGEIGFLEDLRRLNVSLTRAKRKLIATGDSSTLSVH--PTYRRFVEFVKKKGTYVIF
D69085     574  SNRNGSIGFLKDLRRLNVSLTRARRKLIIIGDSRTLSAH--PSYRRLTEFCRKRGFLDEA
C69423     579  SNRKEIGFLDDLRRLNVSLTRARRKLIIMVGDSETLSVN--GTYARLIDHVKRKGVYVEL
P40694     581  SNRKGEVGFLAEDRRINVAVTRARRHVAVICDSHTVNNH--AFLETLVDYFTEHGEVRTA
L24544     582  SNRKGEVGFLAEDRRINVAVTRARRHVAVICDSRTVNNH--AFLKTLVEYFTQHGEVRTA
AJ300306   573  SNSKKEVGFLKDQRRMNVAVTRSRRQCCIVCDTETVSSD--AFLKRMIEYFEEHGEYLSA
Q57568     604  TKN---FGFLKDLRRLNVAITRARKLILIGNENLLKQD--KVYNEMIKWAKSVEEEHKN
T41580     596  SNDVREVGFLAEKRRLNVAITRPKRHLCVIGDSNTVKWAS-EFFHQWVDFLEENAIVMDI
P34243     618  SNEKFEVGFLKEERRLNVAMTRPRRQLVVVGNIEVLQRCGNKYLKSWSEWCEENADVRYP
ABO26643   673  SNNLGAVGFLGDSRRMNVAITRARKHVAVVCDSSTICHN--TFLARLLRHIRYFGRVKHA
D86988     855  ANEHQGIGFLNDPRRLNVALTRARYGVIIVGNPKALSKQ--PLWNHLLNYYKEQKVLVEG
                -SERVED BOX------------------<

B75105     651  PFS
E71080     651  TKDGIR
D69085     632  GLDDVRKWGAS
```

```
C69423      637  DKNGKLGGNPKG
P40694      639  FEYLDDIVPENYTHEGSQGHSRVPK-PKCPSTSIRKPASDQESGQETRAAPRHGRRKPSE
L24544      640  FEYLDDIVPENYSHENSQGSSHAATKPQGPATSTRTGSQRQEGGQEAAAPARQGRKKPAG
AJ300306    631  SEYTN
Q57568      659  KIIQK
T41580      655  DATMFE
P34243      678  NIDDYL
AB026643    731  DPGSLGGSGLGLDPMLPYLG
D86988      913  PLNNLRESLMQFSKPRKLVNTINPGARFMTTAMYDAREAIIPGSVYDRSSQGRPSSMYFQ
                 *****V*****                      ****VI*****

P40694      698  KPPGSHVQSQHSSSANGSDRTGGPDRTEHFRATIEEFVASKESQLEFPTSLSSHDRLRVH
L24544      700  KSLASEAPSQPSLNGGSPEGVESQDGVDHFRAMIVEFMASKKMQLEFPPSLNSHDRLRVH
D86988      973  THDQIGMISAGPSHVAAMNIPIPFNLVMPPMPPPGYFGQANGPAAGRGTPKGKTGRGGRQ

P40694      758  QLAEEFGLRHDSTGEGKARHITVSRRSPASSGSVAPQPSSP-PSPAQAEPEPRAEEPVTV
L24544      760  QIAEEHGLRHDSSGEGKRRFITVSKRAPRPRAALGPPAGTGGPAPLQPVPPTPAQTEQPP
D86988     1033  KNRFGLPGPSQTNLPNSQASQDVASQPFSQGALTQGYISMSQPSQMSQPGLSQPELSQDS

P40694      817  VQAHCPVQLDLKALHLERLQRQQSSQAQTAKGQPGGDSRPQKASQKKKKKEPKGPVMALP
L24544      820  REQRGPDQPDLRTLHLERLQRVRSAQGQPASKEQQASGQ-QKLPEKKKKKAKGHPATDLP
D86988     1093  YLGDEFKSQIDVALSQDSTYQGERAYQHGGVTGLSQY

P40694      877  CEEDFDALVSAVVKADNTCSFSKCSVSTTTLGQFCMHCSHRYYLSHHLPEIHGCGEKARA
L24544      879  TEEDFEALVSAAVKADNTCGFAKCTAGVTTLGQFCQLCSRRYCLSHHLPEIHGCGERARA

P40694      937  HARQRISREGVLYAGSGTKDRALDPAKRAQLQRRLDKKLGELSSQRTSRKKEKERGT
L24544      939  HARQRISREGVLYAGSGTKNGSLDPAKRAQLQRRLDKKLSELSNQRTSRRKERGT
```

Figure 4. Sequence comparison of AtpC-2 with representative polypeptides in the Databanks with amino acid sequence identities >30%. The Clustal W1,8 program at BCM Search Launcher (www.hggsc.bcm.tmc.edu/search/laungher) was used for these alignments. Identical and conserved amino acid residues are presented in gray. The helicase boxes I, Ia, II-VI are indicated (xxIxx). Note the segment between the boxes IV and V with the highest degree of sequence similarities among all proteins (>Conserved Box<). Numbering of the amino acid residues on the left begins at the initiator methionine. The Databank accession numbers (left) refer to the following proteins: B75105, putative DNA helicase from *Pyrococcus abyssi*; E71080, putative DNA helicase from *Pyrococcus horikoshii*; G72429, hypothetical protein TM0005 from *Thermotoga maritima*; D69085, transcription control factor enhancer-binding protein from *Methanobacterium thermoautotrophicum*; C69423, a putative DNA helicase from *Archaeoglobus fulgidus*; P40694, DNA-binding protein SMUBP-2 (immunoglobulin MU binding protein) from mouse; L24544, a DNA helicase from human; AJ300306, the *Arabidopsis* protein AtpC-2 analysed in this study; Q57568, hypothetical ATP-binding protein from *Methanococcus jannaschii*; T41580, putative DNA-binding protein from yeast; P34243, helicase A from yeast; AB026643, DNA helicase-like protein from *Arabidopsis thaliana*; D86988, protein with unknown function from human.

(b) Regulated expression of the gene for the subunit III of photosystem I (*PsaF*) involves phospholipids.

Phospholipids are involved in many signaling pathways in animal systems. Although all components have also been identified in plants, little is know about their role in signal transduction. Many examples demonstrate that phospholipids play an essential role in plant growth and development (summarised in Sopory *et al* 1999). By analysing transgenic tobacco seedlings with promoter::*uidA* fusions from photosystem I genes from spinach, we discovered that the promoter region from *PsaF*, different from those of other photosystem I genes, can be stimulated by pertussis toxin, an activator of heterotrimeric G proteins in plants, serotonin, a stimulator of the turnover of phosphatidyl inositols, phorbol myristate acetate (PMA) and Ca^{2+} (Fig. 7 Top). PMA is a homolog of diacylglycerol, one of the two messengers generated by the PI cycle, which activates protein kinase C. The two signaling pathways initiated by the PI cycle operate at different promoter regions and require different signaling components. Our present understanding of the signaling system is summarized in Fig. 7 (Bottom). Diacylglycerol activates a so far unidentified protein kinase, which in turn requires phosphatase activity to stimulate gene expression *via* a promoter region located upstream of -220. In contrast, the IP_3-dependent pathway releases Ca^{2+} from internal stores and requires kinase activity to promote *PsaF* gene expression *via* a 42 bp segment located at position -220/-178. The involvement of heterotrimeric G proteins, the PI cycle as well as Ca^{2+} in this signaling pathway could be demonstrated both by uptake experiments of the agonists through the roots and by electroporating them into cotyledons of transgenic tobacco plants.

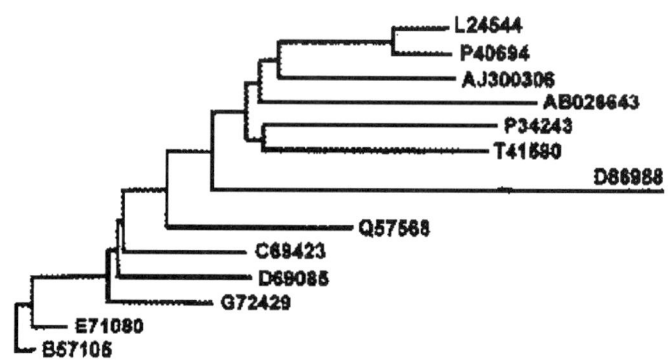

Figure 5. A phylogenetic tree of proteins with more than 30% amino acid identity to AtpC-2 available in the Databanks. Multiple sequence alignments were performed with the Clustal W1,8 program. The tree was generated in TREECON for Windows. The accession numbers correspond to those given in Fig. 4.

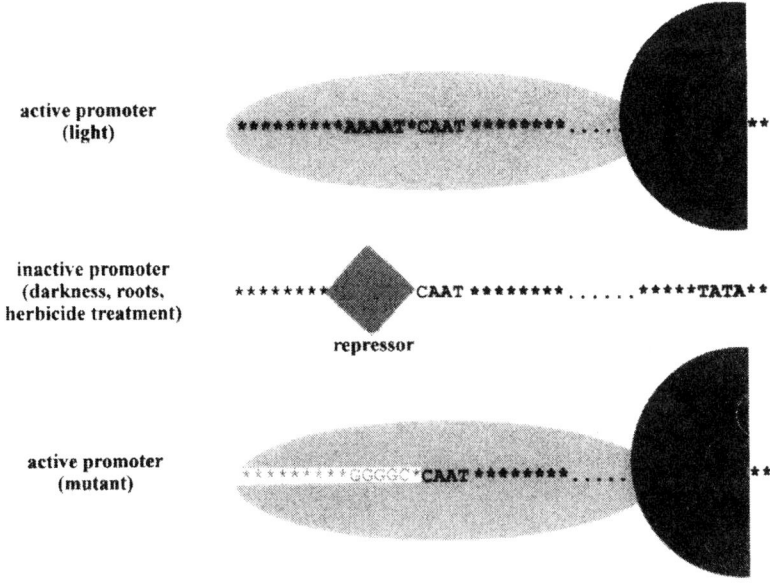

Figure 6. A model describing *AtpC* gene regulation. If the promoter is active the CAAT-box binding complex assembles at the CAAT box and covers an approximately 22 bp region with the central CAAT motif. This allows an efficient assembly of the TATA box complex. Binding of the repressor immediately upstream of the CAAT sequence prevents CAAT box complex assembly. A mutant promoter sequence which prevents the repressor binding results in a constitutive binding of the CAAT box binding protein and thus a constitutively active promoter. Nucleotides located immediately upstream of the CAAT box for which we have generated site-directed mutants and which direct a high constitutive expression in transgenic tobacco are indicated.

Interestingly, the proposed pathway operates only in tissue with functional plastids. If the plastids are impaired by herbicide treatments, application of the above mentioned agonists does not lead to the expression of the transgene. However, modulation of phosphatase and kinase activities activate the responsive *PsaF* promoter regions even in photobleached material. This suggests that activation of kinase and phosphatase activities can by-pass the plastid-mediated block in *PsaF* gene expression (Chandok *et al* 2001).

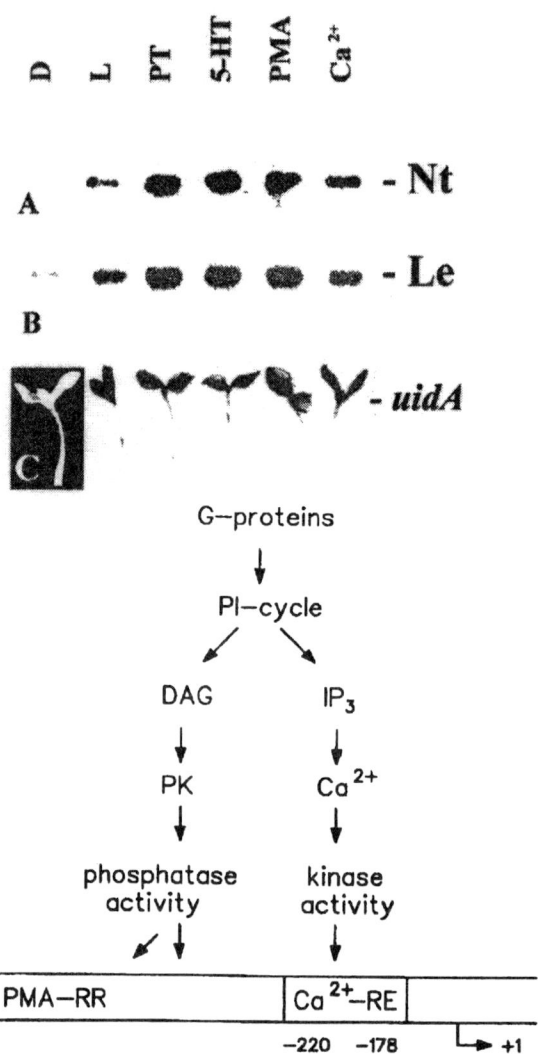

Figure 7 (Top). Northern analysis of etiolated (D) or light-grown (L) tobacco (Nt, panel A) and tomato (Le, panel B) seedlings or of etiolated seedlings treated with PT, 5-HT, PMA or Ca^{2+}. Panel C: transgenic tobacco seedlings in which the *uidA* reporter gene is expressed under the control of the -687/+163 *PsaF* promoter from spinach. (Bottom) A scheme representing the signaling pathways activating *PsaF* gene expression. PI, phosphoinositol; DAG, diacylglycerol; IP_3, inositol triphosphate. The bar at the bottom represents the *PsaF* promoter region; +1 is the transcription start site; Ca^{2+}-RE, Ca^{2+} responsive element; PMA-RR, PMA responsive region.

(c) *PsaD* gene expression involves posttranscriptional processes.

Expression of the *PsaD* gene from different organisms does not fit into the normal concept of promoter-mediated gene expression and additional mechanisms such as gene-internal elements are required to explain the expression of these genes. We could show that intron sequences in the *PsaD* gene are required for light-regulated and plastid-dependent gene expression. We found that transcripts from the spinach *PsaD* gene, when expressed under the control of the 35S RNA CaMV promoter in transgenic tobacco plants, exhibit different associations with isolated polyribosomes. Polyribosomes were isolated from etiolated tobacco seedlings, from seedlings which were transfered to white light for 60 h and seedlings which were transfered to white light in the presence of Norflurazon, a herbicide which blocks plastid biosynthesis due to the inhibition of carotenoid biosynthesis, and in the presence of DCMU, an efficient inhibitor of the photosynthetic electron transfer. The overall *PsaD* transcript level in etiolated and photobleached seedlings is approximately two times lower when compared to white light-grown seedlings in spite of the fact that transcription is driven by the 35S RNA CaMV promoter (Bolle *et al* 1996). This indicates that posttranscriptional processes are also involved in *PsaD* gene expression. However, if RNA samples with comparable amounts of *PsaD* messages were loaded on a sucrose gradient, substantial ribosome association was only observed with extracts from light-grown tissue. Most of the transcripts in extracts from etiolated, or photobleached or DCMU treated material do not appear to be associated with ribosomes. Analysis of various site-directed mutants demonstrated that polysome association of *PsaD* transcripts requires sequences within the first 150 nucleotides downstream of the ATG codon. We are currently trying to define the crucial region in greater detail.

ACKNOWLEDGMENTS

Work was supported by the German Research Foundation, by the Alexander-von-Humboldt Stiftung (Bonn) and the Fond der Chemischen Industrie (Köln).

REFERENCES

Abdallah, F., Salamini, F., and Leister, D., 2000, A prediction of the size and the evolutionary origin of the proteome of chloroplasts of *Arabidopsis*. *Trends Plant Sci.* **5**: 141-142.

Allen, J.F., 1993, Redox control of gene expression and the function of chloroplast genomes – an hypothesis. *Photosynth. Res.* **36**: 95-102.

Bezhani, S., Sherameti, I., Pfannschmidt, T., and Oelmüller, R., 2001, A repressor with similarities to prokaryotic and eukaryotic DNA helicases controls the assembly to the CAAT box binding complex at a photosynthesis gene promoter. *J. B. Chem.* (in press).

Bolle, C., Herrmann, R.G., and Oelmüller, R., 1996, Intron sequences are involved in the plastid- and light-dependent expression of the spinach *PsaD* gene. *Plant J.* **10**: 919-924.

Chandok, M.R., Sopory, S.K., and Oelmüller, R., 2001, Cytoplasmic kinase and phosphatase activites can induce *PsaF* gene expression in the absence of functional plastids: evidence that phosphorylation/dephosphorylation events are involved in interorganellar crosstalk. *Mol. Gen. Genet.* **264**: 819-826.

Chang, C., and Stewart, R.C., 1998, The Two-Component-System. Regulation of diverse signaling pathways in prokaryotes and eukaryotes. *Plant Physiol.* **117**: 723-731.

D'Agostino, I.B., and Kieber, J., 1999, Phosphorelay signal transduction: the emerging family of plant response regulators. *TIBS* **24**: 452-456.

Emanuelsson, O., Nielsen, H., and von Heijne, G., 1999, ChloroP, a neural network-based method for predicting chloroplast transit peptides and their cleavage sites. *Protein Science* **8**: 978-984.

Eraso, J.M., and Kaplan, S., 2000, From Redox Flow to Gene Regulation: Role of PrrC Protein of Rhodobacter sphaeroides 2.4.1. *Biochemistry* **39**: 2052-2062.

Genty, B., Briantais, J.-M., und Baker, N.R., 1989, The relationship between the quantum yield of photosynthetic electron transport and quenching of chlorophyll fluorescence. *Biochem. Biophys. Acta* **990**: 87-92.

Kusnetsov, V., Landsberger, M., Meurer, J., and Oelmüller, R., 1999, The assembly of the CAAT-box binding complex at a photosynthesis gene promoter is regulated by light, cytokinin, and the stage of the plastids. *J. Biol. Chem.* **50**: 36009-36014.

Harder, T., and Simons, K., 1997, Caveolae, DIGs, and the dynamics of sphingolipid-cholesterol microdomains. *Curr. Opin. Cell Biol.* **9**: 534-542.

Hooley, R., 1998, Plant hormone perception and action: a role for G-protein signal transduction? *Phil. Trans. R. Soc. Lond. B Biol. Sci.* **353**: 1425-1430.

Imamura, A., Hanaki, N., Nakamura, A., Suzuki, T., Taniguchi, M., Kiba, T., Ueguchi, C., Sugiyama, T., Li, H., and Sherman, L.A., 2000, A redox-responsive regulator of photosynthesis gene expression in the cyanobacterium *Synechocystis* sp. Strain PCC 6803. *J. Bacteriol.* **182**: 4268-4277.

Li, H., and Sherman, L.A., 2000, A redox-responsive regulator of photosynthesis gene expression in the cyanobacterium *Synechocystis sp.* Strain PCC 6803. *J. Bacteriol.* **182**: 4268-4277.

Lohrmann, J., Buchholz, G., Keitel, C., Sweere, U., Kircher, S., Bäurle, I., Kudla, J., Schäfer, E., and Harter, K., 1999, Differential expression and nuclear localization of response-regulator-like proteins from *Arabidopsis thaliana*. *Plant Biol.* **1**: 495-505.

Mizuno, T., 1999, Compilation and characterization of *Arabidopsis thaliana* response regulators implicated in His-Asp phosphorelay signal transduction. *Plant Cell Physiol.* **40**: 733-742.

Pfannschmidt, T., Nilsson, A., and Allen, J.F., 1999, Photosynthetic control of chloroplast gene expression. *Nature* **397**: 625-628.

Race, H.L., Herrmann, R.G., and Martin, W., 1999, Why have organelles retained genomes? *Trends Genet.* **15**: 364-370.

Sakakibara, H., Hayakawa, A., Deji, A., Gawronski, S.W., and Sugiyama, T., 1999, His-Asp phosphotransfer possibly involved in the nitrogen signal transduction mediated by cytokinin in maize: molecular cloning of cDNAs for two-component regulatory factors and demonstration of phosphotransfer activity in vitro. *Plant Mol. Biol.* **41(4)**: 563-73.

Shaul, P.W., and Anderson, R.G.W., 1998, Role of plasmalemmal caveolae in signal transduction. *Am. J. Physiol.* **275**: L843-L851.

Simons, K., and Ikonen, E., 1997, Functional rafts in cell membranes. *Nature* **387**: 569-572.

Sopory, S.K., Sanan, N., and Oelmüller, R., 1999, Light signal transduction and gene expression. In: Singhai, G.S., Rengar, G., Sopory, S.K., Irrgang, K.-H. (eds), pp. 897-929. Concepts in photobiology: photosynthesis and photomorphogenesis. Kluwer Academic Press, Dordrecht.

Urao, T., Yamaguchi-Shinozaki, K., and Shinozaki, K., 2000, Two-component systems in plant signal transduction. *Trends Plant Sci.* **5**: 67-74.

Yang, Z., 1996, Signal transducing proteins in plants: An overview. In: DPS Verma (ed), Signal transduction in plant growth and development, Springer Verlag, Wien, New York, pp. 1-37.

Regulation of rDNA Transcription in Spinach Plastids by Transcription Factor CDF2

SILVA LERBS-MACHE
Laboratoire de Génétique Moléculaire des Plantes, UMR 5575, CNRS and Université J. Fourier, BP 53, F38041 Grenoble cedex9, France

1. INTRODUCTION

A series of recent-years discoveries has shown that transcription of the plastid genome is highly regulated. Nucleus-encoded (NEP) and plastid-encoded (PEP) RNA polymerase(s) in coordination with different types of transcription factors participate in the expression of the about 150 kbp circular DNA molecule. Most of the plastid transcription units and genes are preceded by several types of promoter structures allowing transcription by different RNA polymerases and suggesting a division of labour between these different enzymes (for reviews see Hess and Börner 1999, Liere and Maliga 2000). However, although genes and cDNAs corresponding to potential plastid transcription factors and RNA polymerase subunits are discovered now, the exact function of them and the interaction of the different transcriptional components during genome expression is not at all clear. We have chosen the spinach plastid *rrn* operon to analyse some of these interactions in more detail. Ribosome biosynthesis is one of the first events taking place during the differentiation of proplastids into photosynthetically active chloroplasts (Harrak *et al* 1995) and it is tightly coupled to plastid metabolic activities. To better understand plastid transcription regulation we have analysed rDNA expression in different plant tissues (i. e. corresponding to different types of plastids), during germination (i. e. corresponding to differentiation of proplastids into chloroplasts) and during different climatic and growth conditions (i. e. corresponding to changes in metabolic activities).

2. CONSTITUTIVE AND UP-REGULATED PLASTID rDNA TRANSCRIPTION

Comparison of rDNA transcription in non-photosynthetically (root) and photosynthetically active tissues has shown that transcription initiates *in vivo* from two different sites, named PtRNA and PC. Both start sites are localised several hundred bp apart on the plastid genome. In root tissues, low-level rDNA transcription starts at PtRNA and the primary transcript includes the tRNA^{Val}. In leaf tissues, both promoters are active, i. e. rDNA transcription is enhanced by additional activation of the PC promoter (Iratni *et al* 1997). In root tissues, the PC promoter can be activated by exposure of roots to light (Lerbs-Mache, unpublished result). During germination, transcription is activated at the PC promoter from the third day of germination (Harrak *et al* 1995, Lerbs-Mache, unpublished result). Both promoters, PtRNA and PC, do not have prokaryotic-type consensus sequences and represent probably NEP promoters, i. e. they are recognized by a nucleus-encoded plastid RNA polymerase (NEP).

3. SEQUENCE CONTEXT OF PC

The DNA sequence surrounding the PC promoter contains two other promoter structures, P1 and P2. These two promoters are of the prokaryotic-type and are recognized in *in vitro* transcription assays by *E. coli* RNA polymerase holoenzyme (Lescure *et al* 1985, Iratni *et al* 1994). These two promoters are also recognized by at least one of the six sigma-type transcription initiation factors that have been revealed in *Arabidopsis* by cDNA sequencing and overproduction in *E. coli* (Hakimi *et al* 2000). These results define these two promoters as PEP promoters and suggest a role for them in rDNA expression by the prokaryotic-type plastid-encoded plastid RNA polymerase. However, the *in vivo* significance of the P1 and P2 promoter is not yet clear. Determination of *in vivo* existing precursor rRNA has not revealed transcripts that correspond to one of these two start sites in spinach (Baeza *et al* 1991, Lerbs-Mache *et al* 1997). Therefore, these two promoters might have functions in regulation of rRNA transcription rather than directly in transcription.

4. REGULATION OF TRANSCRIPTION AT PC

In order to understand regulation of rDNA expression by the three overlapping promoters, P1, PC and P2, we have at first analysed the protein/DNA interactions using techniques like gel mobility shift assays (GMS), UV-crosslinking and DNase I footprinting. These experiments showed that a sequence-specific DNA binding factor, that we have named CDF2, binds to the rDNA promoter region upstream of the PC transcription start site (Baeza *et al* 1991). It also showed that protein/DNA interactions change with spinach growth conditions. L complex (CDF2+PEP) formation dominates in plastids of mature plants having stopped growth and S complex (CDF2) formation is representative for plastids obtained from actively growing plants (Bligny *et al* 2000). DNA binding of CDF2 depends on phosphorylation of one of its components (Iratni and Lerbs-Mache, unpublished result).

4.1. CDF2-A Interacts with PEP and Inhibits rDNA Transcription at P2

CDF2 is composed of at least two different polypeptides of 33 and 35 kDa (Baeza *et al* 1991). The CDF2 binding site comprises the transcription start site of the P1 and the "-35" promoter element of the P2 promoter. Thus, fixation of CDF2 to the rDNA promoter region prevents transcription initiation from the P1-PEP promoter. On the other hand, CDF2 interacts with PEP and fixes the enzyme to the P2 promoter in a relatively stable, but transcriptionally inactive form. This CDF2/PEP complex can be analysed by GMS assays as large complex and inhibits transcription of the *rrn* operon (Iratni *et al* 1994). If CDF2 is separated from PEP by Mono-S FPLC chromatography it elutes from the column as CDF2-A. Once separated from PEP, CDF2-A still binds specifically to the rDNA promoter region, but it has lost its capacity to interact with PEP and to inhibit transcription at P2. This indicates that either additional protein(s) or protein modification(s) are necessary for CDF2/PEP interactions (Lerbs-Mache, unpublished result).

4.2. CDF2-B Interacts with NEP-2 and Activates rDNA Transcription at PC

The PC promoter was confirmed to be a NEP promoter by analyses of rDNA transcription start sites by primer extension in ribosome-deficient white plantlets. Enhancement of PC-initiated transcripts in these plants strongly indicates transcription by a nucleus-encoded RNA polymerase. Accordingly, purification and partial characterisation of the PC-transcribing RNA polymerase clearly distinguishes this enzyme from the tagetitoxin-sensitive PEP. However, based on results obtained by antibody cross-reactions, the PC-transcribing enzyme activity seems also to be different from the 110 kDa phage-like RNA polymerase that has been described so far as NEP (Lerbs-Mache 1993, Hedtke et al 1997). To distinguish the two NEP enzymes we name the phage-like enzyme NEP-1 and the PC-recognizing enzyme NEP-2. A second form of CDF2, CDF2-B, co-elutes in the flow-through fraction of the Mono-S column together with NEP-2. The factor can be separated from NEP-2 by subsequent Mono-Q FPLC chromatography. The Mono-Q recovered CDF2-B factor is different from the Mono-S recovered CDF2-A factor as revealed by their different affinities to the two different ion exchange columns. The two forms of CDF2 are also different with respect to their function. Reconstitution of NEP-2 with CDF2-B (but not CDF2-A) results in correct transcription initiation of the *rrn* operon at PC, revealing CDF2-B as sequence-specific initiation factor for rDNA transcription by NEP-2 (Bligny et al 2000).

4.3. Regulation of rDNA Transcription by Ribosomal Protein L4 ?

Regulation of plastid rDNA transcription by an attenuation-like process has been suggested in 1983 (Briat et al) and first evidence has been provided in 1988 (Laboure et al 1988). The L4 protein belongs to the group of ribosomal proteins for which extra-ribosomal functions in the expression of ribosomal components have been demonstrated in prokaryotes (Yates and Nomura 1980, Zengel et al 1980). Ribosomal protein L4 inhibits the transcription of the S10 operon by interfering in an NusA-mediated attenuation mechanism (Zengel and Lindahl 1990). During fractionation of plastid transcriptional extracts, the nucleus-encoded, plastid-localised, L4 protein co-purifies together with CDF2-A and the phage-like RNA polymerase suggesting a function in CDF2 action or in transcriptional regulation of NEP-1 (Trifa *et al* 1998). Therefore, we have cloned and sequenced the cDNA corresponding to the plastid ribosomal protein L4 in order to used it for overproduction of the protein in *E. coli* and for testing its function in rDNA transcription. We could show that the plastid L4 protein can replace the *E. coli* L4 protein in the NusA-dependent attenuation control of the *E. coli* S10 operon by stabilising stalled transcription complexes in a NusA-dependent reaction. In plastids, the L4 protein inhibits *in vitro* PC-initiated transcription of the *rrn* operon (Trifa and Lerbs-Mache 2000). Although the exact mechanism of the L4 protein action is not yet clear, results indicate extra-ribosomal function(s) of the plastid L4 protein in transcriptional regulation and especially in PC-initiated rDNA synthesis.

5. CONCLUSION

During chloroplast differentiation, *i. e.* during transformation of proplastids into mature chloroplasts, the ribosome number per plastid raises considerably. Thus, the number of ribosomes per plastid is a key element in plastid differentiation and its regulation should be intimately integrated into the general developmental programs governing plant development. All components that are known so far to regulate plastid gene expression, are nucleus-encoded. This assures overall regulation of plastid differentiation by the nucleus. The synthesis of ribosomes starts early during plant germination (Harrak *et al* 1995) and the synthesis of rDNA is highest in very young developmental stages and drops down during differentiation into chloroplasts (Rapp *et al* 1992). If we interpret our results on rDNA transcription in this context, we can propose the following model of transcriptional regulation: The *rrn* operon is transcribed during early phases of plastid development by NEP-2/CDF2. Expression is high because the plastid-encoded, prokaryotic-type RNA polymerase (PEP) is not yet made up. During the process of chloroplast differentiation PEP increases and competes with NEP-2 for CDF2 binding thus down-regulating rDNA expression. In mature chloroplasts, transcription is regulated by multiple mechanisms. Modifications of transcription factor CDF2 (CDF2-A↔CDF2-B) determine its competence either for interaction with NEP or PEP and free ribosomal protein L4 might further modulate rDNA expression when PEP is occupied by transcription of photosynthesis-related genes and thus not available for rDNA repression.

ACKNOWLEDGMENTS

The work has received support from the Region Rhône Alpes and from the European Community with the Programs Emergence and the Biotechnology Programs Semences and BIO4-CT97-2245.

REFERENCES

Baeza, L., Bertrand, A., Mache, R. and Lerbs-Mache, S., 1991, Characterization of a protein binding sequence in the promoter region of the 16S rRNA gene of the spinach chloroplast genome. *Nucleic Acids Res.* **19**: 3577-3581.

Bligny, M., Courtois, F., Thaminy, S., Chang, C.-C., Lagrange, T., Baruah-Wolff, J., Stern, D. and Lerbs-Mache, S., 2000, Regulation of plastid rDNA transcription by interaction of CDF2 with two different RNA polymerases. *EMBO J.* **19**: 1851-1860.

Briat, J. F., Dron, M. and Mache, R., 1983, Is transcription of higher plant chloroplast ribosomal operons regulated by premature termination ? *FEBS Lett.* **163**: 1-5.

Hakimi, M.-A., Privat, I., Valay J.-G. and Lerbs-Mache, S., 2000, Evolutionary conservation of C-terminal domains of primary sigma^{70}-type transcription factors between plants and bacteria. *J. Biol. Chem.* **275**: 9215-9221.

Harrak, H., Lagrange, T., Bisanz-Seyer, C., Lerbs-Mache, S., and Mache, R., 1995, The expression of nuclear genes encoding plastid ribosomal proteins precedes the expression of chloroplast genes during early phases of chloroplast development. *Plant Physiol.* **108**: 685-692.

Hedtke, B., Börner, T. and Weihe, A., 1997, Mitochondrial and chloroplast phage-type RNA polymerases in *Arabidopsis*. *Science* **277**: 809-811.

Hess, W. R. and Börner, T., 1999, Organellar RNA polymerase of higher plants. *Internat. Rev. Cytol.* **190**: 1-59.

Iratni, R., Baeza, L., Andreeva, A., Mache, R. and Lerbs-Mache, S., 1994, Regulation of rDNA transcription in chloroplasts: promoter exclusion by constitutive repression. *Genes Dev.* **8**: 2928-2938.

Iratni, R., Diederich, L., Harrak, H., Bligny, M. and Lerbs-Mache, S., 1997, Organ-specific transcription of the *rrn* operon in spinach plastids. *J. Biol. Chem.* **272**: 13676-13682.

Laboure, A. M., Lescure, A. M. and Briat, J. F., 1988 Evidence for a translation-mediated attenuation of a spinach chloroplast rDNA operon. *Biochimie* **70**: 1343-1352.

Lerbs-Mache, S., 1993, The 110-kDa polypeptide of spinach plastid DNA-depenndent RNA polymerase: Single-subunit enzyme or catalytic core of multisubunit enzyme complexes? *Proc. Natl. Acad. Sci. USA* **90**: 5509-5513.

Lerbs-Mache, S., Baeza,, L., Diederich, L. and Iratni, R., 1997, Regulation of rDNA transcription during chloroplast development. In: Proceedings of the Xth international photosynthesis congress (Mathis, P. ed.) Kluwer Academic Publishers, Dordrecht, Netherlands, Vol. **III**, 557-662.

Lescure, A.-M., Bisanz-Seyer, C., Pesey, H. and Mache, R., 1985, *In vitro* transcription initiation of the spinach chloroplast 16S rRNA gene at two tandem promoters. *Nucleic Acids Res.* **13**: 8787-8796.

Liere, K. and Maliga, P., 2000, Plastid RNA polymerase in higher plants. In: *Regulatory aspects of photosynthesis.* (Anderson, B. and Aro, E. M. eds.), Kluwer Academic Publishers, in press.

Rapp, J. C., Baumgartner, B. J. and Mullet, J., 1992, Quantitative analysis of transcription and RNA levels of 15 barley chloroplast genes. *J. Biol. Chem.* **267**: 21404-21411.

Trifa, Y., Privat, I., Gagnon, J., Baeza, L., and Lerbs-Mache, S., 1998, The nuclear *RPL4* gene encodes a chloroplast protein that co-purifies with the T7-like transcription complex as well as plastid ribosomes. *J. Biol. Chem.* **273**: 3980-3985.

Trifa, Y. and Lerbs-Mache, 2000, Extra-ribosomal function(s) of the plastid ribosomal protein L4 in expression of ribosomal components. *Mol. Gen. Genet.*, in press.

Yates, J. L. and Nomura, M., 1980, *E. coli* ribosomal protein L4 is a feedback regulator protein. *Cell* **2**: 517-522.

Zengel, J. M., Mueckl, D. and Lindahl, L., 1980, Protein L4 of the *E. coli* ribosome regulates an eleven gene r protein operon. *Cell* **2**: 523-535.

Zengel, J.M. and Lindahl, L., 1990, Ribosomal protein L4 stimulates *in vitro* termination of transcription at a NusA-dependent terminator in the S10 operon leader. *Proc. Natl. Acad. Sci. USA* **87**: 2675-2679.

Plastid Ribosome Biogenesis During the Early Steps of Chloroplast Differentiation

Elements Controlling the Activation of Nuclear Genes Encoding Plastid Ribosomal Proteins

REGIS MACHE, JEAN-LUC GALLOIS, AND PATRICK ACHARD
Laboratoire de Génétique Moléculaire des Plantes, UMR 5575. CNRS and Université J. Fourier, BP 53, F38041 Grenoble cedex 9, France

1. INTRODUCTION

Chloroplast differentiation is accomplished in several steps. In embryonic and meristematic cells, proplastids are present, characterised by their small size and the absence of internal membranes. They contain a low copy number of plastid DNA and few plastid ribosomes. Then, plastid differentiation follows several ways depending on plant organs or tissues. Because of its importance for photosynthesis, differentiation of proplastids into chloroplasts has been studied in details. It involves the co-ordinated expression of the nuclear genome and of the plastid genome. In the course of differentiation two main steps can be distinguished: firstly an early step during which the complexes necessary for plastid gene expression are synthesised and secondly, the formation of the photosynthetic apparatus. We will focus the present study on the expression of nuclear genes encoding plastid ribosomal proteins (r-proteins) that are important for early plastid differentiation.

2. **EARLY CHLOROPLAST DIFFERENTIATION IS TRIGGERED BY UPRIGHT SYNTHESIS OF RIBOSOMAL PROTEINS**

In bacteria, ribosome synthesis is strongly correlated with cell growth rate. In plants, chloroplasts in mesophyll cells are actively enlarging and dividing in parallel with early stages of leaf development. In these stages, very active plastid r-protein synthesis is observed. In contrast, in non-photosynthetic tissues containing specific types of plastids (amyloplasts, chromoplasts) plastid ribosomes are synthesised at a low level. In dicotyledonous plants, the first step of chloroplast differentiation can be separated from the second step during seed germination. This developmental phase occurs in darkness before the protrusion of the radicle. A burst of synthesis of nuclear encoded plastid r-proteins occurs during seed germination after the first hours of imbibition. This step occurs about two days before transcription of nuclear encoded photosynthetic genes and before the expression of plastid-encoded genes (Bisanz-Seyer *et al* 1989, Harrak *et al* 1995). The results of differential temporal gene expression are schematically represented in Fig. 1. The end of the first step of chloroplast differentiation can be observed in dark-developed etioplasts that contain a large number of ribosomes and no photosynthetic membranes.

3. **ELEMENTS CONTROLLING THE TRANSCRIPTION OF NUCLEAR ENCODED PLASTID RIBOSOMAL PROTEIN GENES**

A tight control links r-protein synthesis with the need of new ribosomes during the first step of plastid differentiation. Indeed, a continuous synthesis of r-proteins is necessary to feed the small-sized pool of free r-proteins present in the stroma of plastids. In addition to the high gene activity required for the building of new ribosomes, synthesis of r-proteins should be coordinated since r-proteins are present in a stoechiometric amount in ribosomes. As the control of r-protein gene expression is mainly at the level of transcription in the nucleus two main questions could be addressed. What are the signalling elements controlling the burst of r-protein gene activity in the nucleus and, if existing, which elements are shared in the promoters of the gene family to coordinate the expression of the individual genes encoding the plastid r-proteins?

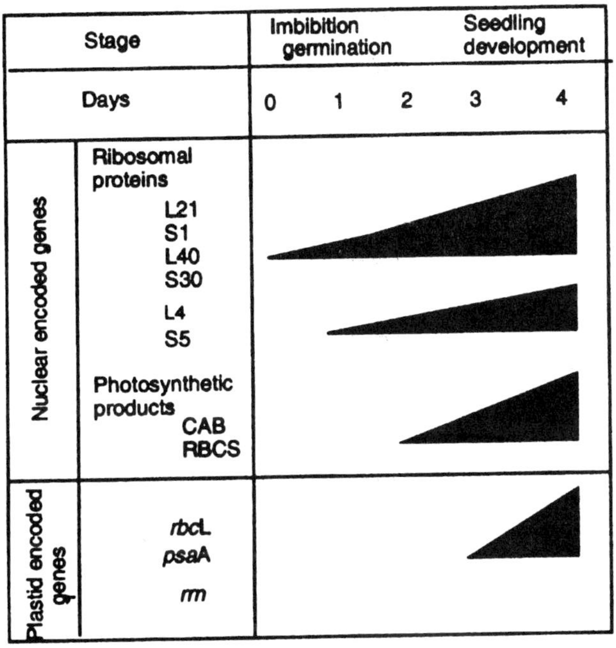

Stage		Imbibition germination		Seedling development		
Days		0	1	2	3	4
Nuclear encoded genes	Ribosomal proteins L21 S1 L40 S30 L4 S5 Photosynthetic products CAB RBCS					
Plastid encoded genes	rbcL psaA rrn					

Figure 1. Early expression of nuclear genes encoding plastid ribosomal protein genes·

3.1. Transcription Regulation at the Initiator Level

Attempts to answer these questions were made by analysing the elements of the promoter of three spinach model r-protein genes encoded in the nucleus. The *RPS1* and *RPL21* genes encode proteins similar to bacterial r-proteins but having their own specificity (Franzetti *et al* 1992a and b, Martin *et al* 1990). The *RPS30* gene encodes a r-protein known for 10 years as a specific chloroplast r-protein (Zhou and Mache 1989) but to whom an *E. coli* protein with a region of similarity and residing at the interface of the bacterial ribosome has recently been discovered (Agafonov *et al* 1999). Besides its ribosomal location, S30 possess the unique ability to accumulate

289

in a free state, at a relatively high level, in plastid stroma (Zhou and Mache 1989). Interestingly, transcription of *RPS1* and *RP L21* genes is differentially regulated in tissues by the usage of two different initiation sites named P1 and P2 (Lagrange *et al* 1993, see Fig. 2). P1 allows a high transcriptional activity and is used specifically in mesophyll cells, or in cotyledons, during the differentiation of plastids into chloroplasts. P2 is used constitutively and allows a low transcriptional expression in non-photosynthetic cells. These two initiation sites are probably used in many other nuclear-encoded, prokaryotic-like, r-protein genes. The usage of a specific promoter for high gene expression is unusual. It probably corresponds to the participation of a specific set of proteins *trans*-activating the transcription initiation complex and displacing the start site at P1, relatively to the less activated initiation complex operating at P2. A complex of 5 proteins has been found in the region surrounding P1 (unpublished results), called activator in Fig. 2, and supporting our hypothesis. In contrast, only one initiation site has been found in the case of *RPS30* although this gene is expressed at a high level. This difference in initiation site possibly reflects the difference in the properties of the S30 protein as indicated above.

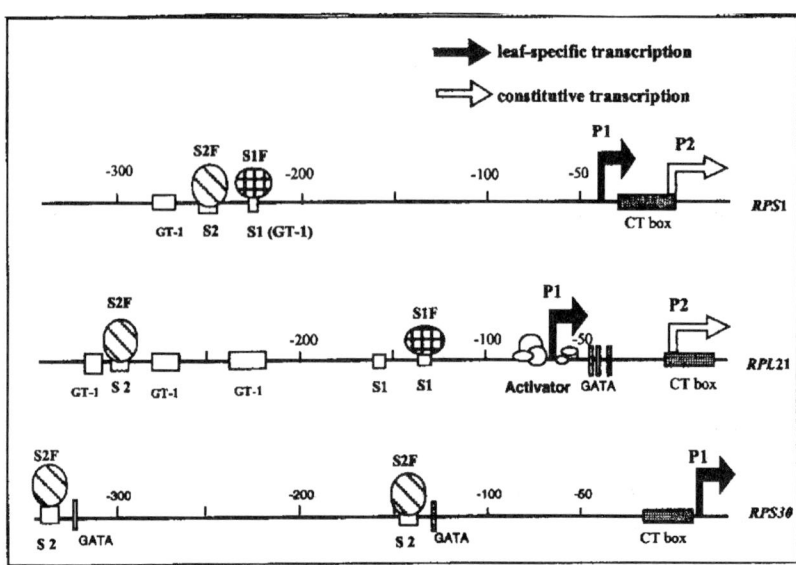

Figure 2. Schematic representation of the promoter of two prokaryotic-like r-protein genes, *RPS1* and *RPL2l,* and of one chloroplast-specific r-protein gene, *RPS30.* Identified controlling elements are indicated.

3.2. Relative Strength of Promoters of Nuclear-Encoded r-Protein Genes

We were interested to compare the strength of the three promoters in transcription activation as they all belong to the same nuclear encoded gene family. The measurements of promoter activities have been made in several organs but only results concerning roots and leaves will be reported here as tissues in these two organs are characterised by the presence of the two main types of plastids, amyloplasts and chloroplasts, respectively. Northern analysis have shown that the *RPS1* and *RPL21* genes are transcriptionally expressed at a comparable level. Both genes are highly expressed in mesophyll cells and are expressed at a low level in root cells. Such a differential expression has also been found for other genes of the same family, the *RPS5* (unpublished results) and the *RPL4* genes (Trifa *et al* 1998). The *RPS30* gene follows the same spatial pattern of expression than the prokaryotic-like genes but to our surprise, the use of transgenic plants has revealed that the expression of the *RPS30* gene in mesophyll tissue is multiplied by a factor of 4 to 5. The study was made in *Arabidopsis* transgenic plants containing the spinach promoter of either the *RPS30* or the *RPL21* gene, coupled with a luciferase reporter gene. The method allows the measurement of the promoter activity *in vivo*, after spraying plants with the substrate luciferin, without injuring plants. It has to be noted that the luciferase protein is transiently accumulated as its half-life is of about 3 hr. Thus, luciferase activity accurately reflects the accumulation of transcripts in tissues. The method can be easily applied to measurements of promoter activity at different developmental stages. Results obtained by using this method are illustrated in Fig. 3. The control plant has no luciferase activity and therefore is not emitting light (*i.e.* is not visible in Fig. 3A). In leaves, the activity of the *RPS30* promoter is comparable to that of the strong CaMV 35S promoter. In apparent contradiction with the results of northern analyses (Bisanz-Seyer *et al* 1989), the luciferase test does not allow the detection of promoter activity in roots. Histochemical detection of a GUS reporter gene in root tissue has shown that the expression is probably restricted to growing cells, at the tip of the root. Thus, the expression of the r-protein gene family might be undetectable in mature root cells, as observed with the luciferase test, and limited to the root tip, as observed with GUS. Altogether, we interpret these observations as that r-protein gene expression is tightly linked to early plastid differentiation in relation with tissue development. Thus, nuclear genes for plastid r-proteins are good markers of tissue development.

3.3. Organ-Specific Positive and Negative Controlling Elements

To determine the *cis*- and *trans*-acting elements of the promoter of the r-protein genes a series of studies have been made. The promoters have been dissected in several regions by deletion and the activity of the different constructs have been tested by transient expression using protoplasts prepared from cultured cells. The results are schematically represented in Fig. 2. It was possible to identify positive and negative distal regions upstream of the initiation sites. In *RPS1*, three *cis*-elements have been identified which are related to the GT-1 *cis*-element present in the light-activated *RBCS-3A* gene and in other genes (Zhou *et al* 1992, Zhou 1999). Mutation or deletion of one of these element, S1 (for site 1), increases transcription of the gene in transfected non-photosynthetic protoplasts or in roots of transgenic plants (Zhou *et al* 1992, Villain *et al* 1996). The S1 element binds to a *trans*-acting factor S1F in leaves but binds to another factor in root cells, named RGTF (Villain *et al* 1996).

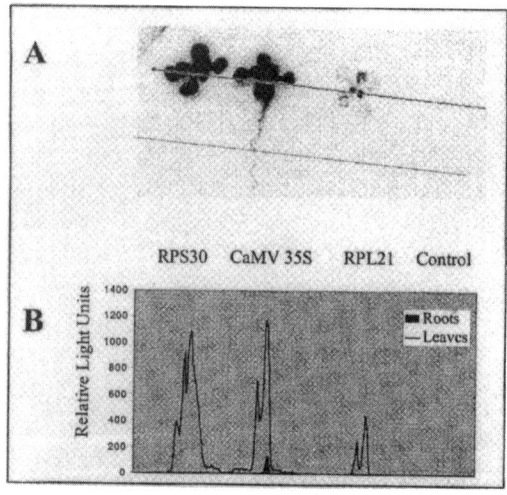

Figure 3. Luminescence of Luciferase-expressing *Arabidopsis*. A, Inverse color video image (taken in the dark) of 25-days-old *Arabidopsis*. As indicated, transgenic plantlets bear Luciferase gene driven respectively by *RPS30*, CaMV 35S and *RPL21* promoters. The control plant has a promoterless Luciferase gene and thus is not visible. The plants were sprayed with 0.5 mM luciferin-0,005% TritonX100. The image was taken by an Intensified CCD camera system (Dual MCP ISIS III, *Photonics Sciences*). B, quantification of light emission in leaves and roots measured along the lines shown in A.

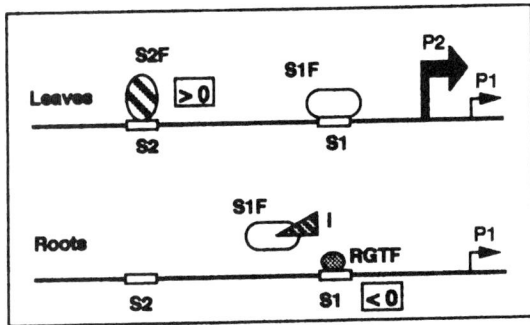

Figure 4. Model for organ specific regulation of r-protein gene activities at the promoter level. S2F is a leaf positive factor not present in roots. S1F is present in both organs but is blocked by the presence of an inhibitor in roots allowing the S1 site to bind to a Root specific factor (RGTF).

Results have shown that an inhibitor present in root cells prevents S1F to interact with S1 and allows the binding of a lower-affinity root factor (Fig. 4). The function of the complex S1/S1F in photosynthetic cells remains obscure as mutation in the S1 element has no effect on transcription of *RPS1* in transgenic leaves. It is worth to note that the S1/S1F complex is also present in the second prokaryotic like gene studied in our laboratory, the *RPL21* gene but is absent in the 900 bp-upstream region of the chloroplast-specific *RPS30* gene.

A new *cis*-acting element, named S2 (for site 2), which sequence is ATACAT, has been discovered recently. The S2/S2F complex has been characterised in the *RP L21* promoter (Lagrange *et al* 1997). S2 is a *cis*-positive element conserved in several promoters of nuclear genes encoding plastid components. It binds a nuclear specific factor, S2F, detected by gel shift assays. Mutation in S2 significantly decreases the activity of the promoter in leaves of transgenic plants but has no effect in roots. Interestingly, root extracts do not form a S2/S2F complex. The same extracts do not prevent the formation of the complex in leaves (Fig. 5, lane L+R) showing that the absence of the complex in roots is not due to the presence of an inhibitor (Fig. 4). The S2/S2F complex is present in the promoter of *RPS1,* and is present in two copies in the promoter of the *RPS30* gene. Thus, the S2/S2F complex activates transcription in a cell specific manner and could be considered as a general early positive factor of chloroplast differentiation.

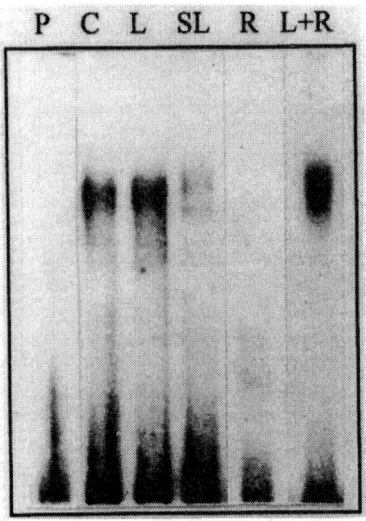

Figure 5. Presence of the S2F factor in different organs, detected by gel shift assays. P, probe; C, dark-grown cotyledon; L, mature leaves; SL, senescent leaves; R, roots; L+R, mixed leaf and root extracts

3.4. Ribosomal protein gene activation during chloroplast differentiation

The prokaryotic-like genes (such as *RPS1* and *RPL21*) have a peak of expression in young leaf tissues that progressively decline with leaf maturation. The luciferase test has shown that the *RPS30* gene, in contrast to prokaryotic-like genes, is evenly expressed during the leaf development (unpublished results). The presence of the positive S2F factor is correlated with the expression of the gene family. As shown in Fig. 5, lane C, S2F is abundant in dark-grown cotyledons confirming that the activation of the nuclear ribosomal protein gene family occurs independently of light and during the first step of plastid development. The factor is present at a low level only in senescent leaves (lane SL) as expected. It is worth to note that the expression of the r-protein genes is very low in meristematic apical cells (primary leaves) as shown in Fig. 6. This observation could be related to the presence of quiescent proplasts which divide later in leaf development.

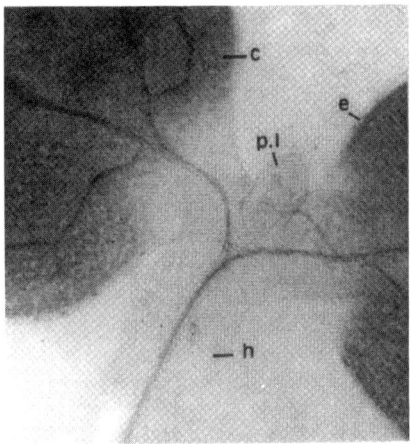

Figure 6. Histochemical detection of the *RPL21* promoter activity in young transgenic tobacco plantlets. The reporter GUS activity is highly detectable in cotyledons (c) but scarcely in primary leaves (p.l) and not in the hypocotyl (h) nor in the epidermis (e).

4. CONCLUSION

Two factors have been identified that control the expression of the nuclear gene family encoding plastid ribosomal proteins of the prokaryotic type. One of these factors, S2F, is a postive factor whereas the function of the other one, S1F, is still unknown. The S2F factor is mesophyll specific. Thus, a specific signal should trigger the expression of this basic factor for chloroplast differentiation and correlatively mesophyll cell differentiation. It is possible that an hormone such as cytokinin participates to the cell-specific signalling. The S30 r-protein, firstly identified as a chloroplast-specific protein, might have a specific function in translation. It is possible that the specific features of the *RPS30* promoter are correlated to the high expression of the gene which is in good correspondance with the observed accumulation of S30 in the plastid stroma.

ACKNOWLEDGMENTS

This work has received support from Region Rhône Alpes, with the Programme Emergence and the Biotechnology Programme Semences.

REFERENCES

Agafonov, D.E., Kolb, v.A., Nazimov, I.V., and Spirin, A.S., 1999, A protein residing at the interface of the bacterial ribosome. *Proc. Natl. Acad. Sci. USA* **96**: 12345-12349.

Bisanz-Seyer, C., Li, Y-F., Seyer, P., and Mache, R, 1989, The components of the plastid ribosome are not accumulated synchronously during the early development of spinach plants. *Plant Mol. Biol.* **12**: 201-211.

Franzetti, B., Carol, P., and Mache, R, 1992a, Characterization and RNA-binding properties of a chloroplast Sl-like ribosomal protein. *J. Biol. Chem.* **267**: 19075-19081.

Franzetti, B., Zhou, D.-X., and Mache, R, 1992b, Structure and expression of the nuclear gene coding for the plastid CSl ribosomal protein from spinach. *Nucl. Acids Res.* **20**: 4153-4157.

Harrak, H., Lagrange, T., Bisanz-Seyer, C., Lerbs-Mache, S., and Mache, R., 1995, The expression of nuclear genes encoding plastid ribosomal proteins precedes the expression of chloroplast genes during early phases of chloroplast development. *Plant Physiol.* **108**: 685-692.

Lagrange, T., Franzetti, B., Axelos, M., Mache, R., and Lerbs-Mache, S., 1993, Structure and expression of the nuclear gene coding for the chloroplast ribosomal protein L21: developmental regulation of a housekeeping gene by alternative promoters. *Mol. Cell. Biol.* **13**: 2614-2622.

Lagrange, T., Gauvin, S., Hye-Jeong, Y., and Mache, R., 1997, S2F, a leaf-specific *trans*-acting factor, binds to a novel *cis*-acting element and differently activates the *rpl21* gene. *Plant Cell* **9**: 1469-1479.

Li, Y-F., Zhou, D.-X., Clabault, G., Bisanz-Seyer, C., and Mache, R., 1995, *Cis*-acting elements and expression pattern of the spinach *rps22* gene coding for a plastid-specific ribosomal protein. *Plant Mol. Biol.* **28**: 595-604.

Martin, W., Lagrange, T., Li, Y -F., Bisanz-Seyer, C., and Mache, R., 1990, Hypothesis for the evolutionary origin of the chloroplast ribosomal protein L21 of spinach. *Curr. Genet.* **18**: 553-556.

Trifa, Y., Privat, I., Gagnon, J., Baeza, L., and Lerbs-Mache, S., 1998, The nuclear *RPL4* gene encodes a chloroplast protein that co-purifies with the T7-like transcription complex as well as plastid ribosomes. *J. Biol. Chem.* **273**: 3980-3985.

Villain, P., Mache, R., and Zhou, D-X., 1996, GT element-mediated cell type-specific transcriptional control. *J. Biol. Chem.* **271**: 32593-32598.

Zhou, D.-X., Li, Y.-F., Rocipon, M., and Mache, R., 1992, Sequence-specific interaction between SIF, a spinach nuclear factor, and a negative *cis*-element conserved in proplastid-related genes. *J. Biol. Chem.* **267**: 23515-23519.

Zhou, D.-X. and Mache, R., 1989, Presence in the stroma of chloroplasts of a large pool of a ribosomal protein not structurally related to any Escherichia coli protein. *Mol. Gen. Genet.* **219**: 204-208. Erratum, *Mol. Gen. Genet* 1990, **223**: 167.

Zhou, D.-X., 1999, Regulatory mechanism of plant gene transcription by GT-elements and GT-factors. *Trends Plant Science* **4**: 210-214.

Mechanism of Regulation of Gene Expression for Chloroplast Proteins

AKHILESH K. TYAGI, JITENDRA P. KHURANA, ARUN K. SHARMA, AMITABH MOHANTY, AMIT DHINGRA, SAURABH RAGHUVANSHI, AND TRIPTI GAUR
Department of Plant Molecular Biology, University of Delhi, South Campus, New Delhi 110021, India

1. INTRODUCTION

Both plastid- and nuclear-encoded proteins are involved in photosynthetic reactions carried out by chloroplasts. While the structure of various plastome-encoded photosynthesis-related genes has been worked out, little is known about the mechanism of regulation of gene expression which is responsible for differentiation of chloroplasts. At the same time, several signal cascades have been worked out in animals and plants for regulation of nuclear gene expression. The aim of the present investigation was to study regulation of plastid and nuclear-encoded photosynthesis-related genes at steady-state mRNA level as influenced by light and development. Earlier, we have reported an interaction of light and developmental (temporal and spatial) cues in establishing steady-state transcript levels of photosynthesis-related genes in rice (Grover *et al* 1998, 1999a, Kapoor *et al* 1993, 1994) and *Arabidopsis* (Grover *et al* 1999b, Jain *et al* 1998, Kochhar *et al* 1996). In this article, a summary of these results along with the new information about the signal transduction, the level of gene regulation and promoter analysis, and the progress made towards developing a plastid transformation system for rice has been presented.

2. PHOTOREGULATION OF PLASTID GENE EXPRESSION

Studies on the expression of photosynthesis-related genes have shown that the steady-state mRNA levels for plastome- encoded genes, *psaA, psbA, psbD and rbcL* in rice are developmentally regulated and are also influenced by light (Kapoor *et al* 1993, 1994). In an attempt to define the role of plant age (temporal development) further, competence of seedlings of varying ages to respond to light was analyzed. The results reflect change in the sensitivity to light signal with age-dependent development (Grover 1996) which is similar to our findings in *Vigna aconitifolia* (Kelkar *et al* 1993). In addition, the transcript levels of these genes were found to fluctuate diurnally (Grover *et al* 1996).

2.1 Role of Plastid-encoded RNA Polymerase

Plastid genes are known to be transcribed by at least two types of RNA polymerases (Tyagi 1999). The plastid-encoded polymerase (PEP) is sensitive to an inhibitor tagetitoxin and nuclear-encoded polymerase (NEP) is insensitive. Interestingly, when seedlings were treated with tagetitoxin in dark, it also blocked accumulation of transcripts for *psbA*, *psaA* and *rbcL* on exposure to light. It shows that most of the light-dependent accumulation of transcripts is a result of new transcription activity by PEP which is sensitive to tagetitoxin. The 16S rRNA gene, however, is transcribed mainly by NEP as its level is not influenced by tagetitoxin and steady-state levels of 16S rRNA do not show light-dependent induction.

2.2 Involvement of Phytochrome

The plant responses triggered by the light stimulus involve perception by sensory photoreceptors (Khurana *et al* 1998, Mustilli and Bowler 1997). To establish the identity of the photoreceptor responsible for light-induced changes in plastid gene expression, 5-day-old dark-grown seedlings of rice were irradiated with red light and red light followed by far-red light. It was observed that red light increased the transcript levels of plastid genes, namely *psbA*, *psaA* and *rbcL*, over a period of 12 h after exposure, and far-red light reversed the effect of red light induction to a significant extent. The red/far-red light reversibility indicates the involvement of phytochrome in the expression of these plastid genes (Grover *et al* 1999a). In some earlier studies too the involvement of phytochromes in plastid gene expression has been reported (Tyagi 1999). In this study, the observation is extended to phytochrome control of *psaA*, *psbA* and *rbcL* expression in rice.

2.3   Role of G-Protein

It has been shown in animals and some plants that stimuli perceived by the cell can be transmitted *via* G-proteins (Bowler and Chua 1994). The agonist of G-protein action, namely cholera toxin (CTX), activates the G-proteins irreversibly by catalyzing the ADP-ribosylation of the α-subunit of stimulatory G-proteins at an arginine residue, thereby activating the signaling pathway downstream. The transcript levels of the plastidic genes showed enhancement over dark level at a concentration of 0.5 $\mu g\ ml^{-1}$ CTX. The second agonist GTPγ-S (a non-hydrolyzable analog of GTP) binds to the γ-subunit and blocks the GTPase activity of the G-proteins. In this state, the α-subunit stays dissociated from the β-subunits of G-proteins thereby maintaining an active state. GTP-βS was used at concentrations of 50, 100 and 200 μM. The enhancement in the transcript levels of *psbA* was more at 50 μM and in case of *psaA* and *rbcL*, both 50 and 100 μM had similar effect. Since the agonists of G-proteins can enhance the steady-state transcript levels in 12 h of treatment in dark itself, it indicates that the G-proteins are involved in light signal transduction pathway.

2.4   Role of Ca^{2+} and Calmodulin in Light Signal Transduction

With the aim of identifying other signaling components acting downsteam to photoreceptors, the role of Ca^{2+} in the regulation of plastid gene expression was investigated by employing agonists and antagonist of Ca^{2+} action (Grover *et al* 1999a). Ca^{2+}, used at concentrations of 1 and 10 mM in the dark, enhanced the transcript levels significantly over the dark control. The Ca^{2+} ionophore, A23187, used at a concentration of 5 µM without external Ca^{2+}, enhanced the steady-state transcript levels of the genes in the dark by as much as the increase elicited by Ca^{2+} alone. The simultaneous addition of Ca^{2+} and ionophore A23187 (10 mM and 5 µM) had no additional effect on plastid transcript levels. The involvement of Ca^{2+} in light-dependent increase in steady-state mRNA levels was also examined by using the well-known Ca^{2+} chelators, EGTA and BAPTA, during light irradiation. Both EGTA and BAPTA reduced the transcript levels of *psbA, psaA* and *rbcL*. Nifedipine and verapamil, wellknown Ca^{2+} channel blockers, also reduced the light-dependent increase in plastid transcript levels. To our knowledge, this is the first report which points towards the involvement of Ca^{2+} in the photoregulation of the transcript levels of the plastid genes. But stimulation of the accumulation of the plastid-encoded proteins has been reported earlier by Neuhaus *et al* (1993) by employing microinjection of Ca^{2+}. To ascertain wheather the Ca^{2+} signal was transmitted through calmodulin, two antagonists of calmodulin action, trifluoperazine (TFP) and W7 were used. Both the antagonists were seen to decrease the steady-state transcript levels in light as compared to the controls. This inhibition indicated that calmodulin is also involved in the plastidic signal transduction pathway. It has been shown earlier that Ca^{2+} and calmodulin are important components of signal transduction for light-stimulated expression of nuclear genes (Bowler and Chua 1994). This has been demonstrated by estimation of transcript levels and reporter gene activity under the control of heterologous promoters of nuclear genes and by the measurement of immunofluorescence specific to *rbcS* and ferredoxin polypeptides in plastids of tomato (*aurea*) hypocotyl cells and soybean cell suspension cultures (Bowler *et al* 1994a, b, Neuhaus *et al* 1993, Romero and Lam 1993).

2.5 Role of Phosphorylation in the Transmission of the Light Signal

To study the possible involvement of protein phosphorylation in the regulation of transcript accumulation of *psbA, psaA* and *rbcL* in rice, inhibitors of protein phosphatases and protein kinases were employed (Grover *et al* 1999a). Okadaic acid, a specific inhibitor of P1 and P2A types of phosphatases,

NaF, a general protein phosphatase inhibitor, and staurosporine, a protein kinase inhibitor were used. Okadaic acid significantly suppressed the accumulation of transcript levels as compared to the irradiated control. NaF also decreased the levels of transcripts and it was more effective than okadaic acid. It is noteworthy that the level of suppression showed gene-specific variation such that 1 mM NaF decreased the *psaA* transcript level to a greater degree as compared to the other genes. Staurosporine treatment resulted in several-fold enhancement of transcript levels in the dark, thereby bringing them to levels comparable to those observed in light. It was also observed that staurosporine enhanced the effect of light and the transcript levels were found to be even higher than in the controls exposed to light alone. Therefore, the dephosphorylated state helps to stimulate plastid gene expression. While the present work was in progress, the role of protein kinases and protein phosphatases has also been reported recently in plastid transcription during greening of barley seedlings by using okadaic acid alone and another inhibitor of protein kinases, K252a (Christopher *et al* 1997).

3. TRANSFORMATION OF CHLOROPLAST

Attempts are also being made to optimize delivery of genes into the plastids by particle bombardment with the ultimate aim of obtaining transplastomic rice for obvious reasons. Vectors have been constructed which would enable stable integration of the transgene into the plastid genome. Gene expression profile of various plastid-encoded genes (*psbA*, *psbB*, ORF216, *16S rRNA*, *psbD-C*) in callus and regenerating tissue of rice has indicated that the 16S rRNA gene is one of the highly expressing genes in these tissues and thus its promoter is suitable for driving the selectable marker and reporter genes of the vector. The vector contains a selectable marker gene *aadA* coding for aminoglycoside-3'-adenyl transferase and a reporter gene coding for β-glucuronidase. Both the genes are driven by 16S ribosomal RNA gene promoter. The cassette containing both of these genes is inserted in between a fragment derived from chloroplast DNA of rice which would facilitate the integration of this cassette in the chloroplast genome through homologous recombination. DNA delivery by particle bombardment resulted in very low transient plastid-specific gene expression.

4. REGULATION OF NUCLEAR GENE EXPRESSION

To understand the influence of light on chloroplast development and function, characterization of nuclear-encoded genes involved in its biogenesis is important. Keeping this in view, we have isolated *psbO, psbP* and *psbQ* genes, encoding three extrinsic proteins of 33, 23 and 16 kDa, respectively, of the oxygen-evolving complex (OEC) from a genomic library of *Arabidopsis thaliana* ecotype Columbia. All the three genes are most likely present in single or low copy number (Grover *et al* 1999b, Jain *et al* 1998, Kochhar *et al* 1996). The proteins are synthesized as precursor polypeptides which could be further cleaved into a transit peptide and a mature peptide. Transcriptional start site was mapped for *psbQ* gene by the primer extension method and a 28 nucleotide long 5' untranslated region was determined for *psbQ* (Grover *et al* 1999b) which is very short in comparison to leader sequence of other photosynthetic genes, such as *petH*, 239 nt (Oelmüller *et al* 1993), *psaF*, 188 nt (Flieger *et al* 1993), and ferredoxin, 95 nt (Elliot *et al* 1989). The transcripts of the genes as influenced by development, light, and organ-specificity have been investigated in detail (Grover *et al* 1999b, Jain *et al* 1998, Kochhar *et al* 1996) and the influence of genetic determinants conferring photoregulation has also been analyzed by employing photomorphogenic mutants (Jain *et al* 1998, Kochhar *et al* 1996).

5. ANALYSIS OF THE ACTIVITY OF PROMOTERS FROM PHOTOSYNTHETIC GENES IN TRANSGENICS

5.1 Analysis of *petH* Promoter of Spinach in Rice

Rice transformation was carried out using *A. tumefaciens* (LBA4404) having pCAMH vector containing g*us* gene driven by promoter from *petH* gene (Oelmüller *et al* 1993) as per the protocol given in Mohanty *et. al* (1999). Southern analysis of transgenic plants using *hph* gene as a probe revealed the presence of at least two copies of the gene integrated in the genome of rice. Histochemical analysis revealed detectable GUS activity only in old yellowish part of the leaves and not in young leaves, roots and anthers. However, fluorometric analysis, which is more sensitive than histochemical assay (Jefferson *et al* 1987), revealed activity in young leaves and young greenish anthers. Roots showed very low activity. In comparison to transgenic tobacco leaves (Oelmüller *et al* 1993), activity of this promoter in transgenic rice leaves was much lower. The low expression of *gus* by this promoter is probably because of the limited compatibility between a dicotyledonous gene promoter and monocotyledonous regulatory mechanisms. Similar observation of low expression of tomato *rbcS* gene promoter in transgenic rice has been reported earlier (Kyozuka *et al* 1993). But, expression of *gus* gene in green tissues and lack of expression in roots suggests that similar organ-specific regulatory mechanisms operate in dicots and monocots.

5.2 Analysis of *psbQ* Promoter in Tobacco

To analyze the *cis*-regulatory elements, promoter fragments of *Arabidopsis psbQ* gene were amplified and hooked to *gus* reporter gene, transformed into *Nicotiana tabacum* var. Samsun, and regulation of expression was studied in the transgenic plants by performing spectroflourometric analysis of GUS activity (Chaudhury *et al* 1995). The transgenic tobacco plants containing *psbQ* promoter::*gus,* growing in the pots under control conditions, were subjected to dark for 5 days and then irradiated with light for 24 h. All the plants tested showed stimulation (3-6 fold) in GUS activity after light treatment (Fig. 1), which is in agreement with the notion of high activity of nuclear-encoded photosynthetic promoters in light as compared to dark (Tyagi *et al* 1999). Transgenic plants were also analyzed for the organ-specific activity of *psbQ* promoter and it was observed that GUS activity in leaves was 6-7 folds higher as compared to roots (Fig. 1) which reflects high activity of the promoter in photosynthesis-competent tissue.

It has been reported previously that elements present in transcribed region are capable of influencing the expression of few nuclear-encoded genes (Bolle *et al* 1994, 1996, Caspar and Quail 1993). The same was observed for *psbQ* where several constructs, which included or excluded 13 bases of leader sequence, and contained some coding regions, were used. Although exclusion of 13 bases from leader sequence and/or a few bases in the vicinity of ATG codon did not influence light regulation, it was found to be very important quantitatively. This region also includes ARL box as reported by Caspar and Quail (1993). Functional significance, either at transcriptional or post-transcriptional level, of such a motif or other sequences incorporated in the constructs remains to be worked out.

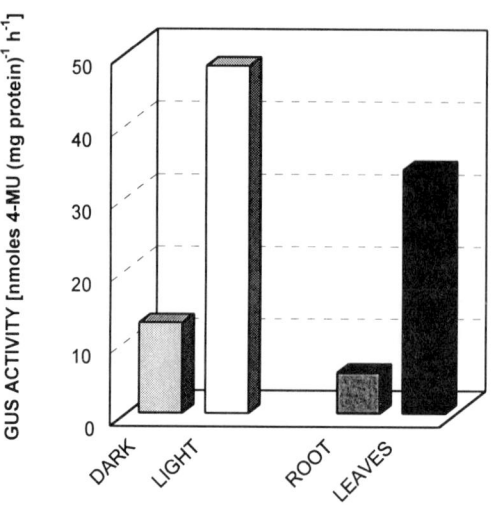

Figure 1. Light-regulated and tissue-specific expression under the control of *Arabidopsis psbQ* promoter in transgenic tobacco plants.

6. CONCLUSIONS

Regulation of photosynthesis-related genes, encoded by nuclear and plastidic genome, in response to light and developmental cues has been studied. Light signal for stimulation of plastid-encoded genes is perceived by phytochrome, and is transduced downstream by signalling intermediates like calcium, calmodulin, G-proteins and protein phosphorylation status, which are known to be involved in regulation of several nuclear genes. A role for calcium and protein phosphorylation during spatial control of plastid gene expression in rice has also been established (Grover *et al* 1998, Kapoor *et al* 1993). Efforts are on to find out if cGMP and phosphoinositide cycle intermediates which are known to be involved in signal transduction during regulation of nuclear-encoded genes in animals, are also involved in regulation of photosynthesis-related nuclear and plastidic genes. It remains to be determined if changes in signalling intermediates are brought about in cytosol and their effects are realised inside or these effects are dependent on changes inside plastids. Experiments with tagetitoxin, the inhibitor of plastid-encoded RNA polymerase, suggest that changes in expression of the plastid-encoded photosynthesis-related genes by light involve activity of plastid-encoded RNA polymerase. Promoters of *petH* and *psbO* were found to be regulated by both light and developmental cues. Though the nucleotides in 5'untranslated region of *psbQ* are important quantitatively, elements controlling qualitative expression reside further upstream. Analysis of nuclear-encoded photosynthesis-related genes in heterologous and homologous plants suggests that though the promoter from dicots do not express well in monocots, organ-specificity of dicot promoter is maintained in a monocot as well as in a heterologous dicot plant. Unfolding the intricacies of several regulatory steps entails more work.

ACKNOWLEDGEMENTS

Our research is funded by DBT, DST and CSIR, India, and the Rockefeller Foundation, USA.

REFERENCES

Bolle, C., Sopory, S., Lübberstedt, T., Herrmann, R.G., and Oelmüller, R., 1994, Segments encoding 5'-untranslated leaders of genes for thylakoid proteins contain *cis*-elements essential for transcription. *Plant J.* **6**: 513-523.

Bolle, C., Herrmann, R.G., and Oelmüller, R., 1996, Different sequences for 5'-untranslated leaders of nuclear genes for plastid proteins affect the expression of the β-glucuronidase gene. *Plant Mol. Biol.* **32**: 861-868.

Bowler, C. and Chua, N.H., 1994, Emerging themes of plant signal transduction. *Plant Cell* **6**: 1529-1541.

Bowler, C., Neuhaus, G., Yamagata, H., and Chua, N.H., 1994a, Cyclic GMP and calcium mediate phytochrome phototransduction. *Cell* **77**: 73-81.

Bowler, C., Yamagata, H., Neuhaus, G., and Chua, N.H., 1994b, Phytochrome signal transduction pathways are regulated by reciprocal control mechanisms. *Genes Dev.* **8**: 2188-2202.

Caspar, T. and Quail, P.H., 1993, Promoter and leader regions involved in the expression of the *Arabidopsis* ferredoxin A gene. *Plant J.* **3**: 161-174.

Chaudhury, A., Maheshwari, S.C., and Tyagi, A.K., 1995, Transient expression of *gus* gene in intact seed embryos of indica rice after electroporation-mediated gene delivery. *Plant Cell Rep.* **14**: 215-220.

Christopher, D.A., Xinli, L., Kim, M., and Mullet, J.E., 1997, Involvement of protein kinase and extraplastidic serine/threonine protein phosphatases in signaling pathways regulating plastid transcription and the *psbD* blue light responsive promoter in barley. *Plant Physiol.* **113**: 1273-1282.

Elliot, R.C., Pedersen, T.J., Fristensky, B., White, M.J., Dicky, L.F. and Thompson, W.F., 1989, Characterization of a single copy gene encoding ferredoxin I from pea. *Plant Cell* **1**: 681-690.

Flieger, K., Tyagi, A., Sopory, S., Cseplö, A., Herrmann, R.G. and Oelmüller, R., 1993, A 42 bp promoter fragment of the gene for subunit III of photosystem I (*psaF*) is crucial for its activity. *Plant J.* **4**: 9-17.

Grover, M., 1996, Ph.D. Thesis *Regulatory aspects of expression of photosynthesis-related genes in rice and characterization of psbQ from Arabidopsis*, University of Delhi.

Grover, M., Maheshwari, S.C. and Tyagi, A.K., 1996, Diurnal fluctuations in steady-state mRNAs of certain chloroplast-encoded photosynthesis-related genes in rice. *J.Plant Biochem. Biotechnol.* **5**: 105-107.

Grover, M., Sharma, A.K., Dhingra, A., Maheshwari, S.C., and Tyagi, A.K., 1998, Regulation of plastid gene expression in rice involves calcium and protein phosphatases/kinases for signal transduction. *Plant Sci.* **137**: 185-190.

Grover, M., Dhingra, A., Sharma, A.K., Maheshwari, S.C. and Tyagi, A.K., 1999a, Involvement of phytochrome(s), Ca^{2+} and phosphorylation in light-dependent control of transcript levels of plastid genes (*psbA, psaA* and *rbcL*) in rice (*Oryza sativa*). *Physiol. Plant.* **105**: 701-707.

Grover, M., Gaur, T., Kochhar, A., Maheshwari, S.C., and Tyagi, A.K., 1999b, Nucleotide sequence of *psbQ* gene for 16-kDa protein of oxygen-evolving complex from *Arabidopsis thaliana* and regulation of its expression. *DNA Res.* **6**: 173-177.

Jain, P.K., Kochhar, A., Khurana, J.P. and Tyagi, A.K., 1998, The *psbO* gene for 33-kDa precursor polypeptide of the oxygen-evolving complex in *Arabidopsis thaliana* – nucleotide sequence and control of its expression. *DNA Res.* **5**: 221-228.

Jefferson, R.A., Kavanagh, T.A. and Bevan, M.W., 1987, GUS fusion: β-*glucuronidase* as a sensitive and versatile gene fusion marker in higher plants. *EMBO J.* **6**: 3901-3907.

Kapoor, S., Maheshwari, S.C., and Tyagi, A.K., 1993, Organ-specific expression of plastid-encoded genes in rice involves both quantitative and qualitative changes in mRNAs. *Plant Cell Physiol.* **34**: 943-947.

Kapoor, S., Maheshwari, S.C., and Tyagi, A.K., 1994, Developmental and light-dependent cues interact to establish steady-state transcripts for photosynthesis-related genes (*psbA, psbD, psaA* and *rbcL*) in rice (*Oryza sativa* L.). *Curr. Genet.* **25**: 362-366.

Kelkar, N.Y,. Maheshwari, S.C., and Tyagi, A.K., 1993, Light–dependent accumulation of mRNAs for chloroplast-encoded genes in *Vigna aconitifolia*. *Plant Sci.* **88**: 55-60.

Khurana, J.P., Kochhar, A., and Tyagi, A.K., 1998, Photosensory perception and signal transduction in higher plants - molecular genetic analysis. *Crit. Rev. Plant Sci.* **17**: 465-539.

Kochhar, A., Khurana, J.P., and Tyagi, A.K., 1996, Nucleotide sequence of the *psbP* gene encoding precursor of 23-kDa polypeptide of oxygen-evolving complex in *Arabidopsis thaliana* and its expression in the wild-type and a constitutively photomorphogenic mutant. *DNA Res.* **3**: 277-285.

Kyozuka, J., McElroy, D., Hayakawa, T., Xie, Y., Wu, R., and Shimamoto, K., 1993, Light-regulated and cell-specific expression of tomato *rbcS-gusA* and rice *rbcS-gusA* fusion genes in transgenic rice. *Plant Physiol.* **102**: 991-1000.

Mohanty, A., Sarma, N.P., and Tyagi. A.K., 1999, *Agrobacterium*-mediated high frequency transformation of an elite *indica* rice variety Pusa Basmati and transmission of the transgene to R2 progeny. *Plant Sci.* **147**: 127-137.

Mustilli, A.C. and Bowler, C., 1997, Tuning into the signals controlling photoregulated gene expression in plants. *EMBO J.* **16**: 5801-5806.

Neuhaus, G., Bowler, C., Kern, R., and Chua, N.H., 1993, Calcium/calmodulin-dependent and -independent phytochrome signal transduction pathways. *Cell* **73**: 937-952.

Oelmüller, R., Bolle, C., Tyagi, A.K., Niekrawietz, N., Briet, S. and Herrmann, R.G., 1993, Characterization of the promoter from the single-copy gene encoding ferredoxin-$NADP^+$-oxidoreductase from spinach. *Mol. Gen. Genet.* **237**: 261-272.

Romero, L.C. and Lam, E., 1993, Guanine nucleotide binding protein involvement in early steps of phytochrome regulated gene expression. *Proc. Natl. Acad. Sci. USA* **90**: 1465-1469.

Tyagi, A.K., 1999, Regulation of plastid gene expression. In *Concepts in Photobiology: Photosynthesis and Photomorphogenesis* (G.S. Singhal, G. Renger, S.K. Sopory, K.D. Irrgang, Govindjee eds.), Narosa Publishing House, New Delhi, pp 739-751.

Tyagi, A.K., Dhingra, A., and Raghuvanshi, S., 1999, Light-regulated expression of photosynthesis-related genes. In *Probing Photosynthesis: Mechanism, Regulation and Adaptation* (M. Yunus, U. Pathre and P. Mohanty eds). In press, Tayler and Francis Publishers Ltd, London.

Functional Analysis of Pea Chloroplast DNA Polymerase and its Accessory Proteins

AMOS GAIKWAD, NASREEN EHTESHAM, D. V. HOP, SHAOJ CHEN AND SUNIL KUMAR MUKHERJEE
Plant Molecular Biology Group, Internationl Centre for Genetic Engineering & Biotechnology (ICGEB), Aruna Asaf Ali Marg, New Delhi - 110 067, India

1. INTRODUCTION

Replication of DNA is a universal process occuring in all the living organisms for the duplicaton of the genetic material. The complex process requires activity of multiprotein complexes for accurate and faithful replication of the genome (Kornberg and Baker 1992, Onrust *et al* 1995). The replicative DNA polymerases are the key enzymes that elongate the DNA chain. These DNA polymerases are assisted by a group of accessory proteins or factors for processive, rapid, accurate and efficient DNA synthesis. A number of such accessory proteins are well characterized from animal, bacterial and yeast systems. PCNA (Proliferating cell nucler antigen), RFC (Replication factor C) and RPA (Replication factor A) and their functional analogues are three of the most extensively studied accessory proteins (Hubscher *et al* 1996, Tsurimoto and Stillman 1989). As eukaryotes have organelles containing their own genetic material, the duplication of this orgenellar DNA becomes essential as these organelles (like mitochondria, chloroplasts *etc.*) are indispensable for any eukaryotic cell. Therefore, accurate DNA replication of the organellar DNA is essential

(Day and Ellis 1984). The study on the organellar DNA replication has remained focussed mainly on the animal mitochondria. Such studies have generated a lot of knowledge in relation to the various DNA polymerases (Ropp and Copeland 1996, Lewis *et al* 1996, Ye *et al* 1996) and their accessory proteins. However, not much of detailed study toward the understanding of the plant DNA replication, especially in the area of chloroplast DNA replication is available (Hernandez *et al* 1993, Van't Hof 1996).

Our laboratory has concentrated its efforts to understand the pea chloroplast DNA replication. Towards this goal, a partially purified pea chloroplast extract that could faithfully replicate the plasmids containing the pea chloroplast origin sequences had been developed (Reddy *et al* 1996). Using the replicative system, a host of replication factors had been identified and some of them, namely the Type 1 topoisomerase, a single strand circle specific nuclease, a DNA polymerase accessory factor *etc.*, had been purified to apparent homogeneity and biochemically characterised (Mukherjee *et al* 1994, Kumar *et.al* 1994, Chen *et. al* 1996). In this paper, we describe the structure and function of a 43 kDa accessory protein (p43) that is required for the processive DNA synthesis by a 70 kDa pea ct-DNA polymerase, purified to homogeneity recently in our laboratory. The processivity and fidelity of the ct-DNA polymerase have been studied. Some of the novel features revealed from the deduced aminoacid sequence of p43 have been highlighted. The major aim of this study was to provide insight for the elongation processes of pea ct-DNA replication.

2, RESULTS AND DISCUSSION

2.1 Purification of the 70 kDa Pea ct-DNA Polymerase Enzyme

Very much like the cellular nucleus and mitochondria, chloroplasts should also possess more than one type of DNA polymerases. To ascertain this, an *in situ* DNA polymerase activity gel assay of the intact pea chloropasts was carried out and we found that at least 3 distinct polymerases were present at around the molecular sizes of Mr. 120, 90 and 70 kDa. However, when purification of the DNA polymerase enzyme from pea chloroplast extracts was attempted, only the 70 kDa form was obtained. The

90 kDa form had been purified earlier and some of the biochemical characteristics were reported (McKown and Tewari 1984). Though some of the chromatographic properties of the 90 kDa and 70 kDa forms were similar, it is difficult to establish the inter-relationship between the above mentioned three forms at the present moment.

2.2 Functions of the 70 kDa Form

To understand the role of the 70 kDa enzyme in pea chloroplast DNA replication, its biochemical properties, processivity and fidelity were studied. Comparison of the properties displayed by the 70 kDa ct-DNA polymerase with the known organellar DNA polymerases (γ type) revealed that it varies from them in some features like its lack of sensitivity to the ddNTPs, Ara CTP and other inhibitors that are known to block the activity of organellar polymerases. Interestingly, the ct-DNA polymerase was highly sensitive to one of the antiviral agents, namely phosphono acetic acid (PAA), which inhibits the viral replicative polymerases. Generally, γ-DNA polymerases prefer RNA templates and show reverse transcriptase like activity whereas the 70 kDa form failed to show such activity and preferred the gapped DNA template. The ct-DNA polymerase lacked the 5'→3' exonuclease activity as revealed by *in situ* activity gel analysis. Ct-DNA polymerase displayed processivity of almost 3.0 kb as compared to 0.1 kb and 1.2 kb processivity of Klenow and sequanase version of T7 DNA polymerase, respectively. The comparatively high rate of processivity under the *in vitro* conditions may qualify the 70 kDa DNA polymerase to be replicative one. However, the role of accessory proteins *in vivo* to further enhance the processivity cannot be ruled out. The 70 kDa DNA polymerase was able to digest single stranded oligonucleotide in the 3'→5' direction. This 3'→5' exonuclease activity was, however, not as efficient as that of *E. coli* Pol. 1. This 3'→5' exonuclease activity was also confirmed by *in situ* activity gel analysis. Majority of the replicative polymerases use the 3'→5'exonuclease activity for the proof-reading function (fidelity). Hence the rate of misincorporation and error-specificity during 70 kDa polymerase mediated DNA synthesis were determined by *in vitro* experiments. A gapped substrate was constructed using the pBSK vector with the gap in the lacZ region. The polymerases were used to fill this gap and such plasmids were transformed in *E. coli* cells. The transformants were examined for the blue/white selection on LB plates containing ampicillin (100 µg/ml), X-gal and IPTG. White colonies on such plates resulted from the cells bearing the plasmid vector that were mutated at the lacZ region. The mutations were detected by DNA sequencing of the mutant plasmids. We found that the ct-DNA polymerase was moderately fidel with the error rate being $1/4 \times 10^5$ bases. The error rate under these *in vitro* conditions was somewhat high but the role of *in vivo* factors (proteins) to decrease the infidelity may not be ruled out. Moreover, the chloroplast DNA is not very large (120 kb) and thus the error rate may further become insignificant. During *in vitro* DNA synthesis, mostly nucleotide substitutions and some replication slippage errors were detected. In 70% the mutagenic events, errors were of GC→AT type. Since the chloroplasts are rich in AT composition, it is interesting to speculate whether this kind of error specificity contributed to the 'AT' richness during evolution.

2.3 Purification and Characterisation of p43

Replicative DNA polymerases, generally, are (a) processive in DNA synthesis (b) proficient in 3'→5' exonuclease activity to show proof reading function (c) assisted by accessory factors for enhanced processivity. The 70 kDa form displayed all the three above mentioned features. We searched for candidate pea chloroplast protein(s) that could act as the accessory factor(s) for the 70 kDa ct-DNA polymerase. In this course of study, a novel 43 kDa protein (p43) was identifed, purified (Fig. 1) and characterised (Chen *et al* 1996). DNA binding properties of p43 were investigated by South Western blotting, gel-retardation, filter-binding and electron-microscopic approaches. The p43 protein bound to DNA more effectively than to RNA suggesting its role only in DNA related processes. The DNA-binding was non-cooperative and preferentially with single stranded DNA. p43 also bound to cognate 70 kDa DNA polymerase in a specific manner and this binding was demonstrated conclusively by using various methods (Chen *et al* 1996). The specific interaction of p43 with the cognate DNA polymerase led to a 10-20 fold of stimulation of DNA synthesis activity of ct-DNA polymerase. By measuring the average size of the nascent DNA made by 70 kDa form both in presence and absence of p43, it was observed that p43 upregulated the processivity of 70 kDa DNA polymerase by a factor of 2. Thus p43 acted as an accessory factor of the ct-DNA polymerase.

2.4 Cloning the Gene Encoding p43

To gain insight into the mechanism of activation of ct DNA polymerase by p43, it was essential to isolate the gene encoding p43. The N-terminal and few of the internal sequences of p43 were obtained by micro sequencing the protein. Oligonucleotides were generated using the available internal aminoacid sequences of p43. Nuclear cDNA library obtained from the leaves of 7-8 day old pea seedlings was screened using these primers and was also processed for immunoscreening using antibodies that were generated against the native p43. By using these two approaches, a full length cDNA clone encoding p43 protein was isolated. Many interesting features were revealed from the deduced amino acid sequence of p43 (Gaikwad et al 1999). The protein was nuclear encoded but chloroplast-targetted and its N-terminal region is very rich in contiguous serine-hydroxyproline residues (S-HyP) that are known to be present in the hydroxyproline glycoproteins (HRGPs). These proteins are highly glycosylated (arabinogalactans), known to be extracellularly localised, involved in various stages of cellular development and known to provide resistance towards pathogens (Kieliszewski and Lamport 1994, Kreuger and van Holst 1996). p43, however, is an exception to this feature in the sense that it is an intracellular protein unlike other HRGPs. It is probably identified as the first of the HRGP like proteins with chloroplastic localisation.

Posttranslational modification of the p43 was evident from its full length cDNA. The deduced amino acid sequence encoded for a protein of about 36 kDa (including the 6.5 kDa of the transit peptide that is required for its transport into chloroplasts). Thus the mature protein was expected to be only of 29 kDa size as adjudged from the cDNA sequence. So the extra 14 kDa observed with the native p43 may be due to the post translational modifications, especially O-glycosylation at the S-HyP residues present at the N-terminal region of the mature protein. Immunoblot analysis using the standard glycan differentation kit confirmed that the protein was indeed glycosylated with sugars like mannose, N-glucosamine. Solvolysis of the p43 followed by gas chromatography of the sugars revealed that the protein is glycosylated, especially with residues like arabinose (87%), xylose (2%) and mannose (1%). Mild chemical deglycosylation of the native p43 with TFMS reduced the apparent size from 43 kDa to about 29 kDa, the size predicted from the cDNA sequence.

To understand the role of the sugar residues in p43 protein, the 29 kDa protein (p29 = deglycosylated p43) was examined for various activities / properties that are associated with p43 protein. Two of the three main features of the p43, namely its binding to the DNA and the 70 kDa cognate DNA polymerase remained unperturbed. However, unlike p43, p29 not only lost its potential to stimulate but also repressed the ct-DNA polymerase activity. This indicated that the sugars might play a very important role in the stimulation of the cognate DNA polymerase.

2.5 Role Played by p43 and p22 in Stimulation

The three activities of p43 namely DNA-binding, 70 kDa DNA polymerase-binding and stimulating the 70 kDa form are perhaps located in separate domains of p43 since various reagents (namely, anti p43 antibodies, deglycosylation *etc.*) affected the three biochemical activities quite differently. To carry out the domain analysis, recombinant expression of p43 was attempted in bacterial hosts. But over expression of p43 was not successful probably due to the N-terminal S-HyP rich region. Therefore a C-terminal region of p43 that lacked the S-HyP domain (corresponding to about 22 kDa size = p22) was overexpressed as a GST fusion protein. The over expressed p22 was tested for the three main features that were demonstrated by the native p43. We observed from Southwestern experiments that p22 did not bind to DNA. Incidentally, the estimated pI of the N-terminal domain (7 kDa S-HyP rich domain = p7) is about 13 because of high abundance of basic amino acid residues, namely arginine. Therefore this N-terminal region (p7) may act as a putative DNA binding site and as p22 lacks this region, it is unable to bind to DNA. But p22 stimulated the cognate DNA polymerase activity several fold and much more efficiently than the native p43. The ability of the p22 to bind to the ct-DNA polymerase was tested by the coimmunoprecipitation assay (Chen *et al* 1996) and the results are summarized in Fig. 2. The p22 protein and the ct-DNA polymerase directly interacted with each other in a coimmunoprecipitation assay. It is worthwhile to mention that deglycosylated p43 (or p29) was more efficient in binding to DNA and 70 kDa DNA polymerase than the native p43.

Thus it is likely that the DNA binding domain of the p43 protein resides in the N-terminal region (p7) and the C-terminal (p22) domain is responsible for the stimulatory activity of 70 kDa DNA polymerase. It is also tempting to speculate that the DNA binding domain alone (i.e. p7) retards the full stimulatory activity of p43. The physiological significance for this repression is being currently investigated. We are also in the process of expressing the

N-terminal S-HyP rich region in a suitable bacterial or yeast system. Once the S-HyP region (p7) is overexpressed, we could obtain clear explanations to some of the queries related to domain analysis. Site specific mutations in the recognised domains responsible for the DNA binding or the DNA polymerase stimulatory function would help further identify the amino acids that play a critical role for these important functions.

2.6 Other Accessory Factors

There will certainly be many other pea chloroplast accessory proteins assisting the 70 kDa DNA polymerase. We have found another strong candidate protein that was able to stimulate the cognate DNA polymerase. The 27 kDa pea chloroplast protein (p27) was purified using a pea PCNA affinity chromatography. PCNA itself is a nuclear protein and very essential for the replicative activity of the nuclear polymerases, namely the pol δ and pol ϵ (Hop et al 1999). p27 cross reacted with the anti-pea PCNA antibodies. The purified p27 stimulated the 70 kDa ct-DNA polymerase activity 10-15 fold in a dose dependant manner. Crosslinking studies with gluteraldehyde revealed that p27 was a trimer in solution. The oligomeric character of p27 was important for enhancing (about 2 fold) the processivity of the 70 kDa DNA polymerase. These features tempt us to consider p27 as an organellar PCNA-like factor. Most importantly, p43 and p27 acted synergistically to upregulate the DNA polymerase activity of the 70 kDa form.

3. CONCLUSION

The chloroplast was chosen for our study as it provides an excellent model system to study the harmonious interaction between the prokaryotic and the eukaryotic replicative machinery. As the chloroplasts have evolved from the endosymbiosed bacteria during the course of evolution, one would speculate that the chloroplasts would have the prokaryotic type of the replicative machinery. However, it turns out that the chloroplast DNA replication does not resemble that of the prokaryotic type. The DNA polymerases and all the accessory factors are nuclear encoded and targeted to the chloroplasts. From the studies of the deletion derivatives of the plastids, there are reasons to believe that all the essential factors required for plastid DNA replication are nuclear encoded and targetted to the plastids. The replicative proteins reveal eukaryotic features both in structure and function. However, all these proteins would use the plastomic DNA which is not complexed with histone like proteins, a hallmark for the prokaryotic templates. This unique combination of the pro- and eukaryotic features makes the study of the organellar DNA replication interesting and challenging. Further, the study on the genes of chloroplast DNA polymerases would enable us to make and select chloroplast mutations with higher photosynthetic activities.

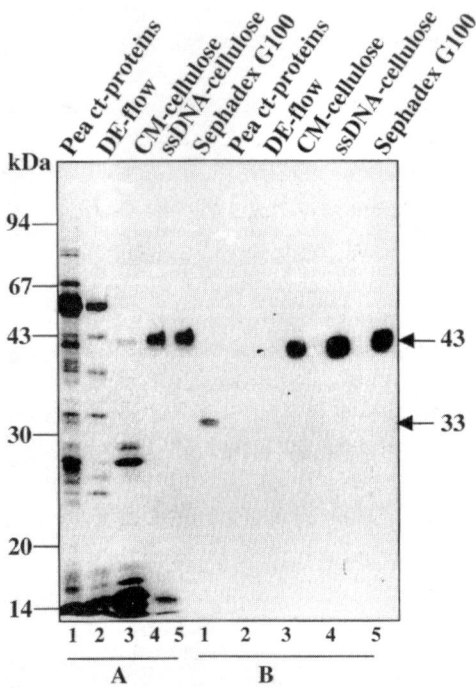

Figure 1. Purification and activity profile of p43: (A) Coomassie stained proteins of various chromatographic fractions, resolved by SDS-PAGE. (B) Autoradiogram of the South Western blot of the gel shown in (A). Denatured radiolabelled DNA containing OriA sequence of pea plastome was used as the probe. About 50, 20, 10, 3 and 2 µg proteins were loaded in the lanes 1, 2, 3, 4 and 5 respectively. The 33 kDa protein is abundantly present in the pea chloroplast and is a RNA-binder.

In this particular study, novelties of p43 are apparent. p43 is an O-arabinogalactan and yet present in the pea chloroplast to help in the pea plastome DNA replication processes. Since deglycosylation of p43 leads to inhibition of 70 kDa DNA polymerase activity, glycosylation seems to play an important role in stimulating the DNA polymerase activity. But this role could be an indirect one. Deglycosylated p43 binds to template DNA more tightly than native p43, and thus might retard DNA polymerase to track down the template. In this model, deglycosylated p7 is a strong DNA binder and a repressor while p22 is the real activator. In the O-glycosylated form of p43, all the basic amino acid residues of p7 may not be available for DNA binding leading to moderate binding of p43 to the template DNA. But this form of DNA binding would help the 70 kDa DNA polymerase latch onto the template DNA for enhanced processivity by virtue of the interaction between p22 and the 70 kDa DNA polymerase. Thus glycosylation of p43 seems to be loosening the DNA binding rather than directly stimulating the 70 kDa DNA polymerase activity.

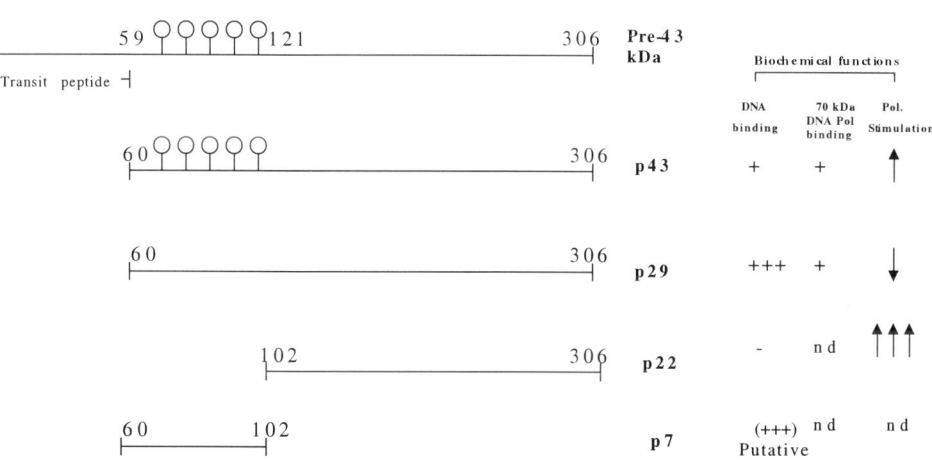

Figure 2. A model showing properties of p43: Pre-43 kDa is a precursor of p43, synthesised in cytosol and targetted to pea choloroplasts using the transit peptide (amino acids 1-59). Ball headed lines are the sites of O-glycosylation. p43 is the mature protein obtained from pea chloroplasts. p29 is derived following mild deglycosylation of p43 using TFMS treatment. p22 is the 22 kDa protein overexpressed in bacteria in the form of GST-p22 fusion protein. The p22 protein is

released from the fusion form following proteolysis with thrombin. p7 is the 7 kDa peptide containing the putative DNA binding and O-glycosylation sites. Arrows pointing up and down represent upregulation and down regulation of DNA polymerase activity respectively. + and - shows the presence and absence of function respectively. The numbers on the top of each line represent the aminoacid coordinates of the depicted sequence.

ACKNOWLEDGEMENT

We would like to thank Ms.Pratibha Chaturvedi for her secretarial assistance.

REFERENCES

Chen, W., Gaikwad, A., Mukherjee, S.K., Choudhury, N.R., Kumar, D., and Tewari, .K., 1996, DNA binding protein from the pea chloroplast interacts with and stimulates the cognate DNA polymerase. *Nucleic Acid Res.* **24**: 3943-3961.

Day, A., and Ellis, T.H.N., 1984, Chloroplast DNA deletions associated with wheat plants regenerated from pollen: possible basis for maternal inheritance of chloroplasts. *Cell* **39**: 359-368.

Gaikwad, A., Tewari, K.K., Kumar, D., Chen, W., and Mukherjee, S.K., 1999, Isolation and characterization of the cDNA encoding a glycosylated accessory protein of pea chloroplast DNA polymerase. *Nucleic Acid. Res.* **27**: 3120-3129.

Hernandez, P., Martin-Parras, L., Martinez-Robles, M.L., Schvartzman, J.B., 1993, Conserved features in the mode of replication of eucaryotic ribosomal RNA genes. *EMBO J.* **12**: 1475-1485.

Hop, D.V., Gaikwad, A., Yadav, B.S., Reddy, M.K., Sopory, S.K., Mukherjee, S.K., 1999, Suppressionof pea nuclear topoisomerase I enzyme activity by pea. *Plant J.* **19**: 153-162.

Hubscher, U., Maga, G., and Podust, V.N., 1996, DNA Replication accessory proteins. In *DNA Replication in Eukaryotic Cells* (M.L. DePamphilis eds), CSHL Press, NY, pp 525-543.

Kieliszewski, M. J.,and Lamport, D.T., 1994, Extensin: repetative motifs, functional sites, post-transcriptional codes and phylogeny. *Plant J.* **5**: 157-172.

Kornberg, A., and Baker, T. (eds), 1992, In *DNA Replication*, 2nd eds, W.H. Freeman & Co., NY.

Kreuger, M., and van Holst, G.J., 1996, Arabinigalactan proteins and plant differentiation. *Plant Mol. Biol.* **30**: 1077-1086.

Kumar, D., Mukherjee, S.K., and Tewari, K.K., 1995, A novel single stranded DNA-specific endonuclease from pea chloroplasts. *J. Exp. Bot.* **46**: 767-776.

Lewis, D.L., Farr, C.L., Wang, Y., Lagina, A.T., and Kaguni, L.S., 1996, Catalytic subunit of mitochondrial DNA polymerase from *Drosophila* embryos: Cloning, bacterial overexpression and biochemical characterization. *J. Biol. Chem.* **271**: 23389-23394.

McKown, R.L., and Tewari, K.K., 1994, Purification and properties of pea chloroplast DNA polymerase. *Proc Natl. Acad. Sci. USA* **81**: 2354-2358.

Mukherjee, S.K., Reddy, M.K., Kumar, D., and Tewari, K.K., 1994, Purification and characterization of an eukaryotic type I topoisomerase from pea chloroplast. *J. Biol. Chem.* **269**: 3793-3801.

Onrust, R., Finkelstein, J., Natkinis, V., Tumer, J., Fang, L., and O'Donnel, M., 1995, Assembly of a chromosomal replication machine: two DNA polymerase, a clamp loader, and sliding clamps in one holoenzyme particle. III. Interface between two polymerases and the clamp loader. *J. Biol. Chem.* **270**: 13348-13391.

Reddy, M.K., Choudhury, N.R., Kumar, D., Mukherjee, S.K., and Tewari, KK., 1994, Characterization and mode of *in vitro* replication of pea chloroplast OriA sequences. *Eur. J. Biochem.* **220**: 933-941.

Ropp, P.A., and Copeland, W.C., 1996, Cloning and characterization of the human mitochondrial DNA polymerase, DNA polymerase gamma. *Genomics* **36**: 449-458.

Tsurimoto, T., and Stillman, B., 1989, Multiple replication factors augment DNA synthesis by the two eukaryotic DNA polymerases, alpha and delta. *EMBO J.* **8**: 3883.

vant Hof, J., 1996, DNA Replication in plants. In *DNA Replication in Eukaryotic Cells* (M.L. De Pamphilis ed.), CSHL Press, NY, pp 1005-1027.

Ye, F., Carrodeguas, J.A., and Bogenhagen, D.F., 1996, The gamma subfamily of DNA polymerases: cloning of a developmentally regulated cDNA encoding Xenopus laevis mitochondrial DNA polymerase gamma. *Nuclic Acid Res.* **24**: 14181-14188.

Signal Transduction by Mitogen-activated Protein Kinases (MAPKs)

ERWIN HEBERLE-BORS, ORNELLA CALDERINI, VIKTOR VORONIN, AND CATHAL WILSON
Vienna Biocenter, Institute of Microbiology and Genetics, University of Vienna, Dr. Bohrgasse 9, A-1030 Vienna, Austria

1. INTRODUCTION

Mitogen-activated protein kinases (MAPKs) are encoded by a large family of serine/threonine protein kinases and are found in all eukaryotes (Kültz 1998). They were initially discovered in association with the cytoskeleton as microtubule-associated protein 2 kinase (MAP-kinase). Later they were found to be involved in the activation of cell division by growth factors, hence the name mitogen-activated protein kinases. Their general role in eukaryotic signal transduction, discovered even later, earned this class of protein kinase the name of extracellular signal-regulated kinases or ERKs. MAPKs are now known to phosphorylate and thus activate or inactivate a variety of substrates, including transcription factors, other components of the protein synthesis machinery, components of the cytoskeleton, and enzymes. Activation of MAPKs is brought about by upstream MAPK kinases (denoted as MAPKKs) through phosphorylation of the conserved threonine and tyrosine residues that are located close to kinase subdomain VIII in all MAPKs. A given dual specificity MAPKK can only activate a specific MAPK and cannot functionally substitute for other MAPKKs. MAPKKs are themselves activated by phosphorylation through upstream kinases that belong to the class of MAPKK kinases (MAPKKKs), such as the Raf and

Mos proteins. A specific set of three functionally interlinked protein kinases (MAPKKK-MAPKK-MAPK) forms the basic module of a MAPK pathway. Several modules may co-exist side by side in a cell and integrate a variety of upstream signals through interactions with other kinases denoted as MAPKKK kinase (MAPKKKK) or G proteins, such as Ras or heterotrimeric complexes. The latter factors often function as coupling agents between a plasma membrane located receptor protein, that senses an extracellular stimulus, and a MAPK module.

MAPKs are today known to be involved in the transduction of very diverse signals and in activating or inactivating a variety of substrates. Not only do they signal information to the basic cell cycle machinery by activating the transcription of cyclin D genes and facilitating the formation of cyclin D-CDK4 complexes, but they also have been shown to activate pathways to recruit ribosomes to nascent transcripts and thereby initiate protein synthesis. MAPKs also phosphorylate histones which, together with MAPK-phosphorylated transcription factors, control gene expression. Finally, nucleotide synthesis is also controlled by MAPKs. ERK MAP kinase regulates the activity of carbamyl phosphate synthetase II which catalyses the initial, rate-limiting step in the *de novo* synthesis of pyrimidine nucleotides. Together with their well-known effect on cytoskeletal proteins, these functions make MAPKs central regulators of growth (Whitmarsh and Davis 2000).

A well-known function of MAPKs is in stress signalling. The p38 and c-Jun N-terminal kinase/stress activated protein kinase (JNK/SAPK) class of MAPKs in animals is activated by ultraviolet radiation, pro-inflammatory cytokines and environmental stress. The p38 kinases are activated in response to endotoxin from gram-negative bacteria, interleukin-1, heat and hyperosmolar stress. In yeast the MPK1 MAPK is required for adaptation to a hypoosmolar environment while the HOG1 MAPK is activated under hyperosmolar conditions and results in an increased biosynthesis of glycerol which acts as an osmotic stabiliser. The SMK1 MAPK is involved in the control of spore formation (Ruis and Schüller 1995).

Beyond growth regulation and stress response, MAPKs have also been found to play important roles in animal development (summarised in Nebreda and Porras 2000). In *Drosophila* a p38 MAPK is involved in *decapentaplegic*-regulated wing development, and an additional role has been proposed in the correct asymmetric development of the *Drosophila* egg which is essential in axis formation of the later embryo. In vertebrates, a role for MAPKs has been identified in cell differentiation of fibroblasts, which are induced by insulin to differentiate into adipocytes, both as a positive regulator of differentiation genes and as a negative regulator of cell cycle genes.

The eukaryotic MAPK family can be divided into three distinct subgroups: the stress-activated protein kinases (SAPKs), the extracellular signal-regulated kinases (ERKs) and the MAPK3 group. In animals and fungi, MAPK members can be found for all these three subgroups. All identified plant MAPKs group together as one subfamily of the ERK subgroup (Kültz 1998). No plant MAPKs are known that group into the SAPK or MAPK3 subgroups.

Sequence comparisons of the predicted amino acid sequences of the plant MAPKs revealed the existence of at least three different subgroups which can be divided into at least five subfamilies, named PERK1-5 (Ligterink 2000). Only the two known MAPKs of monocots (AsPK9 and WCK1) could not be placed into any of the five groups.

The best functional information on PERKs is at present available on the members of groups 1 and 2. The majority of the members of this group signal pathogen attack and/or abiotic stress. Tobacco SIPK and WIPK, *Arabidopsis* ATMPK3, and alfalfa MMK4 are transiently activated by various abiotic and biotic stresses (for review see Meskiene and Hirt 2000).

After isolating MAPK genes from alfalfa (Jonak *et al* 1993) and tobacco (Wilson *et al* 1993) we focused on the tobacco MAP kinases and isolated and characterised three of them (Wilson *et al* 1995). In this review, the present state of knowledge about the two tobacco MAP kinases Ntf4 and Ntf6 is summarised.

2. THE NTF4 MAP KINASE

The amplification of PCR products from pollen cDNA libraries using redundant oligonucleotides designed from conserved regions of MAP kinases suggested that MAP kinases are expressed in pollen (Wilson *et al* 1995). The tobacco ntf4 MAP kinase gene was expressed in pollen, as determined by Northern analysis and the ability to amplify PCR products from pollen cDNA libraries (Wilson *et al* 1993, 1995), and by RT-PCR from pollen-derived mRNA (V. Voronin and C. Wilson, unpublished). Transcription and protein synthesis are developmentally regulated, occurring after pollen mitosis nearly until the mature, dry pollen stage (Wilson *et al* 1997, Prestamo *et al* 1999). This MAP kinase is therefore expressed as a late gene in pollen. The protein accumulates in an inactive form in the mature, dry pollen grain, and no change in protein levels is found in germinating pollen. Hydration of the pollen grain, either in germination medium or in water, results in a rapid and transient activation of the MAP kinase. This relationship of protein abundance on the one hand and kinase activity on the other is quite different from a number of other plant MAP kinase genes whose expression is induced by the application of a stress (Bögre *et al* 1997, Jonak *et al* 1996, Mizoguchi *et al* 1996, Seo *et al* 1995, Zhang and Klessig 1998) and is more similar to the typical situation in animals where also the signal to be transmitted has no effect on the synthesis of the transmitter.

Many late expressed proteins are considered to be necessary for the activation and germination of the mature pollen grain. The kinase is inactivated before pollen tube emergence, suggesting a role in the activation of the pollen grain, i.e. a return to a metabolically active state and preparation of the necessary structures and developmental programme required for tube growth (Wilson *et al* 1999). It is however, not known at present which particular processes are controlled by the NTF4 MAPK. The initial events which occur during rehydration of pollen include the reconstitution of cellular membranes from a gel form to a liquid crystalline state, the repair of damage which has taken place during the dry state and during rehydration, including damage to the DNA of the generative and sperm cells, the reorganisation of the actin cytoskeleton towards the pore where the pollen tube eventually emerges, or the resumption of metabolic activity and of protein synthesis from rapidly reforming polyribosomes utilising the performed mRNAs stored in the dry pollen.

The activation of the tobacco NTF4 kinase by water alone in pollen suggested that hydration is the signal for turning on the MAP kinase cascade. The pollen grains swell rapidly upon hydration, and might activate response mechanisms similar to those identified in hypo-osmotically shocked cells from other organisms. Hypotonic cell swelling activates ion channels and ion transporters, and second messengers such as Ca^{2+} and IP_3 (Strange 1994). Increases in Ca^{2+} concentrations in plant cells after hypoosmotic shock of tobacco cell suspension culture cells may precede the activation of protein kinases (Takahashi *et al* 1997). Some of these responses may be involved in the restoration of the original cell volume, others might function more directly in the response to increases in turgor and changes in cell membrane tension.

The finding that a signal transduction component is active only transiently immediately after pollen rehydration and well before pollen tube emergence is the first indication that specific events operate in the rehydrating pollen which require activation by phosphorylation. No other genes have so far been isolated which are specifically required for pollen activation, as compared to prior pollen development and later pollen tube growth.

As mentioned above, stresses such as cold, drought or salt induce transcription of MAPK genes in alfalfa and *Arabidopsis* while the *ntf4* gene was expressed in a cell-type specific manner before activation of the corresponding protein. Therefore, it can be concluded that transcriptional regulation is an important control mechanism in MAPK signalling cascades in plants. We therefore undertook an analysis of the expression pattern of the tobacco MAP kinase NTF4. The *ntf4* gene promoter region was isolated and a chimerical ntf4 promoter-GUS fusion construct was introduced into tobacco plants. GUS expression was detected in pollen, in developing and mature embryos, and shortly after seed germination, but not in other floral tissues and tissues such as leaf, root, or stem. This expression pattern was confirmed by Northern and Western analyses. In situ hybridisation and immuno-localization studies showed that the expression of the *ntf4* gene and its encoded NTF4 protein occurred in embryos at least from the globular embryo stage until the mature seed, as well as in the seed endosperm. Taken together, the results show that the *ntf4* gene has a very restricted expression pattern, being found only in pollen and seeds.

The NTF4 MAPK is closely related to SIPK. In fact the two are 93% identical, and may originate from the two different ancestral species of tobacco, *Nicotiana sylvestris* and *Nicotiana tomentosiformis* (Zhang and Klessig 2000). In *Arabidopsis* there appears to exist only one member of this class of MAPK, i.e. ATMPK6. Possibly in tobacco the two genes evolved to serve the same biochemical function in different organs. While SIPK seems to be involved in signalling abiotic and biotic stresses in vegetative tissues, the NTF4 MAPK functions in reproductive organs.

Antisense suppression of gene expression is a valuable reverse genetics technique in plants to associate function to a gene. The *ntf4* cDNA was cloned in an antisense orientation behind the pollen-specific and strong *LAT52* promoter and introduced into plants. Western analysis and immunolocalisation of pollen from antisense F1 generation plants showed a complete absence of the NTF4 protein in some of the transformants. However, no phenotype was detected. We expected a phenotype since antisense suppression would probably also knock out the closely related SIPK. The lack of a phenotype was probably due to functional redundancy in the MAPK gene family which was not overcome by the antisense approach.

3. THE NTF6 MAP KINASE

The NTF6 MAP kinase belongs to the PERK4 class of plant MAPKs. This class also harbours the alfalfa MMK3 MAPK. We were in the lucky situation of having antibodies against both MAPK proteins and were searching for a possible function of this class of MAPK. Originally we could not detect the protein or kinase activity in any part of the plant, apart from suspension cultures. Northern blots were also unsuccessful. Turning our attention to proliferation we tested organs rich in dividing cells and could indeed find transcript, protein and kinase activity in very immature organs (Calderini *et al* 1998, Bögre *et al* 1999). A detailed analysis then established that both kinases are associated with proliferation and more specifically with the mitotic phase of the cell cycle. An ultrastructural *in situ* hybridisation study confirmed this conclusion (Prestamo *et al* 1999).

Entry into mitosis was required for the activation of these MAPKs. This was revealed by arresting cells in prophase with a CDK-specific drug, roscovitine. The timing of PERK4 activation in the two species was further dissected by synchronisation with the drug propyzamide, which depolymerises microtubules and blocks cells in metaphase.

Both the alfalfa and tobacco kinases, similar to the animal ERKs, were inactive in metaphase cells with depolymerised microtubules and became active after removing the drug as cells passed through anaphase and telophase. To better define the role of PERK4 in mitosis, suspension cells where released from a metaphase arrest and then the microtubules were again depolymerised with half hour pulse treatments of amiprophos-methyl (APM). A rapid inactivation of MMK3 by the APM-treatment was seen and was further demonstrated by a time course experiment. This allows two interpretations: either intact microtubules are required during ana- and telophase, or the activation of MMK3 and NTF6 is transient and falls within the 10-30 minutes interval of the APM treatments. In another experiment, however, it was not simply the integrity of microtubules which influenced PERK4 activity since taxol, which stabilises microtubules, arrested cells in mitosis. Rather, it seemed that it was passage through anaphase and telophase which was required for NTF6 activation.

This conclusion was supported by an experiment in which the broad specificity kinase inhibitor K252a was added to cells after release from a metaphase arrest imposed by the microtubule-disrupting drug propyzamide. K252a inhibited MMK3 activation and caused a high proportion of cells to proceed to interphase without passing through anaphase/telophase and without forming any microtubule structures, such as the anaphase spindle and phragmoplast.

In *Xenopus*, two distinct MAPKs belonging to the ERK and JNK subclasses were found to control the spindle assembly checkpoint and arrest cells in metaphase until chromosomes are aligned at the cell equator. The localisation of active ERK1 and ERK2 MAPKs to kinetochores during metaphase and the ERK phosphorylation of a phospho-specific epitope recognised by the 3F3/2 monoclonal antibody strongly correlated with the activation of the spindle checkpoint, further substantiating the notion that ERKs play a role in the spindle checkpoint. The two PERK4 kinases, however, are not active in metaphase-arrested cells, making it unlikely that PERK4 is involved in a checkpoint controlling microtubule integrity.

The subcellular localisation of PERK4 is reminiscent of that of ERK1 and ERK2. Both ERKs, as well as MMK3, are found in the cytoplasm in interphase cells and invade the nucleus at the end of G2. Though MMK3 and NTF6 are most active during ana- and telophase, some activation could already be found at prophase. MMK3, NTF6, as well as ERKs, were found at comparable locations in late mitosis. While MMK3 and NTF6 were found associated with the plant-specific phragmoplast in late anaphase and at the midplane of cell division in telophase cells, active ERK1 and ERK2 were detected at the midzone region in late anaphase and at the midbody during telophase and cytokinesis in animal cells. Together, these data strongly support a role for PERK4 in the regulation of plant cytokinesis.

The tobacco MAPKKK, NPK1, has also been implicated in mitosis (Banno *et al* 1993). In contrast to MMK3 and NTF6, NPK1 appears to be activated in cells with severed microtubules. A search for activators of NPK1 led to the identification of kinesin-like proteins (Machida *et al* 1998). ERK1 and ERK2 are also complexed with a microtubule motor protein, CENP-E. The kinesin-like tobacco proteins as well as CENP-E specifically accumulate in mitosis. A hypothetical signal transduction pathway would link the kinesin-like motor protein, the NPK1 MAPKKK and PERK4. The module could operate at several points in mitosis whenever reorganisation of the cytoskeleton takes place.

Cytokinesis has acquired centre stage in plant biology recently. A dynamin-like protein, phragmoplastin, has been isolated from soybean and shown to be associated with cell plate formation (Gu and Verma 1996, 1997). Phosphorylation of animal dynamin by ERK2 inhibits the dynamin-microtubule interaction. Phragmoplastin was found to appear first in the centre of the forming cell plate and, as the cell plate grew outwards, it redistributed to the growing margins of the cell plate. Contrary to this, we have not found a redistribution of MMK3 from the centre to the periphery, indicating that MMK3 is not associated with phragmoplastin during this process. Recently, embryo patterning genes in *Arabidopsis thaliana* such as *knolle* (Lukowitz *et al* 1996) and *keule* (Assaad *et al* 1996) have been identified as being involved in cytokinesis, and the syntaxin-related KNOLLE protein was localised at the cell plate (Lauber *et al* 1997). Similarly to MMK3 in late telophase, when the phragmoplast reached the lateral cortex of the cell, KNOLLE appeared to be present across the entire plane of division. Phragmoplast microtubules might be required to bring these proteins to the cell plate but apparently the microtubules are not required to keep KNOLLE and MMK3 at this location.

Which events of cytokinesis might be regulated by MMK3 and NTF6 is not known at present. These may include the construction of the phragmoplast by regulating microtubule stability or a microtubule-based motor, vesicle transport along the phragmoplast by plus-end-directed motor proteins, or the fusion of these vesicles at the cell plate. However, the link between cytokinesis and pattern formation in the embryo is exciting and warrants more detailed studies on the function of these MAP kinase (Heese *et al* 1998).

4. CONCLUSIONS AND OUTLOOK

The MAP kinases seem to be central regulators in eukaryotes of growth, stress responses and developmental. In plants, we are only beginning to understand their involvement in these processes. Given the large MAPK gene family (more than twenty in *Arabidopsis*), the challenge is not only to identify further functions of plant MAPKs but to identify these functions with the individual members of the gene family. We are attempting to genetically dissect this important gene family using insertion mutants and functional genomics tools.

REFERENCES

Assaad, F.F., Mayer, U., Wanner, G., and Jürgens, G., 1996, The *KEULE* gene is involved in cytokinesis in *Arabidopsis. Mol. Gen. Genet.* **253**: 267-277.

Banno, H., Hirano, K., Nakamura, T., Irie, K., Nomoto, S., Matsumoto, K., and Machida, Y., 1993, *NPK1*, a tobacco gene that encodes a protein with a domain homologous to yeast BCK1, STE11, and BY2 protein kinases. *Mol. Cell Biol.* **13**: 4745-4752.

Bögre, L., Calderini, O., Binarova, P., Mattauch, M., Till, S., Kiegerl, S., Jonak, C., Pollaschek, C., Baker, P., Huskisson, N.S., Hirt, H., and Heberle-Bors, E., 1999, A MAP kinase is activated late in mitosis and becomes localised to the plane of cell division. *Plant Cell* **11**: 101-113.

Bögre, .L, Ligterink, W., Meskiene, I., Barker, P.J., Heberle-Bors, E., Huskisson, N.S., and Hirt, H., 1997, Wounding induces the rapid and transient activation of a specific MAP kinase pathway. *Plant Cell* **9**: 75-83.

Calderini, O., Bögre, L., Vicente, O., Binarova, P., Heberle-Bors, E., Wilson, C., 1998, A cell cycle regulated MAP kinase with a possible role in cytokinesis in tobacco cells. *J. Cell Sci.* **111**: 3091-3100.

Gu, X., and Verma, D.P.S., 1996, Phragmoplastin, a dynamin-like protein associated with cell plate formation in plants. *EMBO J.* **15**: 695-704.

Gu, X., and Verma, D.P.S., 1997, Dynamics of phragmoplastin in living cells during cell plate formation and uncoupling of cell elongation from the plane of cell division. *Plant Cell* **9**: 157-169.

Heese, M., Mayer, U,. and Jürgens, G., 1998, Cytokinesis in flowering plants: cellular process and developmental integration. *Curr. Opin. Plant Biol.* **1**: 486-491.

Jonak, C., Kiegerl, S., Ligterink, W., Barker, P.J., Huskisson, N.S., and Hirt, H., 1996, Stress signalling in plants: A MAP kinase pathway is activated by cold and drought. *Proc. Natl. Acad. Sci. USA* **93**: 11274-11279.

Jonak, C., Páy, A., Bögre, L., Hirt, H., and Heberle-Bors, E., 1993 The plant homologue of MAP kinase is expressed in a cell cycle-dependent and organ-specific manner. *Plant J.* **3**: 611-617.

Kültz, D., 1998, Phylogenetic and functional classification of mitogen- and stress-activated protein kinases. *J. Mol. Evol.* **46**: 571-588.

Lauber, M.H., Waizenegger, I., Steinmann, T., Schwarz, H., Mayer, U., Hwang, I., Lukowitz, W., and Jürgens, G., 1997, The *Arabidopsis* KNOLLE protein is a cytokinesis-specific syntaxin. *J. Cell Biol.* **139**: 1485-1493.

Lukowitz, W., Mayer, U., Jürgens, G., 1996, Cytokinesis in the *Arabidopsis* embryo involves the syntaxin-related KNOLLE gene product. *Cell* **84**: 61-71.

Machida, Y., Nakashima, M., Morikiyo, K., Banno, H., Ishikawa, M., Soyano, T., and Nishihama, 1998, MAPKKK-related protein kinase NPK1: regulation of the M phase of plant cell cycle. *J. Plant Res.* **111**: 243-246.

Meskiene, I., and Hirt, H., 2000, MAP kinase pathways: molecular plug-and-play chips for the cell. *Plant Mol. Biol.* **42**: 791-806.

Mizoguchi, T., Irie, K., Hirayama, T., Hayashida, N., Yamaguchi-Shinozaki, K., Matsumoto, K., and Shinozaki, K., 1996, A gene encoding a mitogen-activated protein kinase kinase kinase is induced simultaneously with genes for a mitogen-activated protein kinase and S6 ribosomal protein kinase by touch, cold, and water stress in *Arabidopsis thaliana*. *Proc. Natl. Acad. Sci. USA* **93**: 765-769.

Nebreda, A.R., and Porras, A., 2000, p38 MAP kinases: beyond the stress response. *TIBS* **25**: 257-260.

Prestamo, P., Testillano, P.S., Vicente, O., Gonzalez-Melendi, P., Coronado, M.J., Wilson, C., Heberle-Bors, E., and Risueno, M.C., 1999, Ultrastructural distribution of a MAP kinase homologue and their transcripts in quiescent and cycling plant cells, and in pollen grains. *J. Cell Sci.* **112**: 1065-1076.

Ruis, H., and Schüller, C., 1995, Stress signalling in yeast. *Bioessays* **17**: 959-965.

Seo, S., Okamoto, M., Seto, H., Ishizuka, K., Sano, H., and Ohashi, Y., 1995, Tobacco MAP kinase: A possible mediator in wound signal transduction pathways. *Science* **270**: 1988-1992.

Strange, K., 1994, Cellular and molecular physiology of cell volume regulation. CRC Press, Boca Raton, FL.

Takahashi, K., Isobe, M., and Muto, S., 1997, An increase in cytosolic calcium ion concentration precedes hypoosmotic shock-induced activation of protein kinases in tobacco cell suspension culture cells. *FEBS Lett.* **401**: 202-206.

Whitmarsh, A.J., Davis, R.J., 2000, A central control for cell growth. *Nature* **403**: 255-256

Wilson, C., Anglmayer, R., Vicente, O., Heberle-Bors, E., 1995, Molecular cloning, functional expression in Escherichia coli, and characterization of multiple mitogen-activated-protein kinases from tobacco. *Eur. J. Biochem.* **233**: 249-257.

Wilson, C., Eller, N., Gartner, A., Vicente, O., and Heberle-Bors, E., 1993, Isolation and characterization of a tobacco cDNA clone encoding a putative MAP kinase. *Plant Mol. Biol.* **23**: 543-551.

Wilson, C., and Heberle-Bors, E., 2000, MAP kinases in pollen. In: MAP kinases in plant signal transduction (H. Hirt, ed.), Springer-Verlag, Berlin Heidelberg New York, pp. 39-51.

Zhang, S., and Klessig, D.F., 1998, The tobacco wounding-activated protein kinase is encoded by SIPK. *Proc. Natl. Acad. Sci. USA* **95**: 7433-7438.

Zhang, S., and Klessig, D.F., 2000, Pathogen-induced MAP kinases in tobacco. (H. Hirt, ed.), Springer-Verlag, Berlin Heidelberg New York, pp. 65-84

Index

Abscisic acid (ABA). *See also* Moss protonema
  MAP kinase activation, 65
  polypeptides induced by, 66–67
Action spectra, phytochrome response, 2–3
Action spectroscopy, 19
Agrobacterium, 130, 132
Albizzia lophantha, 103, 104; *see also* Leaflet
    movements
Alfalfa, 323
Alkalization, of cytosol, in mesophyll cells,
    PEPC, 41–42
Alternanthera pungens, 42, 43
Amaranthus hypochondriacus, 43
Angelica dahurica, 138
Animal cells
  ectoapyrases, 53
  myoinositol phosphates, 71
Apoplast calmodulin, 137–144; *see also* Calmodulin
  CaMBP binding proteins, 142
  functions of, 138–140
  identification, localization and purification of,
    137–138
  transmembrane mechanism of, 140–142
Apoproteins, photoreception, 9, 19
Arabidopsis, 212
  calcium signaling, 127
  chloroplast protein gene expression regulation,
    297, 302, 303–304
  crinkly4 (cr4) signaling, 159, 163
  cytokinin receptor identification, 196, 197
  ectoapyrase, 51, 53
  heat shock factor (HSF), 221–222, 223–224
  kinesin-like calmodulin-binding protein
    (KCBP), 177, 179, 182, 186, 188–189
  mitogen-activated protein kinases (MAPK),
    323, 325, 328
  photosynthesis gene regulation, 264
  plastid ribosome biogenesis, 291
  protonema cultures, 60
  salicylic acid-mediated signal transductions,
    203–204
Arabidopsis mutants, 25–37
  genetic screens, 26–27
  isolation and genetic analysis, 27–28
  molecular characterization, 30–34
    auxin-responsive genes, up-regulation of, 32
    cloning and sequencing, 32–33

Arabidopsis mutants (*cont.*)
  molecular characterizations (*cont.*)
    de-repression in dark, 30
    photosynthesis-related genes, 31
    terpenoid biosynthesis pathway genes, 31–32
  photoreceptor mechanism of action, 25–26
  physiological characterization of, 28–30
    leaf differentiation and flowering in dark,
      29–30
    precocious flowering, 30
  pleiotropic effects, 33–34
Arabidopsis phytochrome(s), 9–17
  apoproteins, 19
  cellular localization, 21–22
  light environment, 9
  phyA and phyB, seedling
    photomorphogenesis, 10–11
  phyD/phyE, 12–13
  signal transduction, 13–15
    gravity-sensing pathways and, 15
    phyA, 14–15
ATP. *See* Extracellular ATP (xATP)
ATP signal-damping role, ectoapyrases, 57
A. tumefaciens, 302
Autophosphorylation, calmodulin-like domain
    protein kinases (CDPK), catalytic
    properties, 147–148
Auxin, calcium/calmodulin-modulated protein
    kinases (CCaMK), 167–168, 172–174; *see
    also* Calcium/calmodulin-modulated
    protein kinases (CCaMK)
Auxin-binding proteins, moss protonema, 59–68,
    63; *see also* Moss protonema
Auxin-responsive genes, up-regulation of,
    Arabidopsis mutants, 32
Avena, 26
Axenic cell line J-2, moss protonema cultures,
    61–62

Blue/UV-A region
  cryptochrome(s) and phototropin, 9
  photomorphogenesis, 19
Brassica juncea, calcium signaling, 127, 129

Calcium
  leaflet movements, 103–111; *see also* Leaflet
    movements

Calcium (cont.)
  phosphoinositide (PI) pathway, Ca^{2+} and $InsP_3$ as secondary messengers, 84
Calcium activation, calmodulin-like domain protein kinases (CDPK), 148
Calcium/calmodulin. See also Calmodulin
  chloroplast protein gene expression regulation, 299–300
  kinesin-like calmodulin-binding protein (KCBP) microtubule interaction, 187–189
Calcium/calmodulin-modulated protein kinases (CCaMK), 167–176
  auxin-mediated signaling, cross-talk, 172–174
  biochemical characteristics, 169–170
  interacting protein EF-1", 170–171
  overview, 167–168
  structural features, 168
Calcium-dependent protein kinases
  calcium signaling, 125; see also Calcium signaling
  moss protonema, 63–67; see also Moss protonema
Calcium homeostasis, myoinositol phosphates, 71–81; see also Myoinositol phosphates
Calcium signaling, 125–136
  EhCaBP homologue role, 130–132
  EhCaBP homologue(s), 127–128
  EhCaBP stimulated kinase (Brassica juncea), 129
  EhCaBP stimulated kinase detection, 129–130
  glyoxalase I role, 132–133
  regulation, 125–126
  ZmCCaMK homologue (cold and salt regulation), 126–127
Calmodulin. See also Apoplast calmodulin
  calcium signaling, 125; see also Calcium signaling
  leaflet movements, 103–111; see also Leaflet movements
  toxic metals tolerance, 113–123; see also Toxic metals tolerance
Calmodulin antagonist
  nyctinastic closure, leaflet movements, 107
  rhythmic movement, leaflet movements, 108–109
Calmodulin-dependent kinases, calcium signaling, 125
Calmodulin-like domain protein kinases (CDPK), 145–155
  catalytic properties, 147–149
    autophosphorylation, 147–148
    calcium activation, 148
    mechanism of activation, 148–149
    substrate specificity in vitro, 147
  domain organization and size, 146–147
  overview, 145–146
  physiological relevance, 149–152
    endogenous substrate identification, 149–150
    induction by signals, 151
    induction by stress, 151–152

Calmodulin-like domain protein kinases (CDPK) (cont.)
  physiological relevance (cont.)
    sequence analysis, 152
    tissue specific expression and subcellular localization, 150
CaMBP binding proteins, apoplast calmodulin, 142
CAM plants, cystolic pH as secondary messenger during light activation of PEPC, 39–48; see also Phosphoenolpyruvate carboxylase (PEPC)
Caspase-like protease activities, detection of, cell death, 214
Caulonema cells, GTP-binding proteins in, moss protonema, 65–66; see also Moss protonema
CDF2 transcription factor, spinach plastid, 279–285; see also Spinach plastid
CDPK. See Calmodulin-like domain protein kinases (CDPK)
C. elegans, kinesin-like calmodulin-binding protein (KCBP), 182
Cell, phytochrome localization, 20, 21–22
Cell death, 209–216
  caspase-like protease activities, detection of, 214
  overview, 209–210
  peptide inhibitors of proteases, 211–213
  reactive oxygen species (ROS) requirement, 210–211
  TMV-induced, P35 overexpression delays, 213–214
Cell differentiation, moss protonema cultures, 62
Cell division, kinesin-like calmodulin-binding protein (KCBP), 183–186
Cell fate, crinkly4 (cr4) signaling, 157–165; see also Crinkly4 (cr4) signaling
Ceratodon purpureus, 59
Chalcone synthase, light-dependent nuclear transport of CPRF2, photomorphogenesis, 21
Chaperone function, heat shock proteins, 218–219
Chlamydomonas, kinesin-like calmodulin-binding protein (KCBP), 183
Chloronema cells. See also Moss protonema
  calcium-dependent protein kinases from, 63–67
  protonema cultures, 60–61
Chloroplast differentiation, 287–296
  nuclear encoded plastid ribosomal protein gene transcription, 288–295
    gene activation during chloroplast differentiation, 294–295
    initiator level, 289–290
    organ-specific positive and negative controlling elements, 292–294
    relative strength of promoters of r-protein genes, 291
  process of, 287

Chloroplast differentiation (cont.)
  ribosomal proteins synthesis triggers, 288
Chloroplast-localized response regulators,
    photosynthesis gene regulation, 262–266
Chloroplast protein gene expression regulation,
    297–307
  analysis of promoters, 302–304
    PBS promoter in tobacco, 305–306
    petH promoter of spinach in rice, 304–305
  chloroplast transformation, 301
  nuclear gene expression regulation, 301–302
  overview, 297–298
  photoregulation, 298–301
    Ca^{2+} and calmodulin in, 299–300
    G-protein role, 299
    phosphorylation role, 300–301
    phytochrome involvement, 298–299
    plastid-encoded RNA polymerase, 298
Cis-elements, photosynthesis gene regulation,
    phospholipids, 267–275
Cold regulation, ZmCCaMK homologue,
    calcium signaling, 126–127
C_4 plants, cystolic pH as secondary messenger
    during light activation of PEPC, 39–48;
    see also Phosphoenolpyruvate
    carboxylase (PEPC)
CPRF2, light-dependent nuclear transport of,
    photomorphogenesis, 21
Crinkly4 (cr4) signaling, 157–165
  biochemical studies, 160–163
    binding proteins, 161
    MAP kinase signaling, 160–161
    Rho-like GTPase interaction, 163
    thioredoxin-h interaction, 161–162
  cell-autonomous differentiation response, 159
  CR4 protein, 158–159
  model for, 164
  mutants identifying candidates for other
    signaling components, 159–160
  overview, 157–158
Cryptochromes
  photomorphogenesis, 19
  spectrum, 9
Cucumber cotyledon phytochromes, 93–101
  inositol phospholipids, 95–98
    neomycin sulfate and lithium effects on IP_3
      levels, 97–98
    neomycin sulfate and lithium effects on
      NRA, 95–97
  materials and methods, 94–95
  nitrate reductase (NR), 93–94
  polypeptide phosphorylation, 99–100
  protein phosphorylation, 98–99
Cucumis sativus. See Cucumber cotyledon
    phytochromes
Cyanophora paradoxa, photosynthesis gene
    regulation, 264
Cystolic pH, as secondary messenger during
    light activation of PEPC, 39–48; see also
    Phosphoenolpyruvate carboxylase (PEPC)

Cytokinesis, mitogen-activated protein kinases
    (MAPK), 327–328
Cytokinin receptor identification, 193–199
  cytokinin binding proteins, 195–196
  G protein-coupled receptor homologue, 196–
    197
  overview, 193
  structure activity and sites of action, 194–195
  two-component receptor involvement, 197
Cytosol/nuclear partitioning, phytochromes, 20

Digitaria sanguinalis, 44, 45
Disease resistance
  nitric oxide-mediated signal transduction,
    204–206
  salicylic acid-mediated signal transduction,
    201–204; see also Salicylic acid-
mediated signal transduction
DNA replication, pea chloroplast DNA
    polymerase, 309–310; see also Pea
    chloroplast DNA polymerase
Drosophila
  heat shock factor (HSF), 222
  kinesin-like calmodulin-binding protein
    (KCBP), 183
  mitogen-activated protein kinases (MAPK), 322

Ectoapyrases, 49–58
  animal cells, 53
  ATP efflux, MDR1 promotes, 50–51
  ATP signal-damping role, 57
  extracellular ATP
    ECM levels, 51–52
    occurrence and function, 50
  historical perspective, 49
  MDR1 co-function in toxin resistance, 54–57
  plant cells, 53–54
EhCaBP homologue
  calcium signaling, 127–128
  role of, calcium signaling, 130–132
EhCaBP stimulated kinase
  calcium signaling (Brassica juncea), 129
  detection of, 129–130
E. histolytica, calcium signaling, 126, 127, 129
Escherichia coli, 264, 289, 311
Eukaryotes, phosphoinositide (PI) pathway, 83–
    84; see also Phosphoinositide (PI)
    pathway
Eukaryotic elongation factor-1″, calcium/
    calmodulin-modulated protein kinases
    (CCaMK), 170–171
Extracellular ATP (xATP), ectoapyrases, 49–58;
    see also Ectoapyrases

Far-red (FR) region, photoreversible
    phytochrome(s), 9
Ferns, 20
Flavonoid biosynthesis, light-dependent nuclear
    transport of CPRF2, photomorphogenesis,
    21

333

Flowering
 in dark, leaf differentiation and, *Arabidopsis* mutants, 29–30
 precocious, *Arabidopsis* mutants, 30
Fluorescent differential display (FDD) technique, phytochrome-dependent gene expression, 4
Funaria hygrometrica, protonema, 59–68; *see also* Moss protonema

Galdieria sulphuraria, 88
Gene expression, phytochrome signals, 3–4
Glyoxalase I, role of, calcium signaling, 132–133
G-protein, chloroplast protein gene expression regulation, 299
G protein-coupled receptor homologue, cytokinin receptor identification, 196–197
Gravity responses, phosphoinositide (PI) pathway and, 84–87
Gravity-sensing pathways, *Arabidopsis* phytochrome(s), signal transduction, 15
Green fluorescent protein (GFP), photomorphogenesis, 22
GTP-binding proteins, moss protonema, 59–68; *see also* Moss protonema

Haemanthus, kinesin-like calmodulin-binding protein (KCBP), 182
Heat shock factor (HSF), 217–226
 downstream effects, 221–222
 growth and development, 223
 heat shock proteins and chaperone function, 218–219
 molecular aspects of manipulation, 223–224
 stress sensing, signaling and feedback regulation, 219–221
 stress tolerance and stress injuries, 217–218
Heavy metals. *See* Toxic metals tolerance
High energy reactions, irradiation, 1
High irradiance response (HIR)
 action spectra, 2–3
 arabidopsis phytochrome(s), phyA/phyB, 10–11
 irradiation, 1
Histidine kinase, 26
Hypersensitive response (HR) cell death, 210; *see also* Cell death

Inositol, phosphoinositide (PI) pathway, 83–84; *see also* Phosphoinositide (PI) pathway
Inositol phospholipids, cucumber cotyledon phytochromes, 95–98; *see also* Cucumber cotyledon phytochromes
Intracellular partitioning, photomorphogenesis, signal transduction in, 19–24; *see also* Photomorphogenesis

Kinesin-like calmodulin-binding protein (KCBP), 177–192
 calmodulin-binding domain location, 179–180

Kinesin-like calmodulin-binding protein (KCBP) (*cont.*)
 cell division and, 183–186
 coding and expression of, 182–183
 kinesin family, novel member of, 178–179
 microtubule binding sites, 181
 microtubule interaction, calcium/calmodulin, 187–189
 minus-ended directed microtubule motor, 181
 myosin- and kinesin-like features, 180–181
 overview, 177–178
 protein interactions, 189
 trichomes development, 186–187
 ubiquity of, 181–182
KNOLLE protein, mitogen-activated protein kinases (MAPK), 328

Lead, toxic metals tolerance, 117–119
Leaf differentiation, flowering in dark and, *Arabidopsis* mutants, 29–30
Leaflet movements, 103–111
 calmodulin antagonist effects on nyctinastic closure, 107
 calmodulin antagonist effects on rhythmic movement, 108–109
 lithium effects on nyctinastic closure, 105–106
 materials and methods, 104–105
 overview, 103–104
Light-dependent nuclear transport, of CPRF2, photomorphogenesis, 21
Lilium, kinesin-like calmodulin-binding protein (KCBP), 182
Lithium
 cucumber cotyledon phytochromes, 95–98
 nyctinastic closure, leaflet movements, 105–106
Low energy reactions, irradiation, 1
Low fluence response (LFR)
 arabidopsis phytochrome(s), seedling photomorphogenesis, 11
 irradiation, 1

Magnaporthe grisea, 55
Maize, 45, 157–165; *see also* Crinkly4 (cr4) signaling
MDR1
 ATP efflux promoted by, 50–51
 ectoapyrase co-function in toxin resistance, 54–57
Mesophyll cells, alkalization of cytosol in, PEPC, 41–42
Metabolic signaling, myoinositol phosphates, 71–81; *see also* Myoinositol phosphates
Microtubule binding sites, kinesin-like calmodulin-binding protein (KCBP), 181
Microtubule interaction, calcium/calmodulin, kinesin-like calmodulin-binding protein (KCBP), 187–189
Microtubule motor, minus-ended directed, kinesin-like calmodulin-binding protein (KCBP), 181

Minus-ended directed microtubule motor, kinesin-like calmodulin-binding protein (KCBP), 181
Mitogen-activated protein kinases (MAPK), 321–329
 crinkly4 (cr4) signaling, 160–161
 NTF4 MAP kinase, 323–325
 NTF6 MAP kinase, 326–328
 overview, 321–323
 tobacco, salicylic acid-mediated signal transduction, 202–203
Mitosis, kinesin-like calmodulin-binding protein (KCBP), 183–186
Molecular genetic analysis, of *Arabidopsis*, 25–37; *see also Arabidopsis* mutants
Mosses, 20
Moss protonema, 59–68
 auxin-binding proteins in, 63
 axenic cell line J-2, 61–62
 calcium-dependent protein kinases, 63–67
 calcium-dependent protein kinases from ABA- and stress-induced polypeptides, 66–67
 abscisic acid (ABA) action, 65
 GTP-binding proteins, 65–66
 cell differentiation regulation, 62
 cultures, 60–61
 paradigm development, 59–60
Mougeotia, 20
Myoinositol phosphates, 71–81
 alternative pathway of $InsP_3$, significance of, 78
 animal cells, 71
 calcium homeostasis, 78
 differential inhibition of high and low affinity sites, 75–77
 phytase binding site, 74
 phytase interaction, 72–73
 plant cells, 71–72
 separation of high and low affinity sites after phytase cleavage by trypsin, 77–78
 ternary complex formation involving phytase, 74–75
Myosin, kinesin-like calmodulin-binding protein (KCBP), 180–181

Neomycin sulfate, cucumber cotyledon phytochromes, 95–98
Nickel, toxic metals tolerance, 117
Nicotiana tabacum. *See* Tobacco
Nitrate reductase (NR), cucumber cotyledon phytochromes, 93–94; *see also* Cucumber cotyledon phytochromes
Nitric oxide-mediated signal transduction, 204–206; *see also* Salicylic acid-mediated signal transduction
NTF4 MAP kinase, mitogen-activated protein kinases (MAPK), 323–325
NTF6 MAP kinase, mitogen-activated protein kinases (MAPK), 326–328
Nuclear gene expression regulation, chloroplast protein gene expression regulation, 301–302

Nyctinastic closure
 calmodulin antagonist effects, leaflet movements, 107
 lithium effects, leaflet movements, 105–106

Osmosignaling, *Galdieria sulphuraria*, phosphoinositide (PI) pathway, 88

Parsley, 20, 21
Pathogens, cell death, 209–216; *see also* Cell death
Pea
 EhCaBP stimulated kinase, detection of, 129–130
 ZmCCaMK homologue, calcium signaling, 126–127
Pea chloroplast DNA polymerase, 309–319
 accessory factors, 315
 DNA replication, 309–310
 70 kDa ct-DNA polymerase enzyme purification, 310–311
 70 kDa form functions, 311–312
 model system, 315–318
 p43 and p22 role in stimulation, 314–315
 p43 cloning, 313–314
 p43 purification and characterization, 312
Peptide inhibitors, of proteases, cell death, 211–213
Phorbol myristate acetate, nitrate reductase (NR), 93–94
Phosphoenolpyruvate carboxylase (PEPC), 39–48
 cytosol alkalization in mesophyll cells, 41–42
 cytosolic pH
  consequences of rise in, 45–46
   PEPC activity correlated with, 42–45
    acid-base modulation, 43–45
    cell sap, 43
 light activation in leaves, 40–41
 model of regulation, 46, 47
 overview, 39–40
Phosphoinositide (PI) pathway, 83–92
 Ca^{2+} and $InsP_3$ as secondary messengers, 84
 eukaryotes, 83–84
 gravity responses and, 84–87
 myoinositol phosphates, 71; *see also* Myoinositol phosphates
 osmosignaling in *Galdieria sulphuraria*, 88
Phosphoinositides, leaflet movements, 103–111; *see also* Leaflet movements
Phospholipid(s), photosynthesis gene regulation, 267–275
Phospholipid-dependent kinases, calcium signaling, 125; *see also* Calcium signaling
Phosphorylation
 calcium/calmodulin-modulated protein kinases (CCaMK), 167–176; *see also* Calcium/calmodulin-modulated protein kinases (CCaMK)
 chloroplast protein gene expression regulation, 300–301

335

Phosphorylation (*cont.*)
  heat shock factor (HSF), 223
  mitogen-activated protein kinases (MAPK), 327–328
  thylakoid membrane dynamics, 241–245; *see also* Thylakoid membrane dynamics
Photomorphogenesis, 19–24
  intracellular localization of phytochromes, 21–22
  light-dependent nuclear transport of CPRF2, 21
  phytochromes, 19–20
  signal transduction in, 22–23
Photomorphogenic mutants, of *Arabidopsis*, 25–37; *see also Arabidopsis* mutants
Photoregulation, chloroplast protein gene expression regulation, 298–301; *see also* Chloroplast protein gene expression regulation
Photoreversible phytochrome(s)
  photomorphogenesis, 19
  spectrum, 9
Photosynthesis
  chloroplast regulation, 297
  thylakoid membrane dynamics, 241–257; *see also* Thylakoid membrane dynamics
Photosynthesis gene regulation, 259–277
  overview, 259–260
  phospholipids involved in, 267–275
  plasma membrane perception, 260–262
  plastid control of gene expression, 262–266
Photosynthesis-related genes, *Arabidopsis* mutants, 31
Phototropin
  photomorphogenesis, 19
  spectrum, 9
PhyA/phyB
  arabidopsis phytochrome(s), seedling photomorphogenesis, 10–11
  phytochrome response, 2–5
PhyD/phyE, *Arabidopsis* phytochrome(s), 12–13
Physcomitrella patens, 59, 60, 67; *see also* Moss protonema
Phytase
  calcium homeostasis, 78
  differential inhibition of high and low affinity sites, 75–77
  myoinositol phosphate interactions, 72–73, 74
  separation of high and low affinity sites after cleavage by trypsin, 77–78
  ternary complex formation involving, 74–75
Phytochrome(s), 1–5
  action spectra, 2–3
  Arabidopsis, 9–17; *see also Arabidopsis* phytochrome(s))
  chloroplast protein gene expression regulation, 298–299
  cucumber cotyledon, 93–101; *see also* Cucumber cotyledon phytochromes
  future prospect, 4–5

Phytochrome(s) (*cont.*)
  gene expression, 3–4
  historical perspective, 1–2
  localization of, 20, 21–22
  photomorphogenesis, 19–20; *see also* Photomorphogenesis
Phytochrome interacting factors (PIF), 26
Phytochrome kinase substrate (PKS), 26
Phytophthora sojae. *See* Soybean-*Phytophthora sojae* interaction
Pigments, photoreception, 1
Pisum sativum, calcium signaling, 127
Plastid-encoded RNA polymerase, chloroplast protein gene expression regulation, 298
Plastid proteins. *See* Chloroplast protein gene expression regulation
Plastid ribosome biogenesis, 287–296
  chloroplast differentiation process, 287
  nuclear encoded gene transcription, 288–295
    gene activation during chloroplast differentiation, 294–295
    initiator level, 289–290
    organ-specific positive and negative controlling elements, 291
    relative strength of promoters of r-protein genes, 291
  ribosomal proteins synthesis triggers chloroplast differentiation, 288
Plastids, gene expression control, photosynthesis gene regulation, 262–266; *see also* Spinach plastid
Pleiotropic effects, *Arabidopsis* mutants, 33–34
Plumular hook open (*pho*) mutants, *Arabidopsis* mutants, 27–34; *see also Arabidopsis* mutants
Polypeptide(s), stress-induced, moss protonema, 66–67
Polypeptide phosphorylation, cucumber cotyledon phytochromes, 99–100
Precocious flowering, *Arabidopsis* mutants, 30
Programmed cell death, 209–210; *See also* Cell death
Protein phosphorylation, cucumber cotyledon phytochromes, 98–99
Protonema. *See* Moss protonema
Pseudomonas syringae pv. phaseolicola, 211, 213

rDNA transcription, spinach plastid, 279–285; *see also* Spinach plastid
Reactive oxygen species (ROS), cell death, 210–211
Red/far-red reversible LFR
  action spectra, 2–3
  photomorphogenesis, 19
  photoreception, 1
Red (R) region, photoreversible phytochrome(s), 9
Rhodobacter sphaeroides, photosynthesis gene regulation, 264
Rho-like GTPase, crinkly4 (cr4) signaling, 163

Rhythmic movement, calmodulin antagonist,
  leaflet movements, 108–109
Rice, 297, 298–299, 304–305
RNA polymerase, plastid-encoded, chloroplast
  protein gene expression regulation, 298
Robinia pseudoacacia, 103, 104; *see also* Leaflet
  movements

Salicylic acid-mediated signal transduction, 201–
  204; *see also* Nitric oxide-mediated
  signal transduction
  Arabidopsis NPR1 interaction, 203–204
  tobacco MAP kinases, 202–203
  tobacco proteins, 201–202
Salt regulation, ZmCCaMK homologue, calcium
  signaling, 126–127
S. cerevisiae, kinesin-like calmodulin-binding
  protein (KCBP), 182
Seedling photomorphogenesis, *Arabidopsis*
  phytochrome(s), 10–11
Serine/threonine kinases, 26
Sorghum, 45
Soybean, 327
Soybean-*Phytophthora sojae* interaction, 227–
  239
  overview, 227–229
  signal transduction process, 230–235
Spinach, 304–305
Spinach plastid, 279–285
  constitutive and up-regulated rDNA
    transcription, 280
  overview, 279–280
  regulation of transcription at PC, 281–283
  sequence context of PC, 280–281
Stress-induced polypeptides, moss protonema,
  66–67
Synechocystis, 26, 264

Ternary complex formation, myoinositol
  phosphates, 74–75
Terpenoid biosynthesis pathway genes,
  Arabidopsis mutants, 31–32
Thermotolerance, heat shock factor (HSF), 217–
  218; *see also* Heat shock factor (HSF)
Thioredoxin-h, crinkly4 (cr4) signaling, 161–162
Thylakoid membrane dynamics, 241–257
  compounds of signal transduction processes,
    249–250
  kinase activity and cDNAs, 245–248
  perspectives on, 252–253
  photosynthesis, 241–245
  subthylakoid location of compounds, 250–252

Thylakoid membrane dynamics (*cont.*)
  thylakoid phosphatases, 248–249
Tobacco, 22
  calcium signaling, 127
  cell death, 213–214
  heat shock factor (HSF), 224
  mitogen-activated protein kinases (MAPK),
    325, 327
  photosynthesis gene regulation, 261–262
  salicylic acid-mediated signal transduction,
    201–203
  toxic metals tolerance, 113–123; *see also*
    Toxic metals tolerance
Tomato, 303
Toxic metals tolerance, 113–123
  lead, 117–119
  materials and methods, 114–116
  model for function and regulation, 119–120
  nickel, 117
  overview, 113–114
  plant sensitivity, 116–117
  plasma membrane, 116
Toxin resistance, ectoapyrase/MDR1 co-function
  in, 54–57
Tradescantia, kinesin-like calmodulin-binding
  protein (KCBP), 182
Transmembrane mechanism, apoplast
  calmodulin, 140–142
Trichomes, development of, kinesin-like
  calmodulin-binding protein (KCBP), 186–
  187
Trypsin, separation of high and low affinity sites
  after phytase cleavage by, 77–78

Up-regulation, of auxin-responsive genes,
  Arabidopsis mutants, 32
UV-A/UV-B region, 9
UVB photoreceptors, photomorphogenesis, 19

Very low fluence response (VLFR)
  arabidopsis phytochrome(s), seedling
    photomorphogenesis, 10
  irradiation, 1

Vigna aconitifolia, 298

Yeast, ATP efflux, MDR1 promotes, 50–51
Yeast two-hybrid screens, 26

ZmCCaMK homologue, calcium signaling, cold
  and salt regulation, 126–127